S0-AZQ-497

ELECTRICAL RACEWAYS
and
OTHER WIRING METHODS

Third Edition

Richard E. Loyd

Delmar Publishers

an International Thomson Publishing company I(T)P®

Albany • Bonn • Boston • Cincinnati • Detroit • London • Madrid
Melbourne • Mexico City • New York • Pacific Grove • Paris • San Francisco
Singapore • Tokyo • Toronto • Washington

NOTICE TO THE READER

Publisher does not warrant or guarantee any of the products described herein or perform any independent analysis in connection with any of the product information contained herein. Publisher does not assume, and expressly disclaims, any obligation to obtain and include information other than that provided to it by the manufacturer.

The reader is expressly warned to consider and adopt all safety precautions that might be indicated by the activities described herein and to avoid all potential hazards. By following the instructions contained herein, the reader willingly assumes all risks in connection with such instructions.

The publisher makes no representations or warranties of any kind, including but not limited to the warranties of fitness for particular purpose or merchantability, nor are any such representations implied with respect to the material set forth herein, and the publisher takes no responsibility with respect to such material. The publisher shall not be liable for any special, consequential or exemplary damages resulting, in whole or in part, from the readers' use of, or reliance upon, this material.

Cover Design: Courtesy of Brucie Rosch

Delmar Staff
Publisher: Alar Elken
Acquisitions Editor: Mark Huth
Project Editor: Megeen Mulholland
Production Coordinator: Toni Hansen
Art/Design Coordinator: Cheri Plasse
Editorial Assistant: Dawn Daugherty
Production Manager: Mary Ellen Black

COPYRIGHT © 1999
By Delmar Publishers
an International Thomson Publishing company I(T)P®

The ITP logo is a trademark under license
Printed in the United States of America

For more information contact:

Delmar Publishers
3 Columbia Circle, Box 15015
Albany, New York 12212-5015

International Thomson Publishing Europe
Berkshire House
168-173 High Holborn
London WC1V 7AA
United Kingdom

Nelson ITP, Australia
102 Dodds Street
South Melbourne,
Victoria, 3205 Australia

Nelson Canada
1120 Birchmont Road
Scarborough, Ontario
M1K 5G4, Canada

International Thomson Publishing France
Tour Maine-Montparnasse
33 Avenue du Maine
75755 Paris Cedex 15, France

International Thomson Editores
Seneca 53
Colonia Polanco
11560 Mexico D. F. Mexico

International Thomson Publishing GmbH
Königswinterer Straße 418
53227 Bonn
Germany

International Thomson Publishing Asia
60 Albert Street
#15-01 Albert Complex
Singapore 189969

International Thomson Publishing Japan
Hirakawa-cho Kyowa Building, 3F
2-2-1 Hirakawa-cho, Chiyoda-ku,
Tokyo 102, Japan

ITE Spain/Paraninfo
Calle Magallanes, 25
28015-Madrid, Espana

Online Services

Delmar Online
To access a wide variety of Delmar products and services on the World Wide Web, point your browser to:
    http://www.delmar.com
    or email: info@delmar.com

A service of I(T)P®

All rights reserved. No part of this work covered by the copyright hereon may be reproduced or used in any form or by any means—graphic, electronic, or mechanical, including photocopying, recording, taping, or information storage and retrieval systems—without the written permission of the publisher.

2 3 4 5 6 7 8 9 10 XXX 02 01 00 99

Library of Congress Cataloging-in-Publication Data
Loyd, Richard E.
    Electrical raceways and other wiring methods / Richard E. Loyd. — 3rd ed.
      p. cm.
    Includes index.
    ISBN 0-7668-0266-3 (alk. paper)
    1. Electric wiring, Interior. 2. Electric conduits. I. Title.
TK3271.L68 1999
  621.319'24—dc21
                       98-30003
                          CIP

CONTENTS

iii

APPENDIX

Codes and Standards   215

Index   *251*

DEDICATION

I dedicate this book to my wife, Nancy, and to the associations that have supported my efforts for so many years: The National Electrical Manufacturers Association Section 5RN (NEMA), The American Iron & Steel Institute (AISI), and the Steel Tube Institute of North America (STI). Without Nancy's tremendous help this book would never have been possible. She has worked as hard as I have to bring this material into book form.

FOREWORD

ELECTRIC PER SE

Electricity! In ancient times, it was believed to be an act of the gods. Not even Greco-Roman civilization could understand this thing we call electricity. It was not until about A.D. 1600 that any scientific theory was recorded, when after seventeen years of research William Gilbert wrote a book on the subject titled *De Magnete*. It was almost another 150 years before major gains were made in understanding electricity so that people might someday be able to control it. At this time Benjamin Franklin, whom we often refer to as the grandfather of electricity or electrical science, and his close friend Joseph Priestley, a great historian, began to gather the works of the many worldwide scientists who had been working independently. (One has to wonder where this industry would be today if the great minds of yesteryear had had the benefit of the great communication networks.) Benjamin Franklin traveled to Europe to gather information, and with that trip our industry began to surge forward for the first time. Franklin's famous kite experiment took place about 1752. Coupling the results of that with other scientific data, Franklin decided he could sell installations of lightning protection to every building owner in the city of Philadelphia. He was very successful in doing just that!

We credit Franklin for the terms *conductor* and *nonconductor*, replacing *electric* per se and *nonelectric*.

With the accumulation of knowledge gained from these many scientists and inventors of the past many other great minds emerged, and lent their names to even more discoveries. Their names are familiar as terms related to electricity. Alessandro Volta, James Watt, Andre Marie Ampere, George Ohm, and Heinrich Hertz, are names of just a few of the great minds that enabled our industry to come together.

Figure F-1A Electricity is here! (Courtesy of Underwriters Laboratories)

The next major breakthrough came only a little more than 100 years ago. Thomas Edison, the holder of hundreds of patents promoting DC voltage, and Nikola Tesla, inventor of the three-phase motor and promoter of AC voltage, began a bitter competition in the late 1800s. Edison methodically made inroads with DC current along the eastern seaboard, and Tesla exhausted his finances for an experimental project with AC current in Colorado. Then the city of Chicago requested bids to light the great Columbian Exposition of 1892. Lighting the world's fair in Chicago, celebrating 300 years since Columbus discovered America, required building generators and installing nearly a quarter of a million lights. Edison teamed up with General Electric and submitted a bid based on the Edison light bulb powered by DC current, but George Westinghouse (already successful in the electrified railroad, the elevator, and approximately 400 other inventions) and Nikola Tesla submitted the lowest bid and were awarded the contract. They based their bid on the Westinghouse Stopper Light powered by alternating current. The installation was made by Guarantee Electric of Saint Louis, Missouri, one of the oldest and largest electrical contractors. Thousands of people came out to see this great lighting display. The arcing, sparking, and fires started by the display were almost as spectacular as the lighting with accidents occurring daily. It is said that Edison embellished the dangers of AC current by electrocuting dogs with it while promoting DC current as being much safer, but the public knew better, because many fires and accidents also were occurring with DC current.

About this time three important events took place almost simultaneously. The Chicago Board of Fire Underwriters hired an electrician from Boston, William Henry Merrill, as an electrical inspector for the exposition. Merrill saw the need for safety inspections and started Underwriters Laboratories in 1894. In 1895 the first nationally recommended code was published by the National Board of Fire Underwriters (now the America Insurance Association). The 1895 National Electrical Code was drafted through the combined efforts of architectural, electrical, insurance, and other allied interests. This code was based on a document that resulted from actions taken in 1892 by Underwriters National Electrical Association, which met and consolidated various codes that already existed. The first *National Electrical Code*® was presented to the National Conference on Electrical Rules composed of delegates from var-

Figure F-1B 1893 Great Columbian Exposition at the Chicago World's Fair. The Palace of Electricity astonished crowds—it astonished electricians too, by repeatedly setting fire to itself. (Courtesy of Underwriters Laboratories)

ious national associations who voted to unanimously recommend it to their respective associations for adoption or approval. After attending the Chicago World's Fair a group from the city of Buffalo, New York, hired George Westinghouse and Nikola Tesla to build AC hydrogenerators to be powered by Niagara Falls. Alternating current (AC) had triumphed over direct current (DC) as this country's major power source. The rest is history.

Figure F-1C Distinctive early labels helped consumers and inspection authorities identify UL certified products.

PREFACE

This book replaces the once popular American Iron & Steel Institute (AISI) *Electrical Steel Raceway Design Manual*, which went out of print in the 1970s. The book has been expanded to cover all raceway types both metallic and nonmetallic, other wiring methods such as all of the cable types, and specific design criteria. Relevant *National Electrical Code®* requirements, basic electricity theory, including the application of electrical formulas, are covered in this book. The exercises found in this book require the user to have the electrical knowledge to apply the requirements of the current edition of the *National Electrical Code®* and Ohm's law.

The primary purpose of this book is to provide students who have a background in basic electricity and a minimum of two years' experience as an apprentice electrician with a concise, easily understood study guide. However, the interests of the electrical designer, the consulting engineer, the installing electrician, the electrical inspector, and especially the student, were carefully considered in the text's preparation. It provides a ready source of basic information on the design of most locations for industrial, institutional, commercial, and residential occupancies. This text may be used for reference, as an aid in selecting the appropriate wiring method, and as study material for advanced students and electricians.

This book includes electrical wiring methods and basic electrical design considerations for all locations. This text is written recognizing that the electrical field has many facets and the user may have diverse interests and varying levels of experience. The text is brief and concise for easy application to everyday practical problems and is supplemented with pictures, drawings, examples, and tables. The information contained in this text facilitates efficient, safe, and economical application of the various types of available wiring methods.

For more detailed information on specific phases of the subject, reference should be made to the current edition of the *National Electrical Code®*, engineering textbooks, application handbooks, and/or appropriate manufacturers' catalogs.

All NFPA material is reprinted with permission from NFPA 70-1999, the *National Electrical Code®*, Copyright © 1998, National Fire Protection Association, Quincy Massachusetts 02269. This reprinted material is not the complete and official position of the National Fire Protection Association on the referenced subject, which is represented only by the standard in its entirety.

ACKNOWLEDGMENTS

Allied Tube & Conduit Company

Bussmann Cooper Industries

Carlon, A Lamson & Sessions Company

Raco Inc., subsidiary of Hubbell, Inc.

Robroy Industries, Inc.

American Iron & Steel Institute (AISI)

Greenlee TexTron, Inc.

Square D Company (in particular the technical information provided on pages 26–32).

Southwire Company

Unity Manufacturing

Crouse-Hinds, Division of Cooper Industries

Triangle Wire & Cable Company

B-Line Systems Inc.

Picoma Industries, Inc.

Ocal Inc.

PermaCoat Industries

LTV Steel Tubular Products Company

Wheatland Tube Company

Western Tube

Underwriters Laboratories

Alflex Corporation

AFC/A Monogram Company

M. P. Husky, Inc.

Special thanks to Charles Eldridge of Indianapolis Power and Light for reviewing the final pages and ensuring they are accurate with the 1999 NEC®.

ABOUT THE AUTHOR

Richard E. Loyd is a nationally known author and consultant specializing in the *National Electrical Code®* and the model building codes. He is president of his own firm, R&N Associates Inc., located in Perryville, Arkansas. He and his wife Nancy travel throughout the United States presenting seminars and speaking at industry-related conventions. He also serves as a code expert at 35 to 40 meetings per year at International Association of Electrical Inspectors (IAEI) meetings throughout the country. Loyd represents the STI (The Steel Tube Institute of North America) as a *NEC®* consultant. He represents the American Iron and Steel Institute (AISI) on ANSI/NFPA 70, the *National Electrical Code®*, as a member of Committee Eight (the panel responsible for raceways). Currently he is immediate past chairman of the National Board of Electrical Examiners (NBEE). Loyd is actively involved in forensic inspections and investigations on a consulting basis and serves as an expert witness in matters related to codes and safety. He is member of the Arkansas chapter of Electrical Inspectors. He also is an associate editor of Intertec Publications (*EC&M* magazine).

Loyd served as the chief electrical inspector and administrator for Idaho and for Arkansas. He served as chairman of NFPA 79 "Electrical Standard for Industrial Machinery," as a member of Underwriters Laboratories Advisory Electrical Council, as chairman for Educational Testing Service (ETS) multistate electrical licensing advisory board, and as a master electrician and electrical contractor. He has been accredited to instruct for licensing certification courses in Idaho, Oregon, and Wyoming, and has taught basic electricity and *National Electrical Code®* classes for Boise State University.

Delmar Publishers Is Your Electrical Book Source!

Whether you're a beginning student or a master electrician, Delmar Publishers has the right book for you. Our complete selection of proven best-sellers and all-new titles is designed to bring you the most up-to-date, technically-accurate information available.

NATIONAL ELECTRICAL CODE®

National Electrical Code® 1999/NFPA

Revised every three years, the *National Electrical Code®* is the basis of all U.S. electrical codes.

Order # 0-8776-5432-8

Loose-leaf version in binder

Order # 0-8776-5433-6

National Electrical Code® Handbook 1999/NFPA

This essential resource pulls together all the extra facts, figures, and explanations you need to interpret the 1999 *NEC®*. It includes the entire text of the code, plus expert commentary, real-world examples, diagrams, and illustrations that clarify requirements.

Order # 0-8776-5437-9

Understanding the *National Electrical Code®*, 3E/Holt

This book gives users at every level the ability to understand what the *NEC®* requires, and simplifies this sometimes intimidating and confusing code.

Order # 0-7668-0350-3

Illustrated Guide to the 1999 *NEC®*/Miller

Technically accurate, highly-detailed illustrations offer insight into Code requirements, and are further enhanced through clearly written, concise blocks of text that can be read very quickly and understood with ease. Organized by classes of occupancy.　　Order # 0-7668-0529-8

Illustrated Changes in the 1999 *National Electrical Code®*, 3E/O'Riley

This book provides an abundantly illustrated and easy-to-understand analysis of the changes made to the 1999 *NEC®*.

Order # 0-7668-0763-0

Interpreting the *National Electric Code®*, 5E/Surbrook

This updated resource provides a process for understanding and applying the *National Electrical Code®* to electrical contracting, plan development, and review.

Order # 0-7668-0187-X

Electrical Grounding, 5E/O'Riley

Electrical Grounding is a highly illustrated, systematic approach for understanding grounding principles and their application to the 1999 *NEC®*.

Order # 0-7668-0486-0

ELECTRICAL WIRING

Electrical Raceways and other Wiring Methods, 3E/Loyd

This hands-on resource provides users with a concise, easy-to-understand guide to the specific design criteria and wiring methods and materials required by the 1999 *NEC®*.

Order # 0-7668-0266-3

Electrical Wiring Residential, 13E/Mullin

Users can learn all aspects of residential wiring and how to apply them to the wiring of a typical house from this, the most widely used residential wiring book in the country.

Softcover Order # 0-8273-8607-9

Hardcover Order # 0-8273-8610-9

Housewiring with the *NEC®*/Mullin

The focus of this new book is the application of the *NEC®* to house wiring.

Order # 0-8273-8350-9

Electrical Wiring Commercial, 10E/Mullin and Smith

Users can learn all aspects of commercial wiring from this comprehensive guide to applying the newly revised 1999 *NEC®*.

Order # 0-7668-0179-9

Electrical Wiring Industrial, 10E/Smith

This practical resource has users work their way through an entire industrial building—wiring the branch circuts, feeders, service entrances, and many of the electrical appliances and subsystems found in commercial buildings.

Order # 0-7668-0193-4

Cables and Wiring, 2E/AVO Multi-Amp

This concise, easy-to-use book is your single-source guide to electrical cables—it's a "must-have" reference for journeyman electricians, contractors, inspectors, and designers.

Order # 0-7668-0270-1

ELECTRICAL MACHINES AND CONTROLS

Electric Motor Control, 6E/Alerich and Herman

Fully updated in this new sixth edition, this book has been a longstanding leader in the area of electric motor controls.

Order # 0-8273-8456-4

Introduction to Programmable Logic Controllers/Dunning

This book offers an introduction to programmable logic controllers

Order # 0-8273-7866-1

Industrial Motor Control, 4E/Herman

This newly-revised book, now in full-color, provides easy-to-follow instructions and essential information for controlling industrial motors.

Order # 0-8273-8640-0

Technician's Guide to Programmable Controllers, 3E/Cox

Uses a plain, easy-to-understand approach and covers the basics of programmable controllers.

Order # 0-8273-6238-2

Programmable Controller Circuits/Bertrand

This book is a project manual designed to provide practical laboratory experience for one studying industrial controls.

Order # 0-8273-7066-0

Electronic Variable Speed Drives/Brumbach

Aimed squarely at maintenance and troubleshooting, *Electronic Variable Speed Drives* is the only book devoted exclusively to this topic.

Order # 0-8273-6937-9

Electrical Controls for Machines, 5E/Rexford

State-of-the-art process and machine control devices, circuts, and systems for all types of industries are explained in detail in this comprehensive resource

Order # 0-8273-7644-8

Delmar's Standard Guide to Transformers/Herman

Delmar's Standard Guide to Transformers was developed from the bestseller *Standard Textbook of Electricity* with expanded transformer coverage not found in any other book.

Order # 0-8273-7209-4

Electrical Transformers and Rotating Machines/Herman

This new book is an excellent electrical resource for electrical students and professionals in the electrical trade.

Order # 0-7668-0579-4

DATA AND VOICE COMMUNICATION CABLING AND FIBER OPTICS

Complete Guide to Fiber Optic Cable System Installation/Pearson
This book offers comprehensive, unbiased, state-of-the-art information and procedures for installing fiber optic cable systems.
Order # 0-8273-7318-X

Fiber Optics Technician's Manual/Hayes
Here's an indispensable tool for all technicians and electricians who need to learn about optimal fiber optic design and installation as well as the latest troubleshooting tips and techniques.
Order # 0-8273-7426-7

A Guide for Telecommunications Cable Splicing/Highhouse
A how-to guide for splicing telecommunications cables.
Order # 0-8273-8066-6

Premises Cabling/Sterling
This reference is ideal for electricians, electrical contractors, and inspectors needing specific information on the principles of structured wiring systems.
Order # 0-8273-7244-2

ELECTRICAL THEORY

Delmar's Standard Textbook of Electricity, 2E/Herman
This exciting full-color book is the most comprehensive book on DC/AC circuits and machines for those learning the electrical trades.
Order # 0-8273-8550-1

Industrial Electricity, 6E/Nadon, Gelmine, and Brumbach
This revised, illustrated book offers broad coverage of the basics of electrical theory and is perfect for those who wish to be industrial maintenance technicians.
Order # 0-7668-0101-2

EXAM PREPARATION

Journeyman Electrician's Review, 2E/Holt
This comprehensive exam prep guide includes all of the topics on the journeyman electrician competency exams.
Order # 0-7688-0375-9

Master Electrician's Exam Preparation, 2E/Holt
This comprehensive exam prep guide includes all of the topics on the master electrician's competency exams.
Order # 0-7688-0376-7

REFERENCE

ELECTRICIAN'S TECHNICAL REFERENCE SERIES

This series of technical reference books is written by experts and designed to provide the electrician, electrical contractor, industrial maintenance technician, and other electrical workers with a source of reference information about virtually all of the electrical topics that they encounter.

Electrician's Technical Reference—Motor Control/Carpenter
Electrician's Technical Reference—Motor Controls is a source of comprehensive information on understanding the controls that start, stop, and regulate the speed of motors.
Order # 0-8273-8514-5

Electrician's Technical Reference—Motors/Carpenter
Electrician's Technical Reference—Motors builds an understanding of the operation, theory, and applications of motors.
Order # 0-8273-8513-7

Electrician's Technical Reference—Theory and Calculations/Herman
Electrician's Technical Reference—Theory and Calculations provides detailed examples of problem-solving for different kinds of DC and AC circuits.
Order # 0-8273-7885-8

Electrician's Technical Reference—Transformers/Herman
Electrician's Technical Reference—Transformers focuses on the theoretical and practical aspects of single-phase and 3-phase transformers and transformer connections.
Order # 0-8273-8496-3

Electrician's Technical Reference—Hazardous Locations/Loyd
Electrician's Technical Reference—Hazardous Locations covers electrical wiring methods and basic electrical design considerations for hazardous locations.
Order # 0-8273-8380-0

Electrician's Technical Reference—Wiring Methods/Loyd
Electrician's Technical Reference—Wiring Methods covers electrical wiring methods and basic electrical design considerations for all locations, and shows how to provide efficient, safe, and economical applications of various types of available wiring methods.
Order # 0-8273-8379-7

Electrician's Technical Reference—Industrial Electronics/Herman
Electrician's Technical Reference—Industrial Electronics covers components most used in heavy industry, such as silicon control rectifiers, triacs, and more. It also includes examples of common rectifiers and phase-shifting circuits.
Order # 0-7668-0347-3

RELATED TITLES

Common Sense Conduit Bending and Cable Tray Techniques/Simpson
Now geared especially for students, this newly expanded manual remains the only complete treatment of the topic in the electrical field.
Order # 0-8273-7110-1

Practical Problems in Mathematics for Electricians, 5E/Herman
This book details the mathematics principles needed by electricians.
Order # 0-8273-6708-2

Electrical Estimating/Holt
This book provides a comprehensive look at how to estimate electrical wiring for residential and commercial buildings with extensive discussion of manual versus computer-assisted estimating.
Order # 0-8273-8100-X

Electrical Studies for Trades/Herman
Based on *Delmar's Standard Textbook of Electricity,* this new book provides non-electrical trades students with the basic information they need to understand electrical systems.
Order # 0-8273-7845-9

To request examination copies or a catalog of all of our titles, call or write to:

Delmar Publishers
3 Columbia Circle
P.O. Box 15015
Albany, NY 12212-5015

Phone: 1-800-347-7707   1-518-464-3500   Fax: 1-518-464-0301

Chapter 1

Codes and Standards in Wiring and Building Design

PURPOSE

Safety is paramount in an electrical installation and must be the prime consideration in designing an electrical system. Over the past 100 years various codes and standards have been developed to provide guidelines and regulations to ensure safe installations. These guidelines govern electricity from its development (generation) and transmission, including associated equipment, to the utilization equipment supplied, to the maintenance and repair of the electrical system.

Primary in designing a wiring system, whether it is a raceway or other wiring method, is an understanding of the part played by the codes and standards governing each method. Codes govern what one may or may not do, and standards govern safety and technical aspects of the products used. Standards also are a means of ensuring that a variety of parts needed in an electrical system fit together, even when produced by multiple manufacturers. The electrical code is the most important for our purposes, but there are times when the building codes can take precedence or govern wiring decisions.

In general, nationally utilized codes and standards are developed by the consensus process. Because there can be differing views on how or what

we install, or the acceptable level of safety to be achieved, the consensus method is used, which means the pros and cons are examined and the language becomes what the majority accepts.

Each organization has its own basic rules and regulations, but the primary goal is to assure widespread distribution of the code or standard under development or revision and opportunity for review by users, producers, and regulators. The code or standard may carry a dual designation, for example, ANSI/UL, which means Underwriters Laboratories developed the standard but public review was through the American National Standards Institute.

NATIONAL FIRE PROTECTION ASSOCIATION STANDARDS (NFPA)

The National Fire Protection Association has acted as the sponsor of the *National Electrical Code®* (*NEC®*) as well as many other safety standards.

The most widely used electrical code in the world is the *NEC®*. The official designation for the *National Electrical Code®* is ANSI/NFPA 70.

NATIONAL ELECTRICAL CODE® (NEC®)

The *NEC®* was first developed about 100 years ago by interested industry and governmental authorities to provide a standard for the essentially safe use of electricity. It is now revised and updated every three years. ANSI NFPA 70-1999 is the latest edition of the *National Electrical Code®*. The Code was first developed to provide the regulations necessary for installations that would safeguard persons and property from the hazards of electricity use. Regular updating continues to provide that assurance to the industry today. The *NEC®* is developed by twenty different code-making panels (CMPs) composed entirely of volunteers. These volunteers come from contractors, installers, manufacturers, testing laboratories, inspection agencies, engineering users, government, and electrical utilities.

Suggestions for the content come from an array of sources, including individuals. Anyone can submit a proposal or comment on a submitted proposal. For more specifics on the *NEC®* process you may contact the National Fire Protection Association, Batterymarch Park, Quincy, MA 02269. Request their free booklet, "The NFPA Standards-Making System."

The *National Electrical Code®* is purely advisory as far as NFPA and ANSI are concerned but is offered for use in law and regulations in the interest of life and property protection. The name "*The National Electrical Code®*" might lead one to believe that this document is developed by the federal government. This is not so; the *NEC®* only recognizes use of products in a particular way. It approves nothing. The *NEC®* has no legal standing until it has been adopted by the authority having jurisdiction (see *NEC®* definition, *Article 100* and *Section 90-4*), usually a governmental entity. Therefore, we must first check with the local electrical inspection department to see which edition, if any, of the *NEC®* has been adopted. Compliance with the *NEC®* coupled with proper maintenance should result in an installation essentially free from hazard, but not necessarily efficient, convenient, or adequate for good service or future expansion of electrical use. The *National Electrical Code®* is not an instruction manual for the untrained, nor is it a design specification. However, it does offer design guidelines.

Section 90-2 defines the scope of the *NEC®*. It is intended to cover all electrical conductors and equipment within or on public and private buildings and structures, including mobile homes, recreational vehicles, floating buildings, and premises such as yards, carnival, parking, and other lots, and industrial substations. The Code also applies to installations of conductors and equipment that connect to the supply of electricity, other installations of outside conductors on the premises, and installations of optical fiber cables. The *NEC®* is not intended to cover installations in ships, watercraft, railway rolling stock, aircraft, automotive vehicles, underground mines, and surface mobile mining equipment. It is not intended to cover installations governed by utilities, such as communication equipment, or transmission, generation, and distribution installations whether in buildings or on rights-of-way.

NOTE: For the complete list of exemptions and coverage, see 1999 *NEC®* Section 90-2.

USING THE NEC®

Mandatory rules in the *NEC®* are characterized by the word *shall,* such as "shall be permitted" or "shall not be permitted." Explanatory information and intent are in the form of fine print notes, which are identified as FPN. All tables and footnotes are a part of the mandatory language. Material identified by the superscript letter *x* includes text extracted from other NFPA documents. A complete list of all referenced NFPA documents can be found in the appendix to the *NEC®*. In revisions to the *NEC®*, new text is indicated by a vertical line on the margin; an enlarged black dot "bullet" indicates deleted text.

To use the *NEC®* one first must have a thorough understanding of *Article 90*, "The Introduction"; *Article 100*, "Definitions"; *Article 110*, "Requirements for Electrical Installations"; and *Article 300*, "Wiring Methods." The next step is to become familiar with *Article 230* on services, *Article 240* on overcurrent protection, and *Article 250* covering grounding. The rest of the Code book can be referred to as needed for a particular type of installation.

One portion of the Code can override another,

sometimes without reference. Therefore, it is imperative to look beyond general wiring methods to the specific type of installation. Some examples are *Articles 300-22* (plenums and environmental air spaces), *517* (health care facilities), and *518* (places of assembly).

> **NOTE:** Other codes and standards may take precedence over the *NEC*®. The NFPA 101 Life Safety Code, applicable building codes, and fire codes may be more restrictive than the *NEC*®. All of these require specific types of wiring methods under certain conditions. Choosing and installing the wrong method before checking the Code thoroughly may prove to be costly when an inspector notifies you that the installation has to come out!

OTHER STANDARDS

Other NFPA standards relating to electrical installations and their safety are listed below.

- NFPA 70A Electrical Code for One- and Two-Family Dwellings
- NFPA 70B Electrical Equipment Maintenance
- NFPA 70E Electrical Safety Requirements for Employee Workplaces
- NFPA 79 Electrical Equipment of Industrial Machinery
- NFPA 99 Health Care Facilities
- NFPA 101 Life Safety Code
- NFPA 497 Classification of Flammable Liquids, Gases, or Vapors and of Hazardous (classified) Locations for Electrical Installations in Chemical Process Areas
- NFPA 499 Classification of Combustible Dusts and of Hazardous (classified) Locations for Electrical Installations in Chemical Process Areas
- NFPA 73 Electrical Systems Maintenance
- NFPA 72 National Fire Alarm Code

> **NOTE**: These are only a few examples of the many NFPA standards that are applicable to specific installations. Designers and installers should review the applicable standards before designing or making installations.

AMERICAN NATIONAL STANDARDS INSTITUTE (ANSI)

The American National Standards Institute (ANSI) is an umbrella organization for the consensus code-making process. ANSI has some of its own volunteer committees, which develop standards that are published and sold. In addition, ANSI coordinates the efforts of several other organizations that develop codes and standards for the electrical industry. Examples are the National Fire Protection Association (NFPA), the Institute of Electrical and Electronic Engineers (IEEE), and Underwriters Laboratories (UL). After a standard has gone through the ANSI process and received public review, the standard then becomes an "American National Standard." The technical information found in the purely ANSI standards may be identical or similar to that found in the dual-designated UL, *NEC*®, and IEEE standards, which are identified ANSI/UL. The designations ANSI/NFPA 70 (the *National Electrical Code*®) or ANSI/IEEE/ANSI/NEMA indicate the document has been circulated for comment through the ANSI system.

NEMA and IEC Standards

NEMA is the leading trade association in the United States representing the interests of electro-industry manufacturers. Founded in 1926 and headquartered Rosslyn, VA, near Washington, DC, its 550 member companies manufacture products used in the generation, transmission and distribution, control, and end-use of electricity. The National Electrical Manufacturers Association (NEMA) also develops standards for some products. These standards are processed through the ANSI mechanism. Manufacturers' representatives who belong to various NEMA sections develop NEMA standards. They write from intimate knowledge of the products they produce. These standards are to provide product consistency, information, and safety. Some major NEMA standards deal with switchgear and overcurrent protection devices. The major NEMA raceway standard is RN-1, which covers plastic-coated tubular steel; there is no UL or ANSI counterpart for this product. NEMA also coordinates much of the input to documents developed by IEC, which is the International Electro-Technical Commission. This organization, although founded in the United States, is now based in Europe and is engaged in writing electrical standards for use internationally. The U.S. is only one of many countries providing input to these standards and has only one vote. Little by little, IEC standards are making their way into U.S. usage. However, they may have some modifications and generally are issued in a UL version of the IEC standard. IEC is introduced here to familiarize you with the term if it should show up in a specification or as a label on a piece of equipment. It is suggested that you read articles appearing on IEC and EC92 to gain knowledge of overseas activity that may well affect the U.S. electrical industry in the future.

FEDERAL GOVERNMENT STANDARDS

Until the late 1970s the federal government had its own standards for most products. In the interest of economy there was a move toward adopting industry standards and canceling the special government specifications that not only took much time to prepare but created excessive purchase costs, because special manufacturing was unnecessarily required. Because many designer specifications still contain outdated references to federal specifications, the old number will be referenced in this book for correlation. For the most part ANSI and UL standards have now been adopted by the federal government for the wiring systems discussed herein.

LOCAL CODES AND REQUIREMENTS

Although most municipalities, countries, and states adopt the *National Electrical Code®*, they may not have adopted the latest edition, and in some cases the authority having jurisdiction (AHJ) may be several editions of the *NEC®* behind. A majority of the jurisdictions amend the *NEC®* or add local requirements. These may be based on environmental conditions, fire safety concerns, or other local experience. An example of one quite common local amendment is that all commercial buildings be wired in raceways. The *NEC®* generally does not differentiate wiring methods in residential, commercial, or industrial installations; however, many local jurisdictions do. Requiring metal raceway wiring, especially in

fire zones, is another common amendment. Some major cities have developed their own electrical codes (e.g., Los Angeles, New York, Chicago, and several other metropolitan areas). In addition to the *National Electrical Code®* and all local amendments, the designer and installer must also comply with the local electrical utility rules. Most utilities have specific requirements for installing the service to the structure. Many unhappy designers and installers have learned about special jurisdictional requirements after making the installation, thus incurring costly corrections at their own expense.

INSURANCE REQUIREMENTS

Many insurance companies have requirements that affect the insurability and rates of the structure. Those requirements should be determined prior to completing the design stage. The degree of risk to the insurer will be the determining factor. It may be possible to make design changes that will greatly improve the insurability and rates. In some cases the wiring methods used and the electrical wiring design may affect such improvement.

APPROVED TESTING LABORATORIES

Underwriters Laboratories (UL) has long been the major product testing laboratory in the electrical industry. In addition to testing, UL is a long-time developer of product standards. By producing to these standards and contracting for follow-up service after undergoing a listing procedure, manufacturers are authorized to apply the UL label or to mark their product. In the past decade other electrical testing laboratories have come on the scene and have become recognized officially. Underwriters Laboratories is no longer the only testing laboratory evaluating electrical products. Some jurisdictions evaluate and approve laboratories, others accept them by reputation. OSHA evaluates and approves testing laboratories. It is the responsibility of the entity specifying the materials to verify that the product has been evaluated by a testing laboratory acceptable to the authority having jurisdiction (AHJ) where the installation is being made. The installer is responsible for assuring that the product is installed according to the manufacturer's instructions and listing (*Section 110-3(b)*).

Figure 1-1 Well-known testing laboratories trademarks.

On most products you will find a Listing Mark or Label. Some AHJ require listed products in all installations. Some of the better-known testing agencies are Metropolitan Electrical Testing (MET), ETL Testing Laboratories (ETL), Applied Engineering, and Factory Mutual (FM). Examples of these are shown in Figure 1-1. Much of the testing by these companies is performed to UL, ANSI, or ASTM standards. Most have not ventured into standards writing, with the exception of FM.

NEC® Section 110-3(b) requires that products be installed in accordance with the manufacturers' instructions, which are part of the listing, and in accordance with the *NEC®*. The instructions may or may not be included with the product. The testing laboratory evaluates products according to a standard designed to ensure that all manufacturers of a

specific item meet a minimum requirement based on expected use in accordance with the *NEC®*. Many electrical system failures, accidents, and fires originate from products that were installed incorrectly or were not the correct product for the application—a violation of *Section 110-3(b)*. The UL "Green Book" (*Electrical Constructions Material Directory*) or the UL "White Book" (*General Information for Electrical Construction, Hazardous Location, and Electric Heating and Air Conditioning Equipment*) provide valuable information that can help avoid the misapplication of materials. These are the only documents that provide information on limitations and requirements for a product listing and should be a part of every designer and installer's library. These and other UL directories and product standards may be obtained from

Underwriters Laboratories, Inc.
333 Pfingsten Road
Northbrook, IL 60062

OCCUPATIONAL SAFETY AND HEALTH ACT (OSHA)

The Occupational Safety and Health Administration (OSHA) began developing and issuing stan-

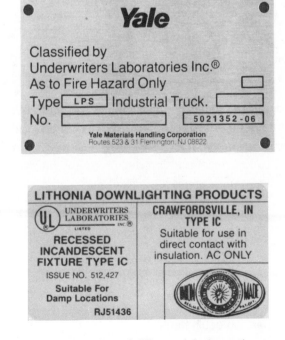

Figure 1-2 Examples of different labels as they appear on equipment.

IDENTIFICATION OF UL LISTED PRODUCTS THAT ARE ALSO UL CLASSIFIED IN ACCORDANCE WITH INTERNATIONAL PUBLICATIONS

Underwriters Laboratories Inc. (UL) provides a service for the Classification of products that not only meet the appropriate requirements of UL but also have been determined to meet appropriate requirements of the applicable international publication(s). For those products which comply with both the UL requirements and those of an international publication(s), the traditional UL Listing Mark and a UL Classification Marking, as described below, may appear on the product as a combination Listing and Classification Marking.

The combination of the UL Listing Mark and Classification Marking may appear as authorized by Underwriters Laboratories Inc.

LISTED
(Product Identity)
(Control Number)

ALSO CLASSIFIED BY UNDERWRITERS LABORATORIES INC.® IN ACCORDANCE WITH IEC PUBLICATION _____

LISTED
(Product Identity)
(Control Number)

ALSO CLASSIFIED BY UNDERWRITERS LABORATORIES INC.® IN ACCORDANCE WITH IEC PUBLICATION _____

UND. LAB. INC. ®   LISTED
(Product Identity)
(Control Number)

ALSO CLASSIFIED BY UNDERWRITERS LABORATORIES INC.® IN ACCORDANCE WITH IEC PUBLICATION _____

Underwriters   LISTED
Lab. Inc. ®   (Product Identity)
(Control Number)

ALSO CLASSIFIED BY UNDERWRITERS LABORATORIES INC.® IN ACCORDANCE WITH IEC PUBLICATION _____

(Product Identity)
CLASSIFIED BY
UNDERWRITERS LABORATORIES INC.®
IN ACCORDANCE WITH
IEC PUBLICATION _____
(Control Number)

OR

(Product Identity)
CLASSIFIED BY
UNDERWRITERS LABORATORIES INC.®
IN ACCORDANCE WITH
IEC PUBLICATION _____
(Control Number)

LOOK FOR THE CLASSIFICATION MARKING
_____

(Product Identity)
(Control Number)

ALSO CLASSIFIED BY UNDERWRITERS LABORATORIES INC.® IN ACCORDANCE WITH IEC PUBLICATION _____

(Product Identity)
(Control Number)

ALSO CLASSIFIED BY UNDERWRITERS LABORATORIES INC.® IN ACCORDANCE WITH IEC PUBLICATION _____

UNDERWRITERS LABORATORIES INC. ®
LISTED
(Product Identity)
(Control Number)

ALSO CLASSIFIED BY UNDERWRITERS LABORATORIES INC.® IN ACCORDANCE WITH IEC PUBLICATION _____

LOOK FOR THE COMBINATION LISTING MARK AND CLASSIFICATION MARKING
_____

IDENTIFICATION OF PRODUCTS CLASSIFIED TO INTERNATIONAL PUBLICATIONS ONLY

Underwriters Laboratories Inc. (UL) provides a service for the classification of products that have been determined to meet the appropriate requirements of the applicable international publication(s). For those products which comply with the requirements of an international publication(s), the Classification Marking may appear in various forms as authorized by Underwriters Laboratories Inc. Typical forms which may be authorized are shown below, and include one of the forms illustrated, the word "Classified", and a control number assigned by UL. The product name as indicated in this Directory under each of the product categories is generally included as part of the Classification Marking text, but may be omitted when in UL's opinion, the use of the name is superfluous and the Classification Marking is directly and permanently applied to the product by stamping, molding, ink-stamping, silk screening or similar processes.

Separable Classification Markings (not part of a name plate and in the form of decals, stickers or labels) will always include the four elements: UL's name and/or symbol, the word "Classified", the product category name, and a control number.

The complete Classification Marking will appear on the smallest unit container in which the product is packaged when the product is of such a size that the complete Classification Marking cannot be applied to the product or when the product size, shape, material or surface texture makes it impossible to apply any legible marking to the product. When the complete Classification Marking cannot be applied to the product, no reference to Underwriters Laboratories Inc. on the product is permitted.

Figure 1-2 *(continued)*   Examples of different labels as they appear on equipment.

dards in 1970 intended to protect employees from hazards in the workplace. In an attempt to eliminate electrical hazards in the workplace, OSHA has continued to develop consensus standards to address specific safety concerns associated with design, installation, maintenance, and repair of electrical systems and components. General safety standards have been developed that cover construction sites and the maintenance and repair of electrically powered machines and equipment. Without qualification, every electrical system, new or existing, must satisfy these requirements, which are retroactive with no time limitations. Any violations constitute OSHA violations and are cause for citations. Electrical systems that are in violation of the OSHA law must be altered or retrofitted to produce full compliance with all OSHA rules. Electrical installations made after March 15, 1972, were the first to come under this OSHA Act of 1970. They were required to comply with the rules in Sections 1910.302 through 1910.308, except for 11 specific exceptions. By April 16, 1981, all the existing electrical installations were required to comply with the rules of the Sections 1910.302 through 1910.308 without exception; every one of the rules was required. Today the new design and safety standards for electrical systems are found in Subpart S, Part 1910 of Title 29 of the Code of Federal Regulations. The entire document is divided into sections and designed numerically as follows:

- 1910.302—Electrical Utilization Systems
- 1910.303—General Requirements
- 1910.304—Wiring Design and Protection
- 1910.305—Wiring Methods, Components, and Equipment for General Use
- 1910.306—Specific Purpose Equipment and Installations
- 1910.307—Hazardous (Classified) Locations
- 1910.308—Special Systems
- 1910.399—Definitions Applicable to This Subpart

The OSHA requirements require all electric products, conductors, and equipment to be approved (listed or labeled by an approved testing laboratory) or to be replaced with products that satisfy the OSHA definition of acceptability. All products must be installed in accordance with the manufacturer's instructions in the listing or labeling for installations requiring the use of approved equipment. The rule clearly states that all products listed in the UL *Electrical Construction Materials Directory* or in other listing directories must be used exactly as described in the application data listed in the book. OSHA is evaluating and approving testing laboratories for the first time. Consult the *Federal Register* to determine which testing laboratories meet OSHA guidelines.

REVIEW QUESTIONS

1. How often is the *National Electrical Code*® revised and updated?

2. Who is the authority having jurisdiction?

3. Name an important thing to determine before designing or installing an electrical system.

4. Name four types of installations *not* covered by the *NEC*®.

5. How is explanatory information characterized in the *NEC*®?

6. What section of the *NEC*® requires equipment to be installed in accordance with the manufacturer's instructions?

7. When does the *NEC*® become a legal document?

8. Are all installations by utilities exempt from the *NEC*®? Explain.

9. What does a vertical line on the margin of the *NEC*® indicate?

10. Is the latest amended edition of the *NEC*® used in all jurisdictions? When was the latest edition of the *NEC*® published?

11. Who invented the three-phase motor?

12. Who set up the first meeting that resulted in the development of the *National Electrical Code*®? When and where was it held?

13. Which organization is engaged in writing electrical standards for use internationally?

14. Who is responsible for enforcing the OSHA Act of 1970 on construction sites? Noncompliance will result in:

Chapter 2

Basic Design Factors

Several factors must be considered together to achieve a well-designed electrical system: safety, capacity, voltage regulation, accessibility, and flexibility. These factors are applicable to the complete distribution or utilization system, or part thereof. The ultimate goal is to provide an efficient electrical system to meet initial as well as future requirements, and to lend itself to easy maintenance and economical alterations.

Before drawing the plans a few things must be known to design a system that will provide adequacy (*NEC® Section 90-1(b)*). The *NEC®* states it will provide a safe and free-from-hazard installation, but it may not provide all the convenience we would like nor provisions for future expansion.

SAFETY

The *NEC®* provides a foundation upon which a safe electrical installation can be designed and installed. The input of a knowledgeable designing engineer is essential to ensure that the electrical system incorporates safeguards for the specific conditions of the location, occupancy, and use; the possibilities of future expansion; and adequate initial and future power capabilities. NFPA 70E, *NEC® Article 305* and OSHA offer guidelines and requirements for safe working conditions and safe temporary electrical systems during the installation (construction) stage.

CAPACITY

It is not uncommon for an electrical system to provide continuous service for 50 years or more. The designed capacity of a new installation must receive comprehensive consideration and often requires close coordination with the serving utility, structure owner, and designer. Oversizing a new system provides good future expansion capabilities at a relatively low cost. However if it is not closely coordinated with all interested parties, it may not be usable when needed.

Special attention should be given to the electrical service's anticipated needs as well as future contingencies. Particular consideration should be given to air conditioning and heating requirements, elevators, data processing and computers, indoor and outdoor lighting, and processing equipment. Attention also should be given to future automation and state-of-the-art upgrading in industrial, commercial, institutional, and residential installations. It is recommended that service equipment, distribution equipment (switchboards and panelboards), feeders, raceways, conductors, junction boxes, and cabinets be specified at least one size larger than the calculated electrical load requires. Branch-circuit panelboards should contain no less than 10% and preferably 25% spare circuit breaker or fuse spaces for expansion capabilities. Working space (physical room) required by *NEC® Sections 110-26* and *110-34* also should provide additional room for future expansion. Spare raceway capacity simplifies rewiring and is a significant factor in minimizing the cost of such work.

9

FLEXIBILITY

A good electrical system provides flexibility. The distribution system should accommodate future physical changes in plant or office layout as production lines are changed, equipment is shifted, or offices are rearranged. The increased popularity of modular office layouts demands an electrical system flexible enough to be rearranged quickly and economically. Raceway types offer the versatility and flexibility to meet today's constantly changing world (see FPN following the definition of "Raceway," *NEC® Article 100*, for a list of raceways).

Steel tubular conduits (*NEC® Article 346*, IMC Article 345, and EMT Article 348) offer almost unlimited expansion and are adaptable to all types of building construction and floor plans. In addition, cellular and underfloor raceways offer creative circuit and outlet arrangement. Surface raceways, busways, and wireways provide additional flexibility in the design of industrial, institutional, and commercial facilities.

Flexibility can be maintained only if adequate additional space is provided for panelboards and switchboards (see dedicated space requirements, *NEC® Section 110-26(f)*). Additional space should be provided for risers and for spare service and feeder conduits serving these areas when run concealed (underfloor, buried in concrete, or in shafts that are either inaccessible or where future installation would be labor-intensive). The importance of flexibility cannot be overemphasized.

ELECTRICAL SYSTEMS AND BUILDING FIRES

The prime objectives of the *National Electrical Code®* are to provide safety from fire and electrical shock. Many of the Code's requirements are based on the premise that an electrical system designed and installed in accordance with the *National Electrical Code®* must not start a fire. In addition, an electrical installation must not contribute to or allow the spread of fire should one originate from other causes. The possible contribution of electrical systems to building fires is undergoing ever more scrutiny as these systems become more extensive and as more uses are

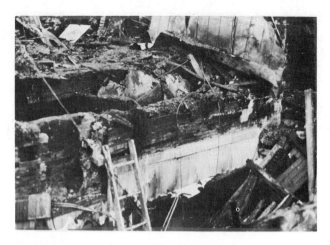

Figure 2-1   Example of fire damage caused by electrical system failure. (Courtesy of Allied Tube & Conduit)

proposed for electrical wiring methods manufactured from a growing variety of materials other than steel.

FIRE RESISTANCE AND SPREAD OF FIRE AND SMOKE

NFPA 101, Fire and Life Safety Code, the three major model building codes, plus a variety of local, state, and city building codes include provisions intended to provide constructions that not only will withstand the effects of fire in a portion of a building, but also will reduce the likelihood of fire spreading within the building. For extremely large or tall structures, building codes usually require fire-resistant construction and noncombustible construction materials. For fire-resistant buildings, building codes require columns, beams, girders, walls, and floor constructions to have a resistance to fire measured by a standard fire test procedure (ASTM E119—Standard Method of Fire Test for Building Construction and Materials).

The fire resistance of a structural assembly is determined from endpoint criteria included in this fire test standard. For load-bearing members such as columns, beams, and girders, maintenance of the load-bearing capability during and after fire exposure is a prime criterion. The unexposed surfaces of wall and floor assemblies must be able to withstand a significant rise in temperature to prevent ignition of combustibles and the spread of fire to the unexposed side of such wall or floor assemblies. It is extremely important in fire-resistant construction that all building components adhere to the approved design of fire-rated construction. Furthermore, during construction every effort should be made to avoid damage to the fire-resistant integrity of the construction by removing protective coatings or coverings or by piercing through entire rated assemblies.

VOLTAGE SELECTION

The choice of voltage selection may depend upon the voltage and capacity that are available from the serving utility at the building site location. This includes the size of the building or facility; nature of the occupancy; and type, size, and character of the electrical loads. Economy and efficiency can be achieved by selecting the voltage most applicable to the electrical loads to be served.

Transmission voltage generally is considered to be the voltage from the generating station to the area substation or in some large industrial and institutional facilities the transmission system. It may feed the facility directly to the customer-owned substations. These systems are above 34.5kV.

Distribution voltage generally is considered as the voltage distributed from one part of the city to another, or from the substations to the premises serving the individual structure service, utilization transformers or customer-owned substations. In some large installations, industrial and high-rise commercial facilities, distribution voltages may deliver current beyond the service to load centers in the building. These systems usually range up to 69kV.

Utilization voltage includes the voltages utilized within a facility to supply lighting, general purpose outlets, and utilization equipment (i.e., motors, machinery, and HVAC). Voltages of 600 volts or less are commonly referred to as utilization voltages. However, voltages through 35kV are covered by the *NEC®* and can be considered utilization voltages.

DISTRIBUTION

Distribution voltage is the nominal voltage used by the serving utility to supply power from the substation to utilization transformers, which step down the voltage to a utilization voltage used to supply individual services. Distribution voltage circuits supply some large facilities that own and maintain their own switching and substations or unit substations. The facility takes power from the serving utility at the "Service point" (*NEC® Definition Article 100* and *Service Conductors Article 230, Part B*), which may be a pole top switch or splice box. Primary metering for these facilities is accomplished with CTs (current transformers). The user is required to locate the service disconnect in accordance with *NEC® Sections 230-200,*

Figure 2-2 Electrical switchboard (Courtesy of Square D Company)

230-205. The distribution circuit then supplies unit substations where the voltage is stepped down to the utilization voltage level and may also supply large motors or other large loads. Distribution voltage is commonly used in large industrial facilities, high-rise buildings, shopping malls, and health care facilities.

Five basic types of distribution systems are:

1. Simple radial-single primary, single transformer, radial secondaries.
2. Primary selective-dual primary, single transformer, radial secondaries.
3. Secondary selective-dual primary, dual transformers, secondary tie circuits.
4. Secondary network-dual primary, secondary loop ties all substations.
5. Looped primary sectionalized primary loop feeds all unit substations.

The choice among these systems depends on the degree of reliability and flexibility demanded by the plant process and considerations of cost. The radial system provides a high degree of reliability and service continuity, but the load requirements may justify a secondary selective system, even though the cost is more. The advantages of all system arrangements must be carefully weighed against load requirements and cost limitations. There is no hard-and-fast rule; each job must be considered individually. Voltages once used for outdoor distribution only are now commonplace in modern interior wiring designs. Lower installation and operating costs, along with favorable use experience, have established higher distribution voltages as acceptable for electrical systems in large-area, one-story industrial plants and in multistory, high-rise office and commercial buildings where long feeders are needed.

The range of voltages from over 600 volts nominal to about 69,000 volts is identified as high or medium voltage when used in buildings. Voltages of higher than 69,000 volts are rarely specified for use in buildings except for special purposes. *The National Electrical Code®* gives no maximum voltage limitations. Service-entrance conductors exceeding 600 volts must be installed in accordance with *Article 230, part H,* with acceptable wiring methods in accordance with *Section 230-202(b).* Where the voltage exceeds 15,000 volts, service-entrance conduc-

tors must enter either metal-enclosed switchgear or a transformer vault conforming to *NEC® Article 450, Sections 450-41* through *450-48. NEC® Section 110-34* prohibits the location of pipes and ducts foreign to the electrical installation in the vicinity of service equipment, metal-enclosed power switchgear, or industrial control equipment, except if provided for fire protection of the electrical equipment. The general requirements for equipment over 600 volts must also comply with *NEC® Article 490 as well as other applicable NEC®* articles.

NOTE: The *National Electrical Code®* defines High, Medium, and Low voltages differently from the rest of the electrical industry. A number of proposals were made to the 1990 *NEC®* based on the work of an NFPA ad-hoc subcommittee that would have made the definitions consistent. (The subcommittee recommended terms consistent with ANSI C84.1, 1982, and IEEE 449: Low voltage = 0 to 1000 volts, Medium voltage = 1001 to 100,000 volts, High voltage = 100,001 to 230,000 volts.) These proposals were rejected and the inconsistency remains.

The *NEC®* itself has less than consistent definitions for these terms (except for high voltage, which is defined in *Section 490-2* as more than 600 volts nominal). *Article 250* "Grounding" calls everything over 1000 volts high voltage; medium voltage is only referenced in *Article 326* "Medium Voltage Cable," as over 2001 volts. Generally, the *NEC®* references nominal voltages and defines them in *Article 100.* Many *NEC®* Articles contain a special part for circuits and equipment over 600 volts. *Article 720* covers circuits and equipment operating at fewer than 50 volts.

Caution: when using the *NEC®* for reference verify the nominal voltage and be sure it is covered in the referenced *NEC®* section.

In the high- or medium-voltage range, the most common voltages are 2400, 4160, 6900, and 13,800 volts (13.8kv) because these provide greater growth possibilities, are more flexible, and for loads exceeding 20,000 kVA are usually more economical than lower voltages.

For loads between 10,000 and 20,000 kVA, 13.8-KV is recommended because of its growth capability and flexibility, although a 13.8-KV system usually is no more economical than a 4160-volt sys-

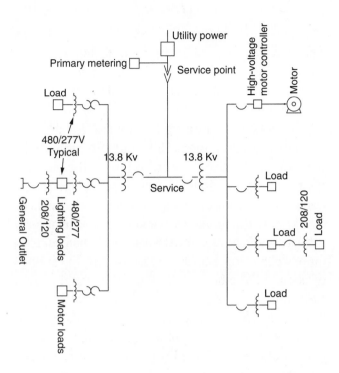

Figure 2-3a Industrial service high voltage offers advantages, but requires one additional level of transformation to usable voltages.

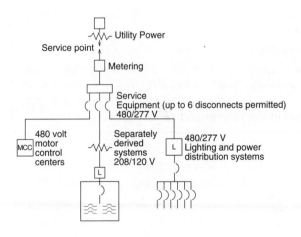

Figure 2-3b Industrial Low-voltage Power System. Distribution sytems, separately derived systems and motor controllers may be located throughout the facility. All must have overcurrent protection and comply with the *NEC®*, but the number of each and location is not limited as to the service.

tem for loads in this range. For loads up to 10,000 kVA and for larger motors, 4160-volt systems should be used. For lighting, 480/277 volts should be used, and 208/120-volt systems should be used for receptacles and appliances.

A comparison between low-voltage and high- or medium-voltage distribution methods is shown in Figure 2-3. It is becoming increasingly common to extend distribution voltages beyond the "Service Point" (defined in *NEC® Article 100*) for customer-owned distribution and utilization. This portion of the electrical system is within the scope of the *National Electrical Code®*. However, the *NEC®* has very limited guidelines for these installations, and it is recommended that the designer as well as the installer also use the ANSI C-2 National Electrical Safety Code (NESC) and the IEEE Green Book for reference to ensure well-designed systems.

UTILIZATION

Utilization voltage is the voltages utilized within a facility to supply lighting, general purpose outlets, and utilization equipment. Voltages of fewer than 600 volts are commonly referred to as utilization voltages and include the voltage ranges referenced in the *National Electrical Code®*. Voltages above 600 volts are generally considered distribution voltages, even though some equipment and motors operate at more than 600 volts. They are generally used in large industrial facilities. Utilization voltages are generally used for branch-circuit supplying equipment, motors, appliances, general-purpose branch circuit, and the lighting loads of a facility.

The most common voltages are 240/120 volt, single phase for one- and two-family dwellings and some multifamily dwellings. The 120-volt single phase is used to serve branch circuits and appliance loads within the facility, and the 240-volt single phase supplies equipment such as ranges, water heaters, and air conditioning within that facility. For other than residential use, such as commercial, office, and industrial facilities, 480-volt, 3-phase line-to-line loads are used to supply the equipment. Equipment such as process equipment, heating, and air conditioning loads are served by the 480-volt, 3-phase system. The 277-volt-to-ground system is used to supply the lighting loads within that building. Generally, facilities supplied by 480/277-volt systems utilize separately derived dry-type transformers to reduce the 480/277-volt, 3-phase system to a 208/120-volt system for supplying general-purpose branch circuits, other in-coincidental loads and some incandescent lighting.

NEC®Section 210-6 covers the requirements for branch-circuit voltage limitations. *NEC®Section 210-6(c)* limits circuits exceeding 120 volts, nominal, between conductors and not exceeding 277 volts, nominal, to ground. These circuits are permitted to supply listed electric-discharge lighting fixtures equipped with medium-base, mogul-base, and other than screw-shell-type lampholders within their voltage ratings. Auxiliary equipment in these circuits can be electric-discharge lamps and cord- and plug-connected and permanently connected utilization equipment.

NEC®Section 210-6(d) 600 Volts Between Conductors. Circuits exceeding 277 volts, nominal, to ground and not exceeding 600 volts, nominal, between conductors are permitted to supply the auxiliary equipment of electric-discharge lamps mounted in permanently installed fixtures on poles at least 22 feet high or similar structures for illumination of outdoor areas, such as highways, roads, bridges, athletic fields, or parking lots, and at least 18 feet on other structures, such as tunnels, cord-and-plug, and permanently connected utilization equipment, with exceptions.

NEC®Section 210-6(b) states that circuits of 120 volts between conductors are permitted to supply the terminals of screw-shell lampholders, medium-base, and other types within their voltage ratings, auxiliary equipment of electric-discharge lamps, and cord-and-plug connected and permanently connected utilization equipment.

NEC®Section 210-6(a) limits the voltage to 120 volts between conductors that supply terminals of lighting fixtures or cord-and-plug connected loads equal to or less than 1440 volt-amperes, nominal, or less than 1/4 horsepower in dwelling units and guest rooms of hotels, motels, and similar occupancies. Motor loads also affect the choice of voltage. For example, a 1-horsepower motor has a full-load current rating for various phase and voltage conditions, as follows:

Motor Type	115V	200V	208V	230V	460V
Single-Phase 1HP	16 Amps	9.2 Amps	8.8 Amps	8 Amps	-
3-Phase 1HP	7.2 Amps	4.1 Amps	4.0 Amps	3.6 Amps	1.8 Amps

The voltages listed are rated motor voltages. The currents listed shall be permitted for system voltage ranges of 110 to 220, 220 to 240, 440 to 480, and 550 to 600 volts.

Motors and controls generally are less costly in the 480-volt rating than at the 240-volt rating. Furthermore, for a given power load, overall conductor losses are less. A 480/277-volt system is more economical than a lower voltage system for demands of 500 kVA (Kilovolt-ampere) and more. For example, a 500-kVA load at 208/120-volt 3-phase would require a conductor ampacity (current-carrying ability) of about 1400 amperes. At 480/277 volts, the same load would require an ampacity of only 600 amperes. The economy in conductor size from the use of higher voltages and 3-phase equipment is obvious and should be considered where load characteristics warrant. All motors rated up to 250 HP, except motors for small single-phase loads and appliances, should be specified for operation at 480 volts.

Larger motors generally are connected to circuits in the distribution voltage range above 600 volts. Utilization voltages being used today are illustrated in Figure 2-4 a–e.

120/240-volt, single-phase, 3-wire (Figure 2-4a) This system is used in residential and small commercial occupancies to provide 120 volts for lighting and small appliance loads and 240-volt single-phase for heavy appliances and small motor loads. The third wire gives this system 100% additional capacity over a 120-volt, 2-wire system, which has been largely superseded in new work.

208/120-volt, 3-phase, 4-wire (Figure 2-4b) This system generally is used for small industrial plants, office buildings, stores, and schools. It provides 120-volt single-phase for lighting and appliances and a 208-volt, 3-phase for moderate power loads. The fourth wire gives this system almost 50% more capacity than a single-phase, 3-wire system.

240/120-volt, 3-phase, 4-wire delta (Figure 2-4c) This system provides single-phase or 3-phase, 240-volt supply for power between phase wires and 120-volt single-phase for lighting between ungrounded phases and the grounded neutral wires. These systems are generally found in commercial buildings and small industrial facilities where motor loads are large compared with lighting loads. This would include the air-conditioning requirements for the facility.

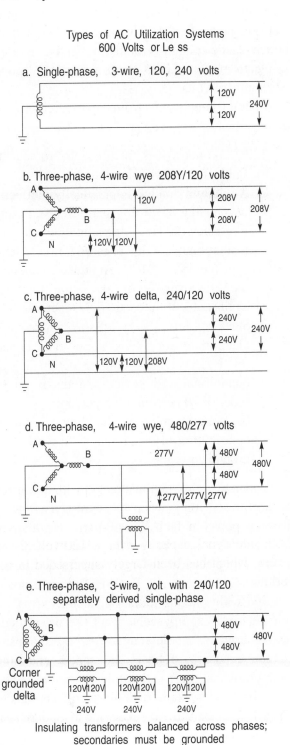

Types of AC Utilization Systems
600 Volts or Le ss

a. Single-phase,   3-wire, 120, 240 volts

b. Three-phase, 4-wire wye 208Y/120 volts

c. Three-phase, 4-wire delta, 240/120 volts

d. Three-phase,   4-wire wye, 480/277 volts

e. Three-phase,   3-wire, volt with 240/120 separately derived single-phase

Corner grounded delta

Insulating transformers balanced across phases; secondaries must be grounded

Figure 2-4 Types of 600-volt or less utilization systems (Courtesy of American Iron & Steel Institute)

The advantage of this system is that if one of the single-phase supply transformers fails, the system may operate at approximately 50% capacity as an open delta system. The disadvantage is that the total load is not balanced over the three phases.

NEC® Sections 384-3(e), 215-8, and *230-56* require that the high-leg, or the leg that is 208 volts to ground, be marked in orange or by other effective means. This high-leg also must be located as the B phase or center phase in panelboards, switchboards, and other equipment within the facility. The exception is equipment within a single section or multi-section switchboard or panelboard; the meter on a 3-phase, 4-wire delta system is permitted to have the same phase configuration as the metering in equipment.

480/277-volt, 3-phase, 4-wire (Figure 2-4d) This system, for circuits in large industrial plants and large commercial buildings, is becoming more common in smaller facilities. It provides 480-volt, 3-phase for motor loads and 277-volt, single-phase for fluorescent lighting systems. Conductor economy is one big advantage of this system. A given amount of electric power can be delivered at 480/277 volts with less than half the conductor metal required at 208/120 volts.

240/480-volt, 3-phase, 3-wire delta (Figure 2-4e) This system provides 3-phase power for 240-volt or 480-volt motors. It is suitable for industrial application with heavy motor loads at voltages listed. Single-phase or 3-phase transformers can be used to secure 120/240-volt single-phase, three-wire or 120/240-volt, 3-phase, 4-wire. Utilization voltages are used for lighting, appliances, and miscellaneous small motor circuits.

Some 600-volt systems may be found in old plants where 550-volt motors are in use. This voltage is not recommended for new installations.

ELECTROMAGNETIC FIELDS THEORY (EMI-EMF)

Electromagnetic fields (EMF) are invisible lines of force that surround the movement of electricity. They are everywhere. They surround power lines, electrical appliances, even the wiring in your house—everywhere that electricity is used. The presence of EMF has been known since the beginning of commercially supplied electricity more than 100 years ago, but there was no scientific interest in its possible health effects until the early 1960's. The communications and electronics industries have been aware of Electromagnetic Interference (EMI) problems from electromagnetic fields for a much longer period of time, and today the concern is growing rapidly as more and more solid-state control devices are being used for common power circuit control. The necessity for shielding from electromagnetic interference (EMI) has become a reality. The question is, how can this effectively be accomplished? Shielding EMI at high frequencies is relatively easy. We can use aluminum foil to shield the conductors and special paints are available for enclosures; there are many ways to effectively shield these circuits. However, 60-hertz power is another matter. Testing by the School of Electrical and Computer Engineering, Georgia Institute of Technology has provided data that indicate aluminum conduit to be practically ineffective and nonconductive plastic raceways to be totally ineffective in reducing electromagnetic field levels in conduit-encased power distribution circuits. The tests clearly show that to eliminate nuisance crosstalk between pwer circuits and control and communication circuits, all must be enclosed in steel raceways or separated by the required distances indicated in this study.

Summary of Georgia Tech Studies

A computer model of conduit-encased power distribution systems has been developed. The model is capable of predicting the level of Electromagnetic Field density at any point around the system. The modeling capability includes electrical metallic tubing (EMT), intermediate metal conduit (IMC), galvanized rigid conduit (GRC), aluminum conduit (ARC) and nonconductive (for example PVC) conduit.

The computer model has been validated with laboratory tests. The laboratory tests consisted of short runs of conduit-encased circuits which were energized with various levels of electric currents. The laboratory tests confirm the model predictions. The results of this work show that galvanized rigid conduit (GRC) is most effective in reducing electromagnetic field levels for encased power distribution

circuits. PVC-coated galvanized rigid conduit (GRC) is equally effective as noncoated GRC of the same steel thickness. IMC and EMT also provide a substantial reduction in the electromagnetic field levels for encased power distribution circuits. The results also indicate that aluminum conduit is practically ineffective in reducing low frequency electromagnetic fields. PVC or any other nonconductive plastic conduit is totally ineffective in reducing electromagnetic field levels for encased power distribution circuits.

Specific conclusions from the research program are:

1. PVC conduit does not affect electromagnetic fields from power circuits.
2. Aluminum conduit is practically ineffective in reducing electromagnetic fields at power frequency (60 Hz). Magnetic field reduction in aluminum conduit encased power systems is on the order of 10%. At higher frequencies the effectiveness of aluminum to shield against electromagnetic fields increases.
3. Steel conduit is very effective in reducing electromagnetic fields at power frequency (60 Hz). Magnetic field reduction in steel conduit-encased power systems is on the order of 70 percent to 95 percent.
4. No conduit provides shielding against the contribution to electromagnetic fields from ground currents.

Magnetic fields vary in strength with the current and are present only when the current is flowing. Electrical power in the U.S. is 60-hertz (Hz) alternating current; that is, it changes direction 120 times per second. The electrical and magnetic fields alternate with the same frequency. EMF-EMI should not be confused with ionizing radiation, such as x-rays, or with microwaves. In recent years the news media have focused on studies suggesting a relationship between exposure to EMF and cancer and other health problems. Electrical utilities are interested because power lines, along with household appliances, are EMF sources. Virtually everyone in modern society is exposed.

Many experts now say that people get more EMF exposure from house wiring, electrical appliances, and distribution lines (which run near homes

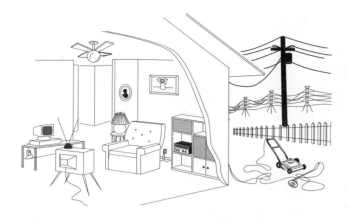

Figure 2-5 We are constantly exposed to EMF by home and utility distribution and transmission lines.

and residential streets) than from the high-voltage transmission lines, but the question of who is exposed to how much EMF or for how long is still under investigation. Assessing the effects of exposure is an important area of research. It is unclear whether EMF exposure is harmful. Research results are highly complex, inconsistent, and inconclusive. There are known biological effects of EMF exposure, but it is uncertain if they are harmful. Most responsible scientists are concerned but not alarmed. The lack of clear, consistent answers demonstrates that more research is in order. It may take a few more years to resolve the EMF question.

Research is being conducted around the world. In the United States research is being conducted by the Department of Energy and by the Navy, and some individual state utilities conduct research. New York has made a large contribution through the New York State Power Lines Project. Significant research has been done in western Europe, Canada and Japan. There are no generally accepted recommendations of action to reduce exposure to electrical and magnetic fields. Some have proposed easy steps to avoid exposure—don't use an electric blanket, move the electric clock away from the bedside. Others recommend more extreme measures, including redesign of electrical products and systems.

At the present time the jury is still out and it is too early to say whether any particular electrical or magnetic fields are harmful. Hopefully in the future, when many research projects are completed and their conclusions accumulated, recommendations will be forthcoming to the public which will resolve these

questions and provide guidence for future electrical installations to assure the safety the public deserves. In the meantime prudent avoidance has been recommended. In instances where there is concern, some power companies have turned to underground lines installed in steel raceways with successful results

Modeling and Evaluation of Conduit Systems

For information on these studies and software regarding "Harmonics and Electromagnetic Fields", contact:

Steel Tube Institute of North America
8500 Station Street, Suite 270
Mentor, Ohio 44060
Phone: (216) 974-6990, Fax: (216) 974-6994
e-mail: sti@apk.net

or

A. P. Sakis Meliopoulos, E. Glytsis or Chien-Hsing Lee at the School of Electrical and Computer Engineering, Georgia Institute of Technology, Atlanta, Georgia 30332

HARMONICS EFFECTS ON THE ELECTRICAL SYSTEM

Harmonics are a phenomenon as old as AC power; however, they have not been a problem until the advent of SCRs, personal computers, data processing equipment, switch mode power supplies, and the like. (Refer to the three parts of Figure 2-6 in this discussion.)

Certain types of nonlinear loads can generate harmonics within a power distribution system. Harmonics cause abnormal heating within transformers, neutral conductors, induction motors, and so on. Harmonics are currents or voltages with frequencies that are integer multiples of the fundamental power frequency. Each has a name associated with the multiplying number (e.g., if the fundamental frequency is 60 Hz, the second harmonic is 120 Hz, the third harmonic is 180 Hz, and so on). Harmonics can be present in voltage or current or both. They occur whenever the wave shape is distorted (i.e., when the wave shape varies from a pure sine wave).

Electric utilities typically generate a voltage that is nearly a sine wave. If the end user connects a linear load such as a resistive heater, the resulting current is a sine wave, and no harmonics are present. If, however, the load is nonlinear and draws short pulses of current within each cycle, the current wave shape is distorted (nonsinusoidal), and harmonic currents flow. In a typical case, each harmonic will have a different amplitude depending on the type of distortion, but in general, the greater the distortion, the larger the harmonic currents. The total current is a combination of the fundamental plus each of the harmonics. Some examples of common equipment that produce harmonic currents include variable frequency drives which control motor speeds, solid-state heating controls, certain types of fluorescent lighting, electronic and medical test equipment, and electronic office machines such as personal computers, printers, and data processing equipment. Particularly worrisome loads are personal computers, owing to their large number, and variable speed motors, because of their occasional large size.

Voltage harmonics are generated on line voltages and are caused by harmonic currents acting in an Ohm's law relationship with the power source impedance (i.e., E = I × Z). For example, a 10-A harmonic current drawn from a source impedance of 0.1 ohm generates a harmonic voltage of 1.0 volt. Line-voltage harmonics can radiate interference in telephone and communication systems. In addition, harmonics on line voltage will generate harmonic

HARMONICS

Harmonic	Frequency (Hz)
1	60
2	120
3	180
5	300
7	420
9	540
11	660

Figure 2-6b

currents in linear loads such as induction motors and power capacitors. Induction motors are particularly vulnerable in higher frequency voltage harmonics for two reasons: (1) the eddy current losses within the motor are proportional to the frequency squared, and (2) certain harmonics, notably the fifth, are negative sequence (backward rotating). Good practice rules for motors dictate that the total harmonic distortion on the line voltage should be less than 5 percent. In accordance with Ohm's law, line-voltage harmonics can be reduced by filters to reduce the harmonic currents or lowering the power source impedance by installing larger transformers and larger, shorter conductors, as well as better connectors. Induction motor installation should be planned so that motor branch circuits do not share loads generating harmonic currents.

Harmonic currents are generated by nonlinear loads that flow through components of the electrical distribution system, including circuit breakers, conductors, bus bars, panels, transformers, and generators. In most cases, these electrical system components were designed to supply utilization equipment calculated and based on 60 Hz. When sizable harmonic currents are present, problems are likely. Transformers don't like harmonics for two reasons. First, the core losses caused by eddy currents and hysteresis tend to increase as the frequency goes up. Copper losses due to skin effect also increase with the frequency. Additionally, harmonics cause certain delta-wye transformer configurations to experience a circulating current in the delta winding. An example is the 208/120-volt transformer that commonly feeds receptacles in commercial buildings. In this case, single-phase, nonlinear loads connected to the receptacles can produce triplen harmonics, which algebraically add in the neutral.

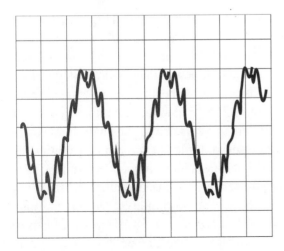

Figure 2-6a   "Dirty power" as seen on an oscilloscope.

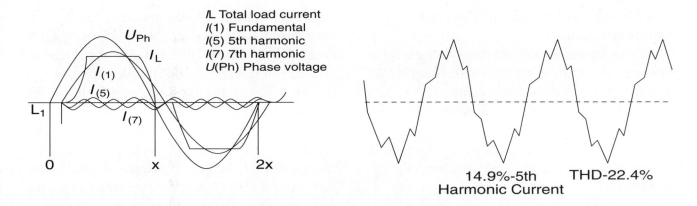

The supplied phase voltage, fundamental load current, 5th & 7th harmonics and the *combined* load current is shown at top left for a typical nonlinear load. Power factor is shown in this figure and the distorted current wave form does not include the damaging effect of the 3rd, 9th, etc., harmonics.

The effects of nonlinear loads on the current waveform may be much worse than the drawing at top right.

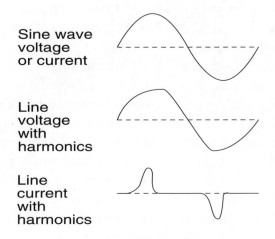

Figure 2-6c

When this neutral current reaches the transformer, it is reflected into the delta primary, where it circulates. The result is that the transformer heating sometimes causes failures.

Electric Panels

Electric panels are designed to carry 60-Hz currents, which then become mechanically resonant to the magnetic fields generated by higher-frequency harmonic currents. Under these conditions, the panel can vibrate and emit an audible buzzing sound at the harmonic frequency.

Circuit Breakers

If the trip mechanism is not designed to respond properly to the higher frequency elements of harmonic currents, the breaker may trip prematurely at low current or fail to trip at the rated trip current. A breaker rated to respond to true rms, or heating value, of the current will have a better chance of protecting against harmonic current loads.

Conductors In 3-phase Systems

Certain harmonic currents can overload neutral conductors. Troublesome harmonics called *triplens* consist of third and odd multiples of the third (i.e., third, ninth, fifteenth.)

These harmonics will add rather than cancel in the neutral of a 3-phase, 4-wire system. Normally the neutral carries only the unbalanced part of the phase currents, but when the triplens are present, it is theoretically possible to have neutral current at the triplen frequency, that is, 1.73 times greater than the phase current, even if the phase currents are perfectly balanced.

Conductors in Single-phase Loads

Personal computers are part of a large class of electronic equipment that employs diode-capacitor power supplies. These power supplies convert the ac line voltage to low voltage dc. The conversion process involves charging large capacitors, each line cycle with narrow pulses of current that is time-coincident, with peaks of line voltage. This process generates odd harmonics, which are mostly third and fifth, with lesser amounts in the seventh and ninth.

In variable speed motor drives, the frequency converters are built in two types, 6-pulse and 12-pulse units. They are named according to the pulses produced for each cycle of the output variable frequency. The primary harmonics produced are equivalent to pulse number plus or minus 1 (i.e., a 6-pulse converter generates fifth and seventh harmonics, and a 12-pulse unit generates eleventh and thirteenth).

The *National Electrical Code®* does not deal effectively or realistically with the harmonics problems of today. Good engineering practice does dictate that an evaluation be made of the anticipated inclusion of equipment that will introduce these harmonics into the system, and appropriate compensation should be made in the transformers and the neutral conductor to provide a safe and efficient electrical system. The *National Electrical Code®* references harmonics (non-linear loads) only in *Articles 220, Sections 220-22* and *Article 310* and *Section 400-5 Note 10(c)* of the *Ampacity Tables of 0-2000 volts*. On a 4-wire, 3-phase wye circuit where the major portion of the load consists of electric-discharge lighting, data processing, or similar equipment, harmonic currents are present in the neutral conductor, and the neutral is considered a current-carrying conductor. This is the only reference and

requires only that a full-size neutral be employed in these conditions. However, good engineering practices tell us that a larger than full-size neutral may be necessary. It is recommended that the electrical system be designed and sized to compensate for the effects of harmonics.

Using type "K"-rated transformers or oversizing the transformer supplying the system are both recommended and effective methods of handling the harmonic problem. It does not cure the problem, but an oversized or "K"-rated transformer will allow a near-normal life expectancy for that equipment.

The following definitions should help one understand this complex problem, which plagues the electrical industry today. It has been estimated that by the year 2000, over 60% of all electrical circuitry will be electronically controlled. This problem is not diminishing, nor will it be solved easily.

Eddy currents. The circulating currents that occur in the core of a transformer are known as eddy currents.

Harmonics. A multiple of the original sine wave frequency, for example, 3rd harmonics, on a 60-Hz system, would be a sine wave at 180 Hz.

Harmonics (zero sequence) filter. A piece of equipment that has a core and coil assembly and is wound to pass harmonics to ground but to block 60-Hz power. Its purpose is to drain most of the harmonics from the system to protect the power transformer and to prevent replacement of the feeder neutral.

Hysteresis losses. Losses in the core of a transformer, owing to the resistance of the steel to the reversing magnetic field.

I^2R losses. Losses due to resistance in a wire (i.e., the coils of a transformer).

K-factor. A rating given to a power transformer for its ability to handle harmonics, that is, a K-13 transformer (see the various definitions of K-factors).

K-1 transformer. A conventional transformer. This transformer does not actually have a K rating, but a K-1 transformer is not constructed to handle any harmonic currents. No transformers are actually rated with a K-1 rating.

K-4 transformer. A transformer designed to handle all the load that a K-1 transformer will handle plus 16% of the fundamental as 3rd harmonic currents, 10% as 5th, 7% as 7th, 5.5% as 9th, and so on, with smaller percentages as the harmonics increase in frequency.

K-13 transformer. This rating handles twice the harmonic loading of a K-4.

K-9, K-20, K-30, etc., transformers. The higher the rating number, the greater the harmonic loading the transformer will handle without overheating.

Linear loads. A load current that follows the sinusoidal wave form of the applied voltage. A poor power factor causes the current to lag behind the voltage but does not distort the current waveform. Examples are motors (without variable frequency drives), incandescent lights, and resistant electric heat.

Nonlinear load. Any load current that does not follow the sinusoidal wave form of the applied voltage. Examples are personal computers, electronic data processing equipment, electronic ballasts, and variable frequency drives.

Nonsinusoidal waveform. A waveform distorted by equipment so as not to follow a sinusoidal shape (see sinusoidal waveform). Normally the current waveform is being referred to when discussing nonsinusoidal waveforms.

Odd harmonics. Any of the odd multiples of the fundamental frequency (i.e., 180 Hz, 300 Hz, 420 Hz, 540 Hz, 660 Hz, etc.).

SCR. Silicon controlled rectifier.

Sinusoidal waveform. A waveform that follows the graphic representation of $y = \sin x$. This is the waveform that is generated by power companies across the nation.

Skin effect. A phenomenon that causes current to flow more on the surface of the conductor as the frequency increases. Skin effect has a negligible effect at 60 Hz, but at the higher frequencies of the harmonics it increases the temperature of transformer windings.

Switch-mode power supply. A power supply that develops a 60-Hz waveform from a 6-pulse (sometimes 12- or 24-pulse) high-frequency switching device, the output of which is filtered. The results are savings in transformer size and waste heat and cost savings to the consumer.

THD. Total harmonic distortion, normally expressed as a percentage.

UPS. Uninterruptable power supply. UPS systems are normally served through switch-mode power supplies and are major contributors to the harmonics problem within a facility.

HARMONICS, THE PROBLEM

Studies indicate that neutral currents may greatly exceed phase currents. Neutral currents are caused by nonlinear loads, which affect the phase currents as well. Transformers have been overheated and destroyed even when a neutral has not been employed. As capacitance has been installed, the capacitors have been destroyed owing to the effect of harmonic currents.

REVIEW QUESTIONS

1. List the three most common utilization voltages systems used in residential and commercial buildings.

2. List two locations where distribution-level voltages are commonly utilized as utilization voltages:

3. Are flexibility and expansion capabilities requirements of the *National Electrical Code*®? If yes or no, which *NEC*® section applies?

4. Which federal regulation and *National Electrical Code*® article regulates temporary electrical installations?

5. List four factors for designers to consider when designing electrical systems.

6. Requirements regulating the amount of clear working space around electrical equipment are found in which article and sections in the *National Electrical Code*®?

7. Which *National Electrical Code*® section regulates the sealing around electrical penetrations through fire resistance rate wall, floors, and ceilings?

8. What are harmonics? Are they a new problem to electrical systems?

9. What are the disadvantages of using a delta system to supply a facility? Are there special marking requirements for the conductors?

Chapter 3

Conductor Considerations

A number of facts are considered when determining the amount of current a conductor can carry safely and economically. Conductor size is basic for determining conductor ampacity, or current-carrying capacity, as are conductor material and type of insulation. (See *Article 100*, "Definitions.") These three facts are used in the NEC® for making the determination. *NEC® Section 310-2* states that conductors shall be insulated, except where bare or covered conductors are permitted elsewhere in the *NEC®*. Conductor material, when referred to in the *NEC®* shall be of aluminum, copper-clad aluminum, or copper unless otherwise specified. Where installed in raceways, conductors of size No. 8 and larger shall be stranded. The minimum size of conductors is listed in *Table 310-5* of the *NEC®*. Conductor material and size and the insulation are three of the determining factors in developing the 0 to 2000-volt tables. These tables are *310-16* through *310-19*.

Table 310-16 covers the ampacity of insulated conductors rated 0-2000 volts, 60° to 90° C, with not more than three conductors in a raceway or cable or in the earth (directly buried), based on an ambient temperature of 30° C (86° F). *Table 310-17* covers the ampacity of single-insulated conductors rated 0-2000 volts in free air based on an ambient air temperature of 30° C (86° F). *Table 310-18* covers the ampacity of three single-insulated conductors rated

0-2000 volts, 150° to 250° C, in a raceway or cable based on an ambient air temperature of 40° C. *Table 310-19* covers the ampacities of single-insulated conductors rated 0-2000 volts, 150° to 250° C in free air based on ambient air temperature of 40° C. In each case, the other determining factors are the ambient temperature in which the conductors are installed. *Sections 310-15(a)* through *310-15(c)* and *Tables 310-15(b)(2)* and *310-15(b)(6)* list other determining factors, which may limit conductor ampacity to values less than those given in these tables. Other factors to consider are the number of conductors in a raceway, whether the load is continuous or intermittent, the physical length of the circuit, and the ambient air temperature where the conductor is located. All these factors are discussed in these sections and must be considered. *Tables 310-61* through *310-84* cover the allowable ampacities for medium- or high-voltage cables up to 35,000 volts.

AMBIENT TEMPERATURE CONSIDERATIONS

NEC® Section 310-10 states that no conductor shall be used in such a manner that its operating temperature will exceed that designated for the type of insulated conductor involved, and in no case shall conductors be associated together in such a way,

Product Data

THW

Thermoplastic Insulated, Heat and Moisture Resistant.
600 volt. Triple E® Aluminum Alloy, Aluminum Alloy
1350 (EC), Copper.

Product Data

XHHW

Crosslinked Polyethylene, High-Heat and Moisture Resistant.
600 volt. Triple E® Aluminum Alloy, Aluminum Alloy
1350 (EC), Copper.

Product Data

THHN

Thermoplastic Insulated, Nylon Sheathed. Heat, Moisture, Oil and
Gasoline Resistant. 600 volt. Copper.
Also rated THWN and MTW in all sizes, 105°C AWM in sizes 14-6 AWG.

Figure 3-1   Common conductor types (Courtesy of Southwire Company)

Ampacities

For ambient temperature other than 30°C (86°F), multiply the ampacities shown above by the appropriate factor shown below.

Ambient Temp. °C	Temperature Rating of Conductor		
	60°C (140°F)	75°C (167°F)	90°C (194°F)
	TYPES †TW, †UF	TYPES †HHW, †THW, †THWN, †XHHW, †USE	TYPES †RHH, †THHN, †XHHW*, †THWN-2, †XHRW-2, †RHW-2, †USE-2
21–25	1.08	1.05	1.04
26–30	1.00	1.00	1.00
31–35	.91	.94	.96
36–40	.82	.88	.91
41–45	.71	.82	.87
46–50	.58	.75	.82
51–55	.41	.67	.76
56–60	....	.58	.71
61–70	....	.33	.58
71–80	....	....	.41

†Unless otherwise specifically permitted in the 1999 *National Electrical Code®* the overcurrent protection for conductor types marked with an obelisk (†) shall not exceed 15 amperes for 14 AWG, 20 amperes for 12 AWG, and 30 amperes for 10 AWG copper; after any correction factors for ambient temperature and number of conductors have been applied.

*For dry and damp locations only. See 75° column for wet locations.

Figure 3-2   (Courtesy of Triangle Wire & Cable Inc.)

with respect to the type of circuit, the wiring method employed, or the number of conductors, that the limiting temperature of any conductor is exceeded. A principal determinant of operating temperature is the ambient temperature, which may vary along the conductor length from time to time and from location to location. The heat generated within the conductor as a result of the load (current flow), the rate at which the generated heat is able to dissipate into the ambient medium, the thermal insulation that covers or surrounds the conductors will all affect the rate of heat dissipation, as will the nature of the soil in the case of direct buried cables, the material in which the conductors are run—steel or nonmetallic conduits—and the adjacent load-carrying conductors.

Adjacent conductors have a dual effect in raising the ambient temperature and impeding heat dissipation. References throughout the *National Electrical Code®* and manufacturers' product literature must be considered when selecting conductors to use on an installation. Several types of conductors are needed for most installations to fulfill the requirements stated. Ampacity correction factors for the carrying capacity of conductors are listed at the bottom of each wiring table in *Article 310. Tables 310-16* through *310-19* include ambient temperature correction factors (see Figure 3-2). When installing conductors of different types in the same raceway, junction box, cabinet box, or cutout box, the temperature rating of the conductors must be considered. More than one calculated or tabulated ampacity could apply. For the conductors within a single enclosure, the lowest value shall be used. For instance, when a Type TW is within the same raceway or enclosure as a Type THWN, the ampacity for both conductors should be calculated based on the TW ampacity column in *NEC® Table 310-16*.

INSULATION

Conductor insulation, as defined in the *NEC®*, *Article 100—Definitions*, consists of a conductor encased in a material of composition and thickness recognized by the Code as electrical insulation. This is not to be confused with "covered," as also defined in *Article 100*. A covered conductor is a conductor encased in a material of composition or thickness not recognized by the Code as electrical insulation. Conductor construction and applications are covered in *Section 310-13* of the Code, and *Table 310-13* covers the application and insulations by wire type. Insulated conductors must comply with the applicable provisions of one or more of these tables: *310-13, 310-61, 310-62, 310-63, or 310-64*. Thermal plastic insulations may stiffen at temperatures colder than -10° C (+14° F) and require care during installation. Thermal plastic insulation may also be deformed at normal temperatures when subjected to pressure, so care must be exercised during installation and at points of support.

Thermal plastic insulation used on DC circuits in wet locations may result in electroendosmosis between the conductor and the insulation. Insulated conductors in wet locations shall be of the lead-covered type, or of types RHW, THW, TW, THHW, THWN, and XHHW, or of one of the types listed for that purpose to be used in wet locations. Some insulations are suitable for dry locations only, but modern technology has permitted wiring manufacturers to produce multi-rated wires, so you may find building wire that is rated THHN/THWN/MTW. These types are permitted in both wet and dry locations. There are conductor insulation types suitable for wet locations, such as submersion in water and corrosive atmospheres. For a complete list of the ratings and characteristics of insulation of conductors, refer to *NEC® Table 310-13*.

As the ambient temperature rises, the allowable ampacity of the conductor diminishes, but at a given temperature a more heat-resistant insulation allows the conductor to be rated closer to its base ampacity. At a room temperature of 60° C (140° F) a Type-TW conductor is not permitted. A THW-insulated conductor is permitted at 58% of its rated ampacity, and a Type-XHHW conductor is permitted at 71% of its basic ampacity at this same temperature. The advantage of using heat-resistant types of insulation is obvious where heat is excessive.

Nevertheless, wires at high-temperature insulations should not be used to obtain high ampacity as a general rule because of possible excessive voltage drop, and because the equipment terminals, and such equipment as circuit breakers and panelboards, may not be suitable for the higher operating temperatures. When calculating allowable ampacity of conductors many things must be considered—*circuit length, equipment, temperature rating, terminal temperature rating*, and *insulation of the conductor used*. (See *Section 110-14C* for temperature limitations.) *NEC® Section 310-15(a)(2)* states that when more than one calculated or tabulated ampacity could apply for a given circuit length, the lowest value shall be used. Additional formula application information can be found in the 1999 *NEC® Appendix B*. However, *Appendix B* is not a part of the requirements and is included for informational purposes only. Therefore, the authority having jurisdiction should be consulted before applying the values calculated using *NEC® Appendix B*.

The effects of temperature associated with the ampacity of conductors has caused a great deal of confusion within the electrical industry for the past several *NEC®* cycles. The 1984 *NEC®* added a requirement for 90° conductors in types NM and NMC cables. Ampacity tables based on the J. H. Neher-M. H. McGrath methods were adopted and inserted into the 1987 *National Electrical Code®*. These methods proved to be difficult and confusing for the average design engineer, installing contractor, and electrician. In the 1990 *NEC®* these tables were removed. However, engineering data were placed in the appendix to the 1990 Code as examples of the J. H. Neher-M. H. McGrath calculation methods for heat and the dissipation of heat within any given raceway or cable system. The 1993 Code has added as a new section, *Section 110-14*, Underwriters Laboratories data to clarify the termination limitations.

Wire Temperature Termination Requirements

Wire temperature ratings and temperature termination requirements for equipment can and do result in rejected installations. Information about this topic can be found in testing agency directories, product testing agency directories, product testing standards, and manufacturers' literature, but most do not consult these sources until it is too late.

Why Are Temperature Ratings Important?

Conductors carry a specific temperature rating based on the type of insulation employed on the conductor. Common insulation types can be found in *Table 310-13* of the *NEC®*, and corresponding ampacities can be found in *Table 310-16*.

Example: A 1/0 copper conductor ampacity based on different conductor insulation types:

Insulation Type	Temperature Rating	Ampacity
TW	60°C	125 amperes
THW	75°C	150 amperes
THHN	90°C	170 amperes

Although the wire size has not changed (1/0 Cu) the ampacity *has* changed due to the temperature rating of the insulation on the conductor. Higher-rated insulation allows a smaller conductor to be used at the same ampacity as a larger conductor with lower-rated insulation, and, as a result, the amount of copper and even the number of conduit runs needed for the job may be reduced.

One common misapplication of conductor temperature ratings occurs when the rating of the equipment termination is ignored. Conductors must be sized by considering where they will terminate and how that termination is rated. If a termination is rated for 75°C, this means the temperature at that termination may rise to 75°C when the equipment is loaded to its ampacity. If 60°C insulated conductors were employed in this example, the additional heat at the connection above the 60°C conductor insulation rating could result in failure of the conductor insulation.

When a conductor is selected to carry a specific load, the user/installer or designer must know termination ratings for the equipment involved in the circuit.

Example: Using a circuit breaker with 75°C terminations and a 150-ampere load, if a THHN (90°C) conductor is selected for the job, from *Table 310-16* select a conductor that will carry the 150 amperes. Although Type THHN has a 90°C ampacity rating, the ampacity from the 75°C column must be selected because the circuit breaker termination is rated at 75°C. Looking at the table, a 1/0 copper conductor is acceptable. The installation would be as shown in Figure 3-3, with proper heat dissipation at the termination and along the conductor length. If the temperature rating of the termination had not been considered, a No. 1 AWG conductor based on the 90°C ampacity may have been selected, which may have led to overheating at the termination or premature opening of the over-current device because of the smaller conductor size. (See Figure 3-4.)

In the example above, a conductor with a 75°C insulation type (THW, RHW, USE, and so forth) also would be acceptable because the termination is rated at 75°C. A 60°C insulation type (TW or perhaps UF) is not acceptable because the temperature at the termination could rise to a value greater than the insulation rating.

General Rules for Application

When applying equipment with conductor terminations, the following two basic rules apply:

- Rule 1 [*NEC® Section 110-14(c)(1)*]—Termination provisions for equipment rated 100A or less or equipment that is marked for No. 14 to No. 1 AWG conductors are rated for use with conductors rated 60°C. (See Figure 3-5.)
- Rule 2 [*NEC® Section 110-14(c)(2)*]—Termination provisions for equipment rated greater than 100A or equipment terminations marked for conductors larger than No. 1 AWG are for use with conductors rated 75°C. (See Figure 3-6.)

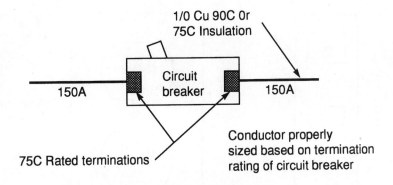

Figure 3-3

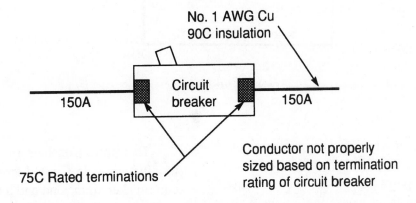

Figure 3-4

There are exceptions to the rules. Two important exceptions are:

- Exception 1—Conductors with higher-temperature insulation can be terminated on lower-temperature rated terminations provided the ampacity of the conductor is based on the lower rating. This is illustrated in the last example, where the THHN (90°C) conductor ampacity is based on the 75°C rating to terminate in a 75°C termination. The table in the next column provides a quick reference of how this exception would apply to common terminations.

- Exception 2—The termination can be rated for a value higher than the value permitted in the general rules if the equipment is listed and marked for the higher temperature rating.

Example: A 30-ampere safety switch could have 75°C rated terminations if the equipment was listed and identified for use at this rating.

Conductor Insulation Versus Equipment Termination Ratings

Termination Rating	60°C	Conductor Insulation Rating 75°C	90°C
60°	OK	OK (at 60°C ampacity)	OK (at 60°C ampacity)
75°C	No	OK	OK (at 75°C ampacity)
60/75°C	OK	OK (at 60°C or 75°C amapcity)	OK (at 60°C or 75°C ampacity)
90°C	No	No	OK*

*The equipment must have a 90°C rating to terminate 90°C wire at its 90°C ampacity.

A Word of Caution

When terminations are inside equipment, such as panelboards, motor control centers, switchboards, enclosed circuit breakers, and safety switches, it is important to note that the temperature rating identified on the equipment labeling should be followed—not the rating of the lug. It is common to use 90°C

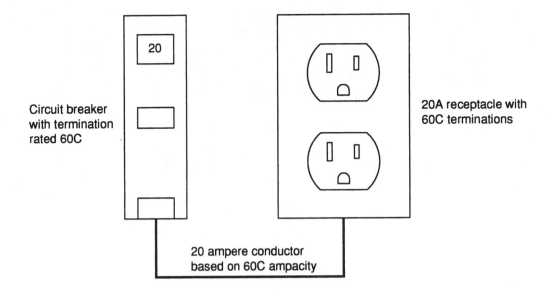

Figure 3-5

rated lugs (i.e., marked AL9CU), but the equipment rating may be only 60°C or 75°C. The use of the 90°C rated lugs in this type of equipment does not give permission for the installer to use 90°C wire at the 90°C ampacity.

The labeling of all devices and equipment should be reviewed for installation guidelines and possible restrictions.

Equipment Terminations Available Today

Remember, a conductor has two ends, and the termination on each end must be considered when applying the sizing rules.

> **Example:** Consider a conductor that will terminate in a 75°C-rated termination on a circuit breaker at one end and a 60°C-rated termination on a receptacle at the other end. This circuit must be wired with a conductor that has an insulation rating of at least 75°C (due to the circuit breaker) and sized based on the ampacity of 60°C (due to the receptacle).

In electrical equipment, terminations are typically rated at 60°C, 75°C, or 60/75°C. There is no listed distribution or utilization equipment that is listed and identified for the use of 90°C wire at its 90°C ampacity. This includes distribution equip-

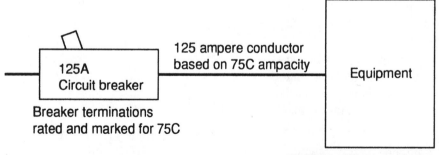

Figure 3-6

ment, wiring devices, transformers motor control devices, and utilization equipment such as HVAC, motors, and light fixtures. Installers and designers who do not know this fact have been faced with jobs that do not comply with the *National Electrical Code®*, and jobs have been turned down by the electrical inspectors.

> **Example:** A 90°C wire can be used at its 90°C ampacity (see Figure 3-7). Note that the No. 2 90°C rated conductor does not terminate directly in the distribution equipment but in a terminal or tap box with 90°C-rated terminations.

Frequently, manufacturers are asked when distribution equipment will be available with terminations that will permit 90°C conductors at the 90°C ampacity. The answer is complex and requires not only significant equipment redesign (to handle the additional heat) but also coordination of the downstream equipment where the other end of the conductor will terminate. Significant changes in the product testing/listing standards also would have to occur.

A final note about equipment is that, generally, equipment requiring the conductors to be terminated in the equipment has an insulation rating of 90°C but has an ampacity based on 75°C or 60°C. This type of equipment might include 100% rated circuit breakers, fluorescent lighting fixtures, and so on, and will include a marking to indicate such a requirement. Check with the equipment manufacturer to see if an special considerations need to be considered.

Higher-Rated Conductors and Derating Factors

One advantage to conductors with higher insulation ratings is noted when derating factors are applied. Derating factors can be required because of the number of conductors in a conduit, higher ambient temperatures, or possibly internal design requirements for a facility. By beginning the derating process at the conductor ampacity based on the higher insulation value, upsizing the conductors to compensate for the derating may not be required.

For the following example of this derating process the two following positions must be considered:

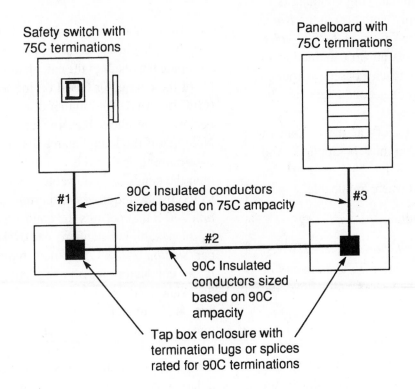

Safety switch with 75C terminations

Panelboard with 75C terminations

#1

#3

90C Insulated conductors sized based on 75C ampacity

#2

90C Insulated conductors sized based on 90C ampacity

Tap box enclosure with termination lugs or splices rated for 90C terminations

Figure 3-7

1. The ampacity value determined after applying the derating factors must be equal to or less than the ampacity of the conductor, based on the temperature limitations at its terminations.
2. The derated ampacity becomes the allowable ampacity of the conductor, and the conductor must be protected against overcurrent in accordance with this allowable ampacity.

Example for the derating process. Assume a 480Y/277 VAC, 3-phase, 4-wire feeder circuit to a panelboard supplying 200 amperes of fluorescent lighting load, and assume that the connectors will be in a 40°C ambient temperature. Also, assume the conductors originate and terminate in equipment with 75°C terminations.

1. Because the phase and neutral conductors will be in the same conduit, the issue of conduit fill must be considered. Note 10(c) in the *Notes To The Ampacity Tables of 0-2000 Volts* states that the neutral must be considered to be a current-carrying conductor because it is supplying electric discharge lighting.
2. With four current-carrying conductors in the raceway, apply Note 8(a) in the *Notes To The Ampacity Tables of 0–2000 Volts*. This note requires an 80% reduction in the conductor ampacity based on four to six current-carrying conductors in the raceway.
3. The correction factors at the bottom of *Table 310-16* also must be applied. And adjustment of .88 for 75°C and .91 for 90°C is required where applicable.

Now the calculations:

Using a 75°C conductor such as THWN:
300 kcmil copper has a 75°C ampacity of 285 amperes. Using the factors from above, the calculations are:

$$285 \times .80 \times .88 = 201 \text{ amperes}$$

Thus, 201 amperes is now allowable ampacity of the 300 kcmil copper conductor for this circuit. If the derating factors for conduit fill and ambient temperatures had not been required, a 3/0 copper conductor would have met these requirements.

Using a 90°C conductor such as THHN:
250 kcmil copper has a 90°C ampacity of 290 amperes. Using the factors from above, the calculations are:

$$290 \times .8 \times .91 = 211 \text{ amperes}$$

211 amperes is less than the 75°C ampacity of a 250 kcmil copper conductor (255 amperes), so the 211 amperes would now be the allowable ampacity of the 250 kcmil conductor. If the calculation resulted in a number larger than the 75°C ampacity, the actual 75°C ampacity would have been required to be used as the allowable ampacity of the conductor. This is critical because the terminations are rated at 75°C.

Note: The primary advantage to using 90°C conductors is exemplified by this example. The conductor is permitted to be reduced by one size (300 kcmil to 250 kcmil) and still accommodate all required derating factors for the circuit.

In summary, when using 90°C wire for derating purposes, begin by derating at the 90°C ampacity. Compare the result of the calculation with the ampacity of the conductor based on the termination rating (60°C or 75°C). The smaller of the two numbers than becomes the allowable ampacity of the conductor. Note that if the load dictates the size of the required over-current device (i.e., continuous load × 125%), then the conductor allowable ampacity must be protected by the required overcurrent device. (See *Section 240-3* that permits the conductor to be protected by the next higher standard size overcurrent device as per *Section 240-6*.) This may require moving to a larger conductor size and beginning the derating. In the example, if the 200 amperes of load were continuous, a 250 ampere overcurrent device would be required. The 201 or 211 amperes from the calculations would not be protected by the 250 ampere overcurrent device and, as such, we would be required to move to a larger conductor meeting these requirements to begin the derating process.

Summary

Several factors affect how the allowable ampacity of a conductor is determined. The key is not to treat the wire as a system but as a component of the total electrical system. The terminations, equipment ratings, and environment affect the ampacity assigned to the conductor. If the designer and the installer remember each rule, the installation will go smoother.

Section 110-14 is especially important in that the requirements for all electrical connections must be made in the proper manner. The 1993 Code has added text clarifications and has identified the termination requirements to meet the listing laboratories recommendations. *Section 110-16* reminds us that we must

---

NEC® 110-14(c) Temperature Limitations

(c) Temperature Limitations. The temperature rating associated with the ampacity of a conductor shall be selected and coordinated so as not to exceed the lowest temperature rating of any connected termination, conductor, or device. Conductors with temperature ratings higher than specified for terminations shall be permitted to be used for ampacity adjustment, correction, or both.

(1) Termination provisions of equipment for circuits rated 100 amperes or less, or marked for Nos. 14 through 1 conductors, shall be used only for one of the following:

(a) Conductors rated 60°C (140°F), or

(b) Conductors with higher temperature ratings, provided the ampacity of such conductors is determined based on the 60°C (140°F) ampacity of the conductor size used, or

(c) Conductors with higher temperature ratings if the equipment is listed and identified for use with such conductors, or

(d) For motors marked with design letters B, C, D, or E, conductors having an insulation rating of 75°C (167°F) or higher shall be permitted to be used provided the ampacity of such conductors does not exceed the 75°C (167°F) ampacity.

(2) Termination provisions of equipment for circuits rated over 100 amperes, or marked for conductors larger than No. I, shall be used only for

(a) Conductors rated 75°C (167°F), or

(b) Conductors with higher temperature ratings provided the ampacity of such conductors does not exceed the 75°C (167°F) ampacity of the conductor size used, or up to their ampacity if the equipment is listed and identified for use with such conductors.

(3) Separately installed pressure connectors shall be used with conductors at the ampacities not exceeding the ampacity at the listed and identified temperature rating of the connector.

FPN: With respect to Sections 110-14(c)(1), (2), and (3), equipment markings or listing information may additionally restrict the sizing and temperature ratings of connected conductors.

(Reprinted with permission from NFPA 70-1999.)

have sufficient working space around electrical equipment of 600 volts nominal or less; that we must have clearances in front and above; we must have headroom to work on the equipment; proper illumination; and that all live parts must be guarded against accidental contact. *Article 110* reminds us that warning signs must be placed and that electrical equipment must be protected from physical damage. *Section 110-22* states that each disconnecting means for motors and appliances, and service, feeders, or branch circuits at the point where it originates must be legibly identified to indicate its purpose and that the identifying markings must be sufficiently durable to withstand the environment involved. Where circuits and fuses are installed with a series combination rating "Caution—Series Rated System," the equipment enclosure shall be legibly marked in the field to indicate the equipment has been applied with a series combination rating. Over 600 volts requirements are found in *Part C* of *Article 110*. These requirements are specific about entrance and access to working space, guarding barriers, enclosures for electrical installations, and separation from other circuitry. Over 600 volts installations must be located in locked rooms or enclosures, except where under the observation of qualified persons at all times. *Table 110-34(e)* covers the elevation of unguarded live parts above working space for nominal voltages 601 and above.

CONDUCTOR AMPACITY ADJUSTMENT FACTORS

The *National Electrical Code®* defines ampacity in *Article 100—Definitions* as "The current in amperes a conductor can carry continuously under the conditions of use without exceeding its temperature rating." Ampacity is used with this meaning throughout this manual in referring to wire and conductor sizes; the *NEC®* abbreviations of cmil for circular mil and kcmil for thousand circular mils are used. These supersede the abbreviations of cm and mcm, respectively, though the latter still can be found on many terminals and conductors made prior to 1990. The ampacity of a conductor does not increase in direct proportion to its size. The ampacity ratings of three copper conductors having different temperature ratings as given in *NEC® Table 310-16* are as follows:

Conductor (Size kcmil)	Temperature Rating			
	60°C (140°F)	75°C (167°F)	85°C (185°F)	90°C (194°F)
250	215	255	275	290
500	320	380	415	430
1000	455	545	590	615

Although 500-kcmil conductors have twice the cross-sectional area of 250 kcmil conductors, their ampacity is only about half again as much. The ampacities of 1000-kcmil conductors are only about 45% more than 500-kcmil conductors. The size of the conductors, particularly for large ampacities, has a definite bearing on the efficient use of both conductor material and raceways. Because large conductors are less flexible than small ones, they are more difficult to handle and pull through raceways, particularly those having offsets and bends. Paralleling of two or more smaller conductors per phase allows higher ampacities than could be obtained with a single large conductor. Even though allowable raceway fill for multiple conductors is less than a single conductor (*NEC® Chapter 9, Table 1*), there is often no need to use a larger raceway to obtain the needed ampacity.

Conductors carry loads continuously; however, *NEC® Article 210 Part B* and *Section 220-19* generally limit the conductors to be loaded to no more than 80% of their rated ampacities. This limitation ensures overcurrent devices are not subjected to loads of more than 80% of their rating. Unlike the conductors, overcurrent devices may open prematurely if fully loaded for long periods, unless specifically designed to carry full current rating continuously. Where overcurrent devices in their enclosures are listed for continuous operation at 100% load, this 80% correction factor does not apply to panelboards, overcurrent devices, or conductors.

ALLOWABLE CONDUCTOR FILL (More Than Three Conductors in a Raceway)

Where the number of conductors in a raceway or cable exceed three, or where single conductors or multiconductor cables are stacked or bundled without maintaining spacing and are not installed in raceways, the maximum allowable load current on each

310-15. Ampacities for Conductors Rated 0-2000 Volts.

(a) General.

(1) Tables or Engineering Supervision. Ampacities for conductors shall be permitted to be determined by tables or under engineering supervision, as provided in (b) and (c).

FPN No. 1: Ampacities provided by this section do not take voltage drop into consideration. See Section 210-19(a), FPN No. 4, for branch circuits and Section 215-2(d), FPN No. 2, for feeders.

FPN No. 2: For the allowable ampacities of Type MTW wire, see Table 11 in the *Electrical Standard for Industrial Machinery,* NFPA 79-1997.

(2) Selection of Ampacity. Where more than one calculated or tabulated ampacity could apply for a given circuit length, the lowest value shall be used.

Exception: Where two different ampacities apply to adjacent portions of a circuit, the higher ampacity shall be permitted ro be used beyond the point of transition, a distance equal to 10 ft (3.05 m) or 10 percent of the circuit length figured at the higher ampacity, whichever is less.

FPN: See Section 110-14(c) for conductor temperature limitations due to termination provisions.

(b) Tables. Ampacities for conductors rated 0 to 2000 volts shall be as specified in the Allowable Ampacity Tables 310-16 through 310-19 and Ampacity Tables 310-20 and 310-21 as modified by (1) through (7).

FPN: Tables 310-16 through 310-19 are application tables for use in determining conductor sizes on loads calculated in accordance with Article 220. Allowable ampacities result from consideration of one or more of the following:

(1) Temperature compatibility with connected equipment, especially at the connection points.

(2) Coordination with circuit and system overcurrent protection.

(3) Compliance with the requirements of product listings or certifications. See Section 110-3(b).

(4) Preservation of the safety benefits of established industry practices and standardized procedures.

(1) General. For explanation of type letters used in tables and for recognized sizes of conductors for the various conductor insulations, see Section 310-13. For installation requirements, see Sections 310-1 through 310-10 and the various articles of this *Code.* For flexible cords, see Tables 400-4, 400-5(A), and 400-5(B).

(2) Adjustment Factors.

(a) *More than Three Current-Carrying Conductors in a Raceway or Cable.* Where the number of current-carrying conductors in a raceway or cable exceeds three, or where single conductors or multiconductor cables are stacked or bundled longer than 24 in. (610 mm) without maintaining spacing and are not installed in raceways, the allowable ampacity of each conductor shall be reduced as shown in Table 310-15(b)(2)(a).

Table 310-15(b)(2)(a). Adjustment Factors for More than Three Current-Carrying Conductors in a Raceway or Cable

Number of Current-Carrying Conductors	Percent of Values in Tables 310-16 through 310-19 as Adjusted for Ambient Temperature if Necessary
4–6	80
7–9	70
10–20	50
21–30	45
31–40	40
41 and above	35

Reprinted with permission from NFPA 70-1999.

FPN: See Appendix B, Table B-310-11, for adjustment factors for more than three current-carrying conductors in a raceway or cable with load diversity.

Exception No. 1: Where conductors of different systems, as provided in Section 300-3, are installed in a common raceway or cable, the derating factors shown in Table 310-15(b)(2)(a) shall apply to the number of power and lighting conductors only (Articles 210, 215, 220, and 230).

Exception No. 2: For conductors installed in cable trays, the provisions of Section 318-11 shall apply.

Exception No. 3: Derating factors shall not apply to conductors in nipples having a length not exceeding 24 in. (610 mm).

Exception No. 4: Derating factors shall not apply to underground conductors entering or leaving an outdoor trench if those conductors have physical protection in the form of rigid metal conduit, intermediate metal conduit, or rigid nonmetallic conduit having a length not exceeding 10 ft (3.05 m) and the number of conductors does not exceed four.

(b) *More than One Conduit, Tube, or Raceway.* Spacing between conduits, tubing, or raceways shall be maintained.

(3) Bare or Covered Conductors.
Where bare or covered conductors are used with insulated conductors, their allowable ampacities shall be limited to those permitted for the adjacent insulated conductors.

(4) Neutral Conductor.

(a) A neutral conductor that carries only the unbalanced current from other conductors of the same circuit shall not be required to be counted when applying the provisions of Section 310-15(b)(2)(a).

(b) In a 3-wire circuit consisting of two phase wires and the neutral of a 4-wire, 3-phase wye-connected system, a common conductor carries approximately the same current as the line-to-neutral load currents of the other conductors and shall be counted when applying the provisions of Section 310-15(b)(2)(a).

(c) On a 4-wire, 3-phase wye circuit where the major portion of the load consists of nonlinear loads, harmonic currents are present in the neutral conductor; the neutral shall therefore be considered a current-carrying conductor.

(5) Grounding or Bonding Conductor. A grounding or bonding conductor shall not be counted when applying the provisions of Section 310-15(b)(2)(a).

(6) 120/240-Volt, 3-Wire, Single-Phase Dwelling Services and Feeders. For dwelling units, conductors, as listed in Table 310-15(b)(6) shall be permitted as 120/240-volt, 3-wire, single-phase service-entrance conductors, service lateral conductors, and feeder conductors that serve as the main power feeder to a dwelling unit and are installed in raceway or cable with or without an equipment grounding conductor. For application of this section, the main power feeder shall be the the feeder(s) between the main disconnect and the lighting and appliance branch-circuit panelboard(s), and the feeder conductors to a dwelling unit shall not be required to be larger than their service-entrance conductors. The grounded conductor shall be permitted to be smaller, than the un-grounded conductors, provided the requirements of Sections 215-2, 220-22, and 230-42 are met.

Table 310-15(b)(6). Conductor Types and Sizes for 120/240-Volt, 3-Wire, Single-Phase Dwelling Services and Feeders. Conductor Types RH, RHH, RHW, RHW-2, THHN, THHW, THW, THW-2, THWN, THWN-2, XHHW, XHHW-2, SE, USE, USE-2

| Conductor (AWG or kcmil) | | Service or Feeder Rating (Amperes) |
Copper	Aluminum or Copper-Clad Aluminum	
4	2	100
3	1	110
2	1/0	125
1	2/0	150
1/0	3/0	175
2/0	4/0	200
3/0	250	225
4/0	300	250
250	350	300
350	500	350
400	600	400

Reprinted with permission from NFPA 70-1999.

(7) Mineral-Insulated, Metal-Sheathed Cable. The temperature limitations on which the ampacities of mineral-insulated, metal-sheathed cable are based shall be determined by the insulating materials used in the end seal. Termination fittings incorporating unimpregnated, organic, insulating materials shall be limited to 90°C (194°F) operation.

(c) Engineering Supervision. Under engineering supervision, conductor ampacities shall be permitted to be calculated by means of the following general formula:

$$I = \sqrt{\frac{TC - (TA + DeltaTD)}{RDC(1 + YC)RCA}}$$

Where:

TC = Conductor temperature in degrees Celsius (°C)

TA = Ambient temperature in degrees Celsius

Delta TD = Dielectric loss temperature rise

RDC = dc resistance of conductor at temperature *TC*

YC = Component ac resistance resulting from skin

effect and proximity effect

RCA = Effective thermal resistance between conductor and surrounding ambient.

FPN: See Appendix B for examples of formula applications.

(Reprinted with permission from NFPA 70-1999.)

---

conductor must be reduced as shown in the box that follows.

The table is based on the following formula:

A_2 = (the square root of) $0.5N/E \times (A_1)$

where A_1 = Table ampacity multiplied by factor from *Table 310-15(b)(2)(a)*, N = Total number of conductors used to obtain factor from *Table 310-15(b)(2)(a)*, E = Desired number of energized conductors, and A_2 = Ampacity limit for energized conductors.

These reductions are to minimize heat buildup in the raceway, stack, bundle, or cable that may tend to damage the insulation over a long period of time. Allowable ampacities of multiple conductors in a raceway after applying these derating factors, as they are called, are shown in the preceding table. Additional correction factors may be necessary to avoid prolonged overheating of conductors and insulation from outside sources.

The following example illustrates the use of the derating and temperature correction factors:

Example: Determine the ampacity of three #1-0, three #2, and six #4 Type-THW conductors in a 3-phase, 3-wire delta system (no neutral wire) for operation at a temperature of 40°C (104°F). The maximum operating temperature for THW wire from *NEC® Table 310-13* is 75°C (104°F). According to *Table 310-15(b)(2)*, ampacities must be reduced to 50% of the values listed in the tables. According to the *Correction Factors* below *Table 310-16,* ampacities also must be reduced by a correction factor of 88% for ambient temperature. Derating factors are as follows:

Wire Size	Ampacity	X	Factor for More than 3 Conductors in a Raceway	X	Factor for Temperature Over 30°C	=	Net Ampacity
1/0	150		50%		.88		66
2	115		50%		.88		50.6
4	85		50%		.88		37.4

NEC® *Section 310-15* states that the ampacity for conductors rated 0–2000 volts shall be as specified in *Tables 310-16* through *310-21* and their accompanying notes. The ampacity for solid dielectric insulated conductors rated 2001 through 35,000 volts shall be as specified in *Tables 310-69* through *310-84* and their accompanying notes. However, under engineering supervision, ampacity may be calculated by the following general formula. Examples of these calculations and the formula applications can be found in Appendix B of the 1999 *NEC®*. However, the Appendix is not a part of the requirements of the *NEC®* and is included for information purposes only. Therefore, the authority having jurisdiction may determine the application of this material for complying with *NEC® Section 310-15*.

VOLTAGE DROP

The resistance, or impedance, of conductors may cause a substantial difference between the voltage measured at the service entrance of a facility and the point of use at the equipment utilizing the said voltage. It is necessary to minimize the voltage drop in conductors to ensure good service. Excessive voltage drop impairs the operation of electrical equipment. Lower-than-rated voltage results not only in decreased light output from fluorescent lighting, but also may cause starting and operating difficulties. The total lumen output for a 2-lamp, rapid-stop fluorescent fixture in general varies about 1% for each percent change in line voltage. A similar voltage drop of 1% causes about 3 1/2% loss in lumen output for filament-type lamps.

Although motors generally are designed for a 10% overall voltage variation, they operate more efficiently at nameplate voltages. Excessive voltage drop (more than 5% of the line voltage) can cause serious motor overheating. If the overheating is serious enough, it could cause a fire if the overcurrent or thermal protective devices fail to operate properly.

Although the *NEC®* does not make mandatory requirements to govern voltage drop, it would be a violation of *Section 110-3B* to supply voltage to equipment at less than the nameplate rating. *NEC® Section 210-19* includes a fine print note (FPN4), which includes guides to the allowable voltage drop in branch-circuit conductors; *NEC® 215-2* includes a similar fine print note for feeder conductors. Adherence to these fine print notes will provide a reasonable efficiency of operation. As a matter of information, the *NEC®* states in FPNs, as listed above, the following: "Conductors for branch circuits or feeders as defined in *Article 100* should be sized to prevent voltage drop exceeding 3% at the farthest outlet of power, heating, and lighting loads or combinations of such loads, and where the maximum total voltage drop on both feeders and branch circuits to the farthest outlet does not exceed 5% will provide reasonable efficiency of operation" (courtesy *NEC®*). This means that the voltage drop must be fairly equally divided between the feeder and the branch-circuit conductors. If the branch-circuit voltage drop is 3% of the rated voltage, the voltage drop in the feeder should not exceed 2%. A voltage drop greater than 5% of line voltage can result in a variation greater than 10% from designed operating voltage, as in the case of a 240-volt motor connected to a 208-volt branch circuit having an excessive voltage drop.

Available heat from an electric heating appliance is proportional to the square of the applied voltage. If a heating unit is rated at 240 volts and the terminal voltage is 5% less, or 228 volts, the heat generated at the appliance is reduced by 10%. At a 10% reduction, or 216 volts, the heat generated at the appliance is reduced by nearly 20%.

It is important, therefore, to take proper precautions to avoid excessive voltage drop in all types of circuits, whether for lighting, power, or heating. The basic formula for determining voltage drop in a 2-wire, direct-current circuit; a 2-wire, single phase; or a 3-wire, single-phase, alternating-current circuit with a balanced load at 100% power factor (where reactance is negligible) is:

$$\text{Voltage Drop} = \frac{2 \times R \times L \times I}{\text{Circular Mils}}$$

where R = Resistivity of conductor material (12 ohms per circular mil foot for copper, 19 ohms per circular mil foot for aluminum); L = one-way length of circuit (feet), and I = current in one conductor (amperes) (see *Table A-15* in appendix).

Impedance Factors for Finding Voltage Drop in Copper Conductors

Wire Size AWG	Single-phase AC, Power Factor					
	100	90	80	70	60	DC
14	.5880	.5360	.4790	.4230	.3650	.5880
12	.3690	.3330	.3030	.2680	.2320	.3690
10	.2320	.2150	.1935	.1718	.1497	.2320
8	.1462	.1373	.1248	.1117	.0981	.1462
6	.0918	.0882	.0812	.0734	.0653	.0918
4	.0578	.0571	.0533	.0489	.0440	.0578
2	.0367	.0379	.0361	.0337	.0309	.0363
1	.0291	.0311	.0299	.0284	.0264	.0288
1/0	.0233	.0257	.0252	.0241	.0227	.0229
2/0	.0187	.0213	.0212	.0206	.0196	.0181
3/0	.0149	.0179	.0181	.0177	.0171	.0144
4/0	.0121	.0152	.0156	.0155	.0151	.0114
Wire Size, kcmil						
250	.0102	.0136	.0143	.0143	.0141	.0097
300	.0086	.0121	.0128	.0131	.0130	.0081
350	.0074	.0109	.0118	.0122	.0122	.0069
400	.0066	.0101	.0111	.0115	.0116	.0060
500	.0054	.0089	.0099	.0105	.0108	.0048
600	.0047	.0083	.0093	.0099	.0103	.0040
700	.0041	.0077	.0088	.0094	.0098	.0034
750	.0039	.0075	.0086	.0093	.0097	.0032
800	.0037	.0073	.0084	.0091	.0095	.0030
900	.0034	.0069	.0081	.0088	.0093	.0027
1000	.0031	.0067	.0079	.0086	.0091	.0021

Voltage drop in two wires of a single-phase AC or DC circuit is equal to current × length of run in hundreds of feet × impedance factor.

Impedance factors in table multiplied by 0.86 will give phase-to-phase voltage drop for 3-phase AC circuits.

Figure 3-8 Impedance factors for voltage drop (Courtesy of American Iron & Steel Institute)

For 3-phase, 4-wire, alternating-current circuits, the voltage drop between 1-phase wire and the neutral is one-half of the value derived from this formula. For 3-phase circuits at 100% power factor, the voltage drop between any 2-phase wires is 0.866 × the voltage drop calculated by the formula.

In circuits where each phase and neutral conductor consists of two or more wires, the voltage drop is determined by using the current in only one wire of a phase conductor and the neutral. For loads of other than 100% (unity) power factor, the preceding formula does not apply. For example, where motors are a major part of a circuit load, the power factor may be considerably less than one. Determining voltage drop in circuits having a load of less than unity power factor requires a more complicated formula. Figure 3-8 lists impedance factors developed from this formula for single-phase circuits.

For 3-phase circuits, the voltage drop may be computed by multiplying the single-phase voltage drop by .86. Should the voltage drop so determined be greater than the suggested desirable percentage of the circuit voltage, a larger-size conductor should be used, the circuit length should be shortened, or the circuit load should be reduced until the voltage drop is within the required or desired limits.

Example: Find voltage drop in a 300-foot circuit using No. 4/0 Type TW copper conductors in a 3-phase circuit. Allowable ampacity (*NEC® Table 310-16*) is 195; power factor, 80%. Voltage drop $= 195 \times 3.00 \times 0.0156 \times 0.86 = 7.9$ volts.

CONDUCTORS RUN IN PARALLEL

When large ampacities are required, the use of parallel conductors should be carefully investigated. Conductors larger than 500 kcmil generally are considered questionable from the standpoint of ease of installation and operating economies. *NEC® Section 310-4* states that conductors of size 1/0 or larger comprising each phase, neutral, or equipment-grounding conductor are permitted to be connected in parallel (electrically joined to form a single conductor) only if the following requirements are met: all parallel conductors are of the same length, are of the same conductor material, have the same circular mil area, have the same insulation type, and are terminated in the same manner. Although paralleling of conductors may prove more practical and economical than using one larger conductor, the voltage drop in parallel conductors in a 3-phase, alternating-current circuit at 100% power factor may be greater than the voltage drop in a single large conductor. In a

Where Conductors are run in parallel in multiple raceways or cables as permitted by *Section 310-4* the equipment grounding conductor when required shall be run in parallel. Each parallel equipment grounding conductor shall be sized based on the overcurrent device.

Example: 1000 ampere overcurrent device protecting a feeder run in three individual nonmetallic conduits or Type MC cable *Table 250–122* requires that a 2/0 copper or 4/0 aluminum equipment grounding conductor be run in each conduit or cable. Note: Most cable manufacturers include a reduced equipment grounding conductor and where cable is specified the cable may have to be special ordered to get the properly sized equipment grounding.

circuit at 90% power factor, the voltage drop is about equal for single and parallel conductors.

If all parallel conductors are to be installed in a single raceway, as in Figure 3-9, it is necessary to derate them as per *NEC® Article 310, Table 310-15(b)(2)*, when more than three conductors are installed in a raceway. Hence, it is better in many cases to use two or more smaller raceways. Therefore, when conductors having different insulation materials are added to conductors already in a raceway, the new conductors must have an equal or higher temperature rating than the existing ones, but the lower ampacity rating of the two insulation types must be used for all conductors, as per *NEC® Section 300-3*.

When more than one raceway is used, *NEC® Section 300-3(b)* applies, with exceptions stating that each raceway shall contain all phase conductors, the neutral, and the equipment grounding conductor if present. There is another exception in *Section 300-5(i)* that permits an isolated phase in nonmetallic raceways. This requirement is clarified in *Section 300-20*: Where conductors carrying alternating current are installed in metal enclosures or metal raceways, they are arranged so as to avoid heating the surrounding metal by induction.

To accomplish this, all phase conductors and, where used, the neutral and all equipment grounding conductors are grouped together. For example, if three wires are in parallel for each phase, three raceways are required. This would result in three wires in each raceway for a single-phase, 3-wire system, and four wires in each raceway for a 3-phase, 4-wire system. Figure 3-11 illustrates a 3-phase, 4-wire system with two wires connected in parallel for each phase and for the neutral, requiring two raceways with four wires in each. Where two or more raceways enclose conductors in parallel, they are required by *Section 310-4* to have the same physical characteristics. (Otherwise, the raceways enclosing the different conductors might have a different electrical characteristic, such as unequal inductive reactance.) Currents in the conductors are in individual reactance, and currents in the conductors in the individual raceways would not be in phase at their common terminal. This condition must be avoided for proper operation of multiphase equipment such as motors. Inductive heating caused by improper installation of conductors in steel raceways and enclosures is discussed in Chapter 5.

Although derating factors will provide for safe installation of large numbers of conductors in a raceway, there are some drawbacks to this practice. The more conductors enclosed in a raceway, the greater the chance of a serious breakdown. Where continuity of service is important, it usually is advisable to install circuit conductors in separate conduits or raceways. Also, derating conductors may not provide the most effective use of the conductor material. It is very important when employing parallel conductor runs that all conductors be installed using the same wiring method throughout the parallel run. For instance, if parallel conductors are installed in individual metallic conduits enclosing each phase conductor and the neutral conductor, the same wiring method should be used from the point of origin to the termination point. Changing wiring methods, such as terminating the raceways in a wireway or auxiliary gutter and extending the grouped parallel conductors to the termination point, may result in excessive inductive heating. Although paralleling of circuit conductors has distinct advantages, great care must be taken to avoid conductor current imbalance and inductive heating.

One 3" rigid or IMC steel conduit

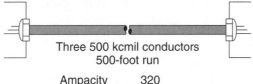

Three 500 kcmil conductors
500-foot run

Ampacity	320
DC resistance	0.0216 ohms/1000 ft
Factor for AC	1.13

Voltage drop = 320 ∞ 500 ∞ 0.0216/100 ∞ 1.13 = 3.9 volts
per conductor (7.8 volts for two conductors)

Two 2" rigid or IMC steel conduit

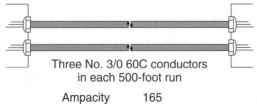

Three No. 3/0 60C conductors
in each 500-foot run

Ampacity	165
DC resistance	0.642 ohms/1000 ft
Factor for AC	1.04

Voltage drop = 320 ∞ 500 ∞ 0.642/100 ∞ 1.04 = 5.5 volts
per conductor (11.0 volts for two conductors)

The voltage drops are computed for loads at 100 per-cent power factor. The differences decrease as power factors become less than 100 percent.

Figure 3-9   Voltage drop in single and paralleled conductors

NEUTRAL CONDUCTOR (GROUNDED CONDUCTOR)

Most neutral systems are divided into two general categories, either solidly grounded or impedance grounded. Impedance-grounded neutral systems can be divided into several subcategories. The system ground point always should be at the power source. An archaic concept of grounding at the load or at other points of the system because of availability or convenience is not recommended because of problems caused by multiple ground paths and the danger that the system could be left ungrounded and unsafe. The *National Electrical Code®* recognizes this danger and prohibits system grounding at any place except the source or service equipment. The neutral conductor is defined in *Section 310-15(b)(4)* of the *NEC®*. The size of the required neutral for an electrical system is found in *Section 220-22*.

The basic rule is that the neutral conductor

a. Single raceway

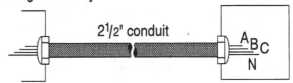

Example: 3-phase, 4-wire feeder, total design load = 300 amperes, and 75C copper conductors, maximum wire size to be No. 3/0.

Phase current is equal to load current; 300 amperes. Minimum size 75C copper conductor for 300 amperes load, from *NEC® Table 310-16,* is 350 Kcmil. Therefore, at least two conductors per phase are needed or six conductors. From *NEC Table 310-15(b)(2)(a),* not over 80% of tabular ampacity values may be used when 4 to 6 conductors are in a raceway. Therefore, conductor ampacity for each of the two conductors must be more than 150 by a factor of 1/0.8:

$$I = \frac{150}{0.8} = 190 \text{ amperes}$$

Ampacity of a No. 3/0, 75C copper conductor is 200. Two conductors of this size will carry the required load.

For a 300-ampere load, the required ampacity for the neutral conductor is 200 amperes plus 70% of excess over 200,

$$200 + (0.7 \times 100) = 270$$

NEC® Section 220-22). For two neutral wires, each must carry 270/2 or 135 amperes (applicable to neutrals which carry only the unbalanced current from other conductors need not be derated).

Minimum conduit size for six No. 3/0 and two No. 1/0 copper conductors

$$6 \times 0.3286 = 1.9728 \text{ sq. in.}$$
$$2 \times 0.2367 = 0.4734 \text{ sq. in.}$$
$$= 2.4462 \text{ sq. in.}$$

For not more than 40% fill, a single 3-inch conduit is required (*NEC® Tables 4* and *5*).

b. Parallel raceways

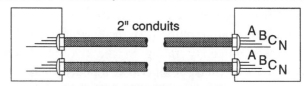

Three No. 3/0 60C conductors
in each 500-foot run

For three current-carrying conductors plus one neutral in each raceway, no derating is necessary. At 150 amperes per conductor, No 1/0 75C conductors may be used for both phase and neutral wires. Conductors for each phase and neutral must be in each raceway. For parallel runs, two 2-inch conduits are required.

Figure 3-10   Multiple conductors in parallel as in *NEC® Section 310-4* (Courtesy of American Iron & Steel Institute)

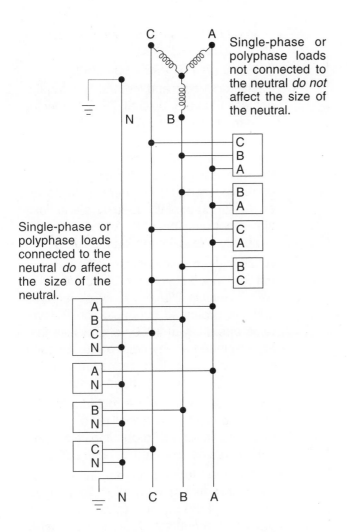

Single-phase or polyphase loads not connected to the neutral *do not* affect the size of the neutral.

Single-phase or polyphase loads connected to the neutral *do* affect the size of the neutral.

Section 310–15(b)(4) a through *c.*

(4) Neutral Conductor.

(a) A neutral conductor that carries only the unbalanced current from other conductors of the same circuit need not be counted when applying the provisions of *Section 310–15(b)(2)a.*

(b) In a 3-wire circuit consisting of 2-phase wires and the neutral of a 4-wire, 3-phase wye-connected system, a common conductor carries approximately the same current as the line-to-neutral load currents of the other conductors and shall be counted when applying the provisions of *Section 310–15(b)(2)a.*

(c) On a 4-wire, 3-phase wye circuit where the major portion of the load consists of nonlinear loads, there are harmonic currents present in the neutral conductor, and the neutral shall be considered to be a current-carrying conductor.

Figure 3-11 Size of neutral conductor as in *NEC®* *Sec. 220-22* (Courtesy of American Iron & Steel Institute)

must have sufficient ampacity to carry the maximum possible imbalance of the connected load between the neutral and any of the ungrounded conductors of the circuit. Although *Section 310-15(b)(4)c* only requires that the neutral be full size, where the effects of harmonics are present good engineering practices dictate that the neutral conductor be sized to the maximum possible current on the electrical system, which may require a neutral 1.5 to 1.75 times the size of the ungrounded conductors of the system. In many cases, on 3-phase, 4-wire wye systems, typically 208/120-volt systems, each ungrounded conductor should have its own full-size neutral when supplying personal computers and other data processing equipment.

In common inductive-resistant systems, such as residential and commercial store buildings, a 3-wire, 120/240-volt circuit should be the total of the 120-volt load between the neutral and the more heavily loaded ungrounded conductor, as the more lightly loaded conductor could be carrying no current at all, because 240-volt loads are not connected to the neutral. They would not affect the commutation of its size. The same applies in general to 3-phase loads on 3-phase, 4-wire circuits. For a feeder supplying household electric ranges

Example: Given a 3-phase load, not connected to neutral, of 200 amperes per phase conductor and single-phase unbalanced loads (phase to neutral) of 50, 100, and 150 amperes. Then, the load in amperes divides as follows:

	Max. current per conductor	
	Phase conductor	Neutral conductor
3-phase load	200	0
Single-phase loads	150	150*
Total	350	150

The minimum size of copper 75° C wire is 650 kcmil for the phase conductor and No. 1/0 for the neutral. Good design, however, calls for larger conductors.

*Maximum possible neutral load is 150 amperes. With all loads connected, resultant current in the neutral would be 87 amperes.

and other cooking appliances, the maximum unbalanced load is considered to be 70% of the load on the ungrounded conductors, as determined in the allowance with *NEC® Table 220-19*. For 3-wire direct current; 3-wire, single-phase alternating current; and 4-wire, 3-phase alternating-current feeds, a further demand factor of 70% may be applied to that portion of the unbalanced load that exceeds 200 amperes.

No demand factor can be applied, and hence no reduction in the size of the neutral can be made, for that portion of the feeder or circuit supplying electric-discharge lamps, such as fluorescent fixtures. Neutral conductors are approximately equal to phase currents on these circuits because of third harmonics developed by the lamp ballast.

Example: a new 3-phase, 4-wire wye connected feeder is connected by an existing junction box to three single-phase plus neutral circuits, each consisting of 2-phase conductors and a neutral. Two of these circuits are in one conduit, the third is in another. Single-phase circuit conductors are all #2 THW copper. Determine the ampacity of each single-phase circuit and the required size of the new service.

- If electric discharge ballasts are in the circuit, the load is considered continuous (3 hours or more continuous operation). See definition in *NEC® Article 100* for "continuous load."
- Neutrals in these circuits must be counted as current-carrying conductors (Refer to *NEC® Section 310-15(b)(4)b, Ampacity Tables of 0–2000 Volts*).
- The ampacity of #2 THW conductors, derived from *NEC® Table 310-16*, is 115 amperes.
- *Circuit ampacities:* In conduits with a single circuit no derating is required for only 3 conductors: the ampacity equals 115.
- In conduits with two circuits 80% derating factors must be used, as all six wires are considered current-carrying conductors.
- Each of these two circuits has a designed ampacity of 0.8 × 115 = 92.
- The sum of the ampacities of all three circuits is the required ampacity of the service conductors and service neutral (*NEC® Section 220-22*).
- The value is 115 + 92 + 92 = 299.
- Each conductor and neutral will require not less than a 350 kcmil, 75˚C conductor with an ampacity of 310.
- The derating factors for more than three conductors per raceway do not apply to the grounding conductor (equipment = grounding conductor) where used as required for nonmetallic raceways and flexible raceways over 6 feet in length, generally. *(See Section 310-15(b)(5))*.
- Additional requirements for sizing the neutral (grounded conductor) can be found in *NEC® Section 250-24 (b)*, which requires that the grounded conductor of AC systems operating at 1000 volts or less that are grounded at any point be run to each service-disconnecting means and bonded to each disconnecting-means enclosure.
- The conductor must be routed with the phase conductors and shall not be smaller than the required grounding electrode conductors as specified in *Table 250-66*.
- In addition, for service-entrance phase conductors larger than 1100 kcmil (copper) or 1750 kcmil (aluminum), the grounded conductor shall not be smaller than 12 1/2% of the area of the largest service-entrance phase conductor.
- Where the service-entrance phase conductors are paralleled, the size of the grounded conductor shall be based on the equivalent area for the parallel conductors as indicated.
- There are some exceptions listed.

EQUIPMENT-GROUNDING CONDUCTORS

The term *equipment grounding* refers to the interconnection and grounding of nonelectrical metallic elements of a system (See *NEC® Article 100—Definitions*). Examples of components of equipment grounding systems are metallic conduit, motor frames, equipment enclosures, and a grounding conductor. Note that a grounding conductor is part of the equipment-grounding system, as distinguished from a grounded conductor, which is part of a power distribution system. The requirements for grounding a system are covered in *Article 250* of the *NEC®*, *Section 250-2*.

> **(a) Grounding of Electrical Systems.**
>
> Electrical systems required to be grounded shall be connected to earth in a manner that will limit the voltage imposed by lightning, line surges, or unintentional contact with higher voltage lines and that will stabilize the voltage to earth during normal operation.

> **(b) Grounding Electrical Equipment.**
>
> Conductive materials enclosing electrical conductors or equipment or forming part of such equipment, shall be connected to earth so

as to limit the voltage to ground on these materials. Where the electrical system is required to be grounded, these materials shall be connected together and to the supply system grounded conductor as specified by this article. Where the electrical system is not solidly grounded, these materials shall be connected together in a manner that establishes an effective path for fault current.

> **(c) Bonding of Electrically Conductive materials and Other Equipment.**
>
> Electrically conductive materials, such as metal water piping, metal gas piping, structural steel members, that are likely to become energized shall be bonded *as specified by this arti-*

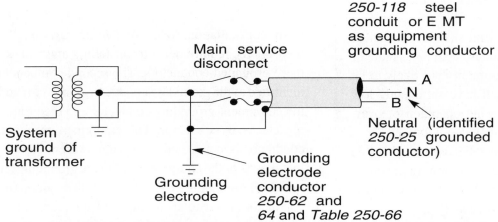

Requirements for grounding electrode conductor, grounded system as per part C *Article 250* grounded system

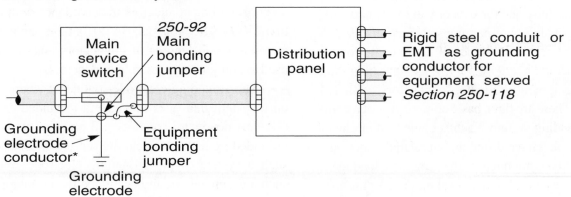

*Grounds both the identified grounded conductor and the equipment grounding conductor. It is connected to the grounded metallic cold water piping on premises, or other available grounding electrode, such as metal building frame if effectively grounded, or a made electrode *Sections 250-50* and *52*

Figure 3-12 Grounding principles (Courtesy of American Iron & Steel Institute)

cle to the supply system grounded conductor, or *in the case of an ungrounded electrical system to the electrical system grounded equipment,* in a manner that establishes an effective path for fault current.

The IEEE document Green Book reminds us that equipment-grounding conductors are installed to

- Reduce shock hazard to personnel
- Provide adequate current-carrying capability, both in magnitude and duration to accept the ground-fault current permitted by the over-current protection system without creating a fire or explosive hazard to building or contents
- Provide a low-impedance return path for ground-fault current necessary for timely operation of the overcurrent protective system

---

(250-2)(d) Performance of Fault Current Path.

The fault current path shall be permanent and electrically continuous, shall be capable of safely carrying the maximum fault likely to be imposed on it, and shall have sufficiently low impedance to facilitate the operation of over-current devices under fault conditions.

The earth shall not be used as the sole equipment grounding conductor or fault path.

(FPN): See *Figure 250-2* for information on the organization of *Article 250.*

Sections 250-90 and *250-96* require bonding to ensure continuity and the capacity to safely carry the fault currents likely to be imposed. *Section 250-118 (b)* lists the acceptable types of equipment-grounding conductors. *Section 250-122* lists the minimum size of equipment-grounding conductors, with exceptions. There are three basic components of a complete grounding system. Each is composed of one or more parts that have different, but related, functions.

The first component is the grounded conductor. This may be a grounded neutral or, on a 3-phase, 3-wire, delta-connected system, a grounded phase conductor. The latter sometimes is called a *corner-grounded* system. In all cases, the grounded conductor must be white, natural gray, or with three continuous white stripes on other than green insulation, or be identified at the terminals and junction

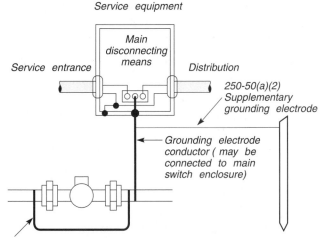

Bonding jumper around water meter filtering equipment, and other similar fittings in water piping likely to be disconnected, such as unions, where grounding connection is not on street side of water meter.

Figure 3-13 Bonding jumpers as in *NEC® Article 250-50(a)* (Courtesy of American Iron & Steel Institute)

boxes at installation. *NEC® Article 200, Section 200-6* covers the requirements for installing and marking these grounded conductors. The second component, identified as the grounding conductor, consists of both the equipment-grounding conductor and the grounding electrode conductor. The equipment-grounding conductor is connected to the non-current-carrying metal parts of the equipment and to the grounding terminals of grounding-type receptacles. The grounding conductor also extends to the service equipment enclosure. Metal raceways, rigid or IMC conduit, and EMT electrical metallic tubing may serve as equipment grounding conductors in accordance with *Section 250-118.* See appendix 16-20 for tables giving the maximum safe lengths of conductors and conduits used as equipment grounding conductors.

The grounding-electrode conductor, where used on a system that does not have a grounded circuit (system) conductor, such as a 3-phase, 3-wire ungrounded system, connects the service enclosure of such a system to the grounding electrode. Where such a grounded circuit is supplied by a transformer secondary that is grounded at any point, the grounded conductor must extend to the service equipment enclosure, even though it will not be used as a circuit conductor on the premises of the facility.

The grounding-electrode conductor in this case connects the service enclosure and the special grounded conductor to the grounding electrode. This special grounded conductor functions as a grounding conductor to carry any fault currents directly to the grounded point on the secondary, so that such fault currents will not have to travel along the high impedance path through the earth. This special grounded-grounding conductor, in furnishing a low-impedance path for fault current, facilitates the operation of the overcurrent device on ungrounded circuits.

On a grounded system using a grounded neutral, or a grounded phase conductor, the common grounding-electrode conductor connects both the equipment grounding conductor and the service enclosure, and the grounded-circuit, or system, conductor to the grounding electrode.

> **Equipment Ground:** The grounding conductor must be bare, green, or green with one or more yellow stripes.

The finish may be, and usually is, covered only and must not be relied on as insulation. The color coding is required by the *NEC®* only for equipment grounding conductors, but in some jurisdictions it is required for grounding conductors as well.

The third component of a grounding or grounded system is the bonding jumper. An equipment-bonding jumper connects two or more equipment grounding conductors or two or more non-current-carrying metal parts where continuity and adequate ampacity of equipment grounding cannot otherwise be ensured or relied upon. Examples of this are a flush box and the grounding terminal of the grounding-type receptacle (unless the receptacle has an integral device designed and ap-

proved for this purpose), or a service-entrance raceway and a service-entrance enclosure. These bonding jumpers must be bare, green, or green with one or more yellow stripes. A main bonding jumper bonds the service enclosure to the grounded bus of a grounded system. A circuit-bonding jumper is a current-carrying conductor, usually a wire or strap, that bonds two or more parts of a circuit, whether grounded or ungrounded. An instance of this is an expansion joint in which the ampacity of the circuit cannot otherwise be ensured around the joint. A circuit-bonding jumper is not part of the grounding system. It is mentioned here only because of its similarity in name with other types of bonding jumpers.

The grounding-electrode conductor is connected to the grounding electrode. The grounding electrode must be an underground, cold water, metal piping system where available, as specified in *NEC® Section 250-50*. Such piping system or portion of piping system that is effectively grounded for a length of at least 10 feet underground must be supplemented by one or more of the grounding means in *NEC® Sections 250-50 or 52*. For example, it could be a metal frame of a building where "effectively grounded" as defined in *NEC® Article 100* means "intentionally connected to earth through a ground connection or connections of sufficiently low impedance and having sufficient current-carrying capacity to prevent the buildup of voltages which may result in undue hazard to connected equipment or to persons."

Another acceptable type of grounding electrode is a concrete-encased electrode (Ufer Ground), which consists of an electrode encased by at least 2 inches of concrete located within or near the bottom of the foundation, or footing, which is in direct contact with the earth, and consisting of at least 20 feet of one or more steel reinforcing bars or rods not less than 1/2 inch in diameter, or consisting of at least 20 feet of bare copper conductor not smaller than No. 4.

Other means as listed in *NEC® Section 250-52*, if no grounding electrode is available on the premise, a ground rod or pipe may be driven, not less than 8 feet in length, and it shall consist of one of the following materials—a pipe or conduit not smaller than 3/4 trade size, of iron or steel, and with an outer surface galvanized or otherwise metal-coated for corrosion protection. A rod of iron or steel at least 5/8 inches in diameter, nonferrous or stainless steel

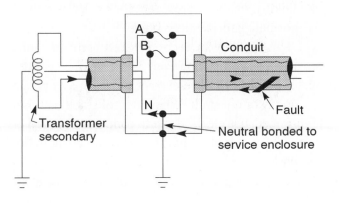

Figure 3-14 Bonding at service disconnect enclosure as required in *Section 250-92* (Courtesy of American Iron & Steel Institute)

rods or their equivalent less than 5/8 inch shall be listed and shall not be less than 1/2 inch in diameter. *Section 250-56* states that a single electrode consisting of a rod, pipe, or plate that does not have a resistance of 25 ohms or less shall be supplemented by an additional electrode of any of the types specified in *Sections 250-50* or *250-52*. Where multiple rod, pipe, or plate electrodes are installed, they shall be not fewer than 6 feet apart.

NEC® *Section 250-162* covers the requirements for grounding direct-current systems. The neutrals of all 3-wire DC systems supplying premises shall be grounded. All 2-wire DC systems supplying premises shall be grounded, with exceptions. *Section 250-20* covers the requirements for alternating-current circuits and systems to be grounded. Alternating-current circuits of fewer than 50 volts shall be grounded under certain conditions. Alternating-current systems of 50–1000 volts supplying premise wiring and premise wiring systems shall be grounded under any of the following conditions: (1) where the system can be so grounded at the maximum voltage to ground or the ungrounded conductors do not exceed 150 volts; (2) where the system is 3-phase, 4-wire wye connected in which the neutral is used as a circuit conductor; (3) where the system is 3-phase, 4-wire delta-connected in which the midpoint of one phase is used as a circuit conductor; (4) where a grounded service conductor is permitted to be uninsulated.

Examples of common distribution voltages and the grounding requirements are shown in Figure 3-15. There are exceptions to these general rules. Alternating current systems of 1 kV and over supplying mobile or portable equipment shall be grounded as specified in *NEC®* *Section 250-188*. Where supplying other than portable equipment, such systems shall be permitted to be grounded. Where such systems are grounded, they shall comply with the applicable provisions of this article. Separately derived systems whose power is derived from generator, transformer, or converter windings and that have no direct electrical connections, including a solidly connected grounded circuit conductor to supply conductors originating in another system, if required to be grounded, as in *NEC®* *Article 250*, shall be grounded as specified in *Section 250-30*, which requires that they be grounded to the nearest building steel or

nearest available other grounding electrodes in accordance to Part C at Article 250.

The grounding rules in the *NEC®* deserve special emphasis. The electrode-grounding connections at the transformer and at the service equipment provide parallel paths to ground for fault currents. Two parallel paths have much less resistance than either separately. Thus, they provide a low-resistance connection, essential for minimum impressed voltage should a fault occur. For example, if a temporary fault current is 100 amperes and the total impedance of the ground fault path is 10 ohms, the voltage drop (difference of potential) across the fault path will reach 1000 volts. If a 2-ohm grounding path is achieved using the underground-water-pipe grounding connection plus the transformer-electrode grounding connection (utility), the voltage drop is held to 200 volts. A fault current of 25 amperes has a voltage drop of 50 volts across the 2-ohm ground. Thus, the grounding connection at the transformer not only lessens the effects of the fault, but also helps to ground a fault on the source side of the service equipment.

A fault to ground may occur on the interior wiring of a grounded system when the grounded conductor from the transformer to the service equipment has a faulty connection, is open circuited, or is otherwise impaired or inadequate. In this event, some of the fault current travels from the service-equipment enclosure via the service grounding electrode, the earth, the transformer grounding electrode, and back along the ungrounded conductor to the overcurrent device protecting the faulted phase. In this way an alternate, or standby, path is provided for fault current if the normal path is interrupted.

This standby protection against ground faults on interior wiring, however, is by no means adequate, because the overcurrent device may operate too slowly or not at all. Where the sole fault path is a high-impedance electrode earth path, an uncleared or delayed clearing of an arcing fault could result in fire and prolonged shock hazard.

The electrical separation between the electrode-earth path over the fault-phase conductor produces a high-inductive reactance, which, when added to the electrode-earth resistance, may result in a large impedance. Thus, a main bonding jumper

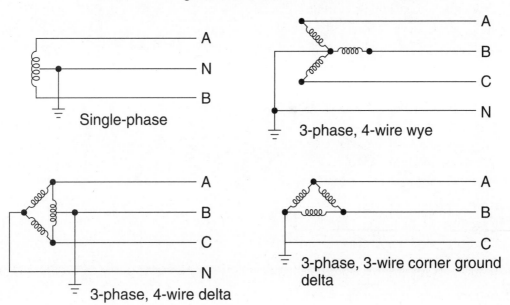

Systems that must be grounded — 150 volts or less to ground
or where bare grounded service conductor is used

Single-phase

3-phase, 4-wire wye

3-phase, 4-wire delta

3-phase, 3-wire corner ground
delta

Figure 3-15 Example of AC systems that are required to be grounded by *NEC®* *Section 250-20 and 250-21* (Courtesy of American Iron & Steel Institute)

is required to bond the service-equipment enclosure to the grounded conductor and to provide a low-impedance path for the fault current. The fault current travels from the point of the fault on the interior wiring along the equipment-grounding conductor (raceway, wire, armor, or bus) to the service-equipment enclosure through the main bonding jumper to the grounded conductor, along the grounded conductor to the transformer winding, and back through the ungrounded conductor to the overcurrent device protecting the fault-phase conductor.

This low-impedance path should result in rapid operation of the overcurrent device and fast clearing of the fault. The grounded conductor of a grounded system is required to be brought to the service equipment from the transformer whether or not it is to be used as a circuit conductor. Hence, on a grounded system, this conductor plus the main bonding jumper at the service and the bonding jumper between the transformer case and the transformer winding are always available as a low-impedance path for fault currents. There is no dependence on the electrode-earth path except where the grounded conductor is inadvertently interrupted.

- Where both the grounded conductor and the main bonding jumper are provided, properly connected, and of adequate capacity, only three conditions can prevent rapid operation of the overcurrent device:

  1. A faulty or improperly rated overcurrent device
  2. A high-resistance fault, such as an arcing fault, or a fault through high-resistance contact with a grounded object
  3. A high-resistance, equipment-grounding conductor connecting the faulted equipment to the service-equipment enclosure.

- The third condition can be prevented by proper grounding installed as specified in the *NEC®*.
- Even where the raceway does not provide the sole equipment-grounding conductor but is supplemented by an equipment-grounding wire or bus, the raceway provides a parallel path with the wire or bus and also protects the conductor. Tubular

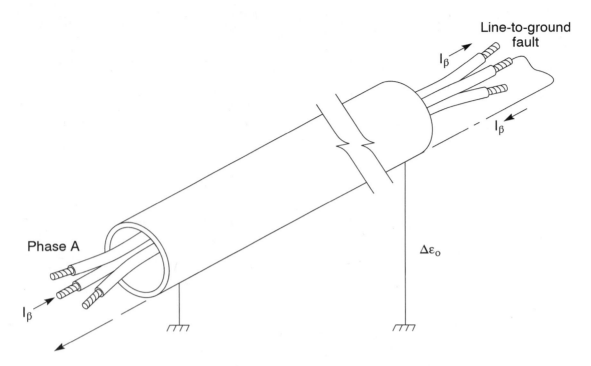

Figure 3-16   Tubular metal raceway used as a grounding conductor

metal raceway such as rigid conduit, IMC, or EMT metal raceways fit the preferred conductor geometry perfectly, within which the phase conductors and neutral conductors are enclosed. In this way effectiveness is markedly improved. The returning ground-fault current distributes itself about the entire enclosing metal shell in such a fashion as to result in a lower round-trip voltage drop. The electrical behavior during the line-to-line fault is that of a coaxial line except for the effects of resistivity in the metal shell. All electric and magnetic fields are contained inside the shell. The external space magnetic field becomes zero.

Even when the raceway does not provide the sole equipment-grounding path but is supplemented by an equipment-grounding conductor wire or bus enclosed within, the raceway provides a parallel path and also protects the wire and bus from physical damage. Where connected as a parallel path, the raceway not only lowers the total impedance of the grounding path, but carries 95% or more of the current. Furthermore, the raceway provides the additional protection of a grounding shield. For instance,

in the unlikely event that a phase conductor becomes damaged within a raceway and an uninsulated spot resulting from the damage comes in contact with a grounded raceway, the resulting fault current would be carried to the grounded service-equipment enclosure through the main bonding jumper to the system and the overcurrent device.

By contrast, a damaged phase conductor in a nonmetallic raceway could remain exposed to accidental contact indefinitely. A fault within a nonmetallic raceway may or may not provide a low-impedance return path to trip the overcurrent device protecting the circuit. Some designers prefer only bare-equipment grounding conductors in nonmetallic raceways to facilitate early detection when conductor insulation damage occurs.

An equipment-grounding conductor and a grounded-circuit conductor are not the same electrically. It is unsafe to connect the grounded-circuit conductor to the raceway or to any other non-current-carrying equipment at any point on the load side of the service equipment.

If the grounded-circuit conductor is not insulated from the non-current-carrying equipment on the load side of the service equipment, then the raceways and possibly the piping and ducts will carry continuously a portion of the neutral current on systems having a grounded neutral, and a portion of the phase current on systems having a grounded phase conductor. If this section of the grounded conductor becomes open circuited, the full circuit-to-ground voltage is impressed upon such non-current-carrying equipment in the event of a fault-to-ground.

The flow of such currents over raceways and other equipment can be a serious fire hazard. If a raceway is in casual contact with a pipe, duct, or building structure member, arcing may present a shock hazard. For these reasons, the grounded circuit conductor must be insulated throughout its length, as would an ungrounded circuit conductor. Since World War II *NEC® Section 250-140* has permitted the grounded conductor to be used for grounding the frames of ranges, wall-mounted cooking units, and clothes dryers under some conditions, but this has been changed in the 1996 *NEC®.* This practice will continue to be permitted for existing installations. However, all new installations of these appliances will be required to be wired in accordance with all the requirements of *Article 250.* The frames and all non-current-carrying parts are now required to be grounded with a separate equipment grounding conductor; the grounded conductor will not be permitted to be used for this purpose. This change will require that all ranges, counter-mounted cooking equipment, and clothes dryers be connected with a four-wire cord and receptable.

See Appendix Figures A-16–A-20 for maximum lengths for raceways and conductors used for equipment grounding.

CAPABLE OF CARRYING THE AVAILABLE FAULT CURRENT

Short-Circuit Calculations

Throughout the *National Electrical Code®*, as well as in industry standards, you will find references to fault currents, short-circuit currents, ground-fault currents, interrupting ratings, impedances of transformers, and the like.

- *Section 110-9.* **Interrupting Rating.** Equipment intended to interrupt current at fault levels shall have an interrupting rating sufficient for the nominal circuit voltage and the current that is available at the line ter-mnals of the equipment."

- *Section 110-10.* **Circuit Impedance and Other Characteristics.** The over-current protective devices, the total impesance, the component short-circuit current ratings, and other characteristics of the circuit to be protected shall be selected and coordinated to permit the circuit-protective devices used to clear a fault to do so without extensive damage to the electrical components of the circuit. This fault shall be assumed to be either between two or more of the circuit conductors, or between any circuit conductor and the grounding conductor or enclosing metal raceway. Listed products applied in accordance with their listing shall be considered to meet the requirements of this section.

These tough code requirements mean that we need to know the available fault currents at the main service equipment as well as at all points in the system. Too often, engineers, electricians, and contractors look for adequate interrupting ratings of the fuses or breakers in the main service equipment only. This is wrong. The Code tells us to observe available fault currents throughout the electrical system. This means panelboards, motor controllers, motor control centers, distribution centers, and so on.

Determining Short-circuit Current at Various Distances From Transformers, Switchboards, Panelboards, and Load Centers Using the Point-to-Point Method

A simple method of determining the available short-circuit current (also called fault current) at various distances from a given location is the "point-to-point method." Reasonable accuracy is obtained when this method is used with 3-phase and single-phase systems.

The following procedure demonstrates the use of the "point-to-point method."

Step 1. Determine the full-load rating of the transformer—in amperes—from the transformer nameplate, tables, or the following formulas:

a) For 3-phase transformers:

$$I_{FLA} = \frac{kVA \times 1\,000}{E_{L-L} \times 1.73}$$

where E_{L-L} = Line-to-line voltage

b) For single-phase transformers:

$$I_{FLA} = \frac{kVA \times 1\,000}{E_{L-L}}$$

Step 2. Find the percent impedance (Z) on nameplate of transformer.
Step 3. Find the transformer multiplier "M_1":

$$M_1 = \frac{100}{\text{transformer \% impedance (Z)} \times 0.9}$$

(**Note:** Because the marked transformer impedance can vary +/– 10% per the UL standard, the 0.9 factor above takes this into consideration to show "worst case" conditions.)
Step 4. Determine the transformer let-through short-circuit current at the secondary terminals of transformer. Use tables or the following formula:

a) For 3-phase transformers (L-L-L):

$$I_{SCA} = \text{transformer}_{FLA} \times \text{multiplier "}M_1\text{"}$$

b) For 1-phase transformers (L-L):

$$I_{SCA} = \text{transformerFLA} \times \text{multiplier "}M_1\text{"}$$

c) For 1-phase transformers (L-N):

$$I_{SCA} = \text{transformer}_{FLA} \times \text{multiplier "}M_1\text{"} \times 1.5$$

The L-N fault current is higher than the L-L fault current at the secondary terminals of a single-phase, center-tapped transformer. At some distance from the terminals, depending on the wire size, the L-N fault current is lower than the L-L fault current. This can vary from 1.33

to 1.67 times. These figures are based on the change in the turns ratio between primary and secondary, infinite-source impedance, a distance of zero feet from the terminals of the transformer, and 1.2 times oX (reactance) and 1.5 times oX (resistance) for the L-N vs. L-L resistance and reactance values.

For simplicity, in Step 4c we used an approximate multiplier of 1.5. First, do the L-N calculation at the transformer secondary terminals, Step 4c, then proceed with the point-to-point method.

Step 5. Determine the "f" factor:

a) For 3-phase faults:

$$f = \frac{1.73 \times L \times I}{N \times C \times E_{L-L}}$$

b) For 1-phase, line-to-line (L-L) faults on 1-phase, center-tapped transformers:

$$f = \frac{2 \times L \times I}{N \times C \times E_{L-L}}$$

c) For 1-phase, line-to-neutral (L-N) faults on 1-phase, center-tapped transformers:

$$f = \frac{2 \times L \times I}{N \times C \times E_{L-N}}$$

where
L = the length of the circuit to the fault, in feet.
I = the available fault-current—in amperes—at the beginning of the circuit.
C = the constant derived from the "C" tables (see page TK) for the specific type of conductors and wiring method.
E = the voltage, line-to-line or line-to-neutral. See step 4a, b, and c to decide which voltage to use.
N = the number of conductors in parallel.
X = the fault current in amperes at the transformer terminals.

Step 6. After finding the "f" factor, refer to Chart M (see page 51) the appropriate value of the multiplier "M_2" for the specific "f" value. Or, calculate as follows: $M_2 = \dfrac{1}{1 + f}$

Step 7. Multiply the available fault current at the beginning of the circuit by the multiplier "M_2" to determine the available symmetrical fault current at the fault.

I_{SCA} at fault = I_{SCA} at beginning of circuit $\times$ "M_2"

"C" VALUES—Copper Conductors

| | Three Single Conductors | | | | | | Three Conductor Cable | | | | | |
| | Steel Conduit | | | Nonmagnetic Conduit | | | Steel Conduit | | | Nonmagnetic Conduit | | |
AWG or kcmil	600V	5KV	15KV	600V	5KV	15KV	600V	5KV	15KV	600V	5KV	15KV
14	389	—	—	389	—	—	389	—	—	389	—	—
12	617	—	—	617	—	—	617	—	—	617	—	—
10	981	—	—	981	—	—	981	—	—	981	—	—
8	1557	1551	1557	1556	1555	1558	1559	1557	1559	1559	1558	1559
6	2425	2406	2389	2430	2417	2406	2431	2424	2414	2433	2428	2420
4	3806	3750	3695	3825	3789	3752	3830	3811	3778	3837	3823	3798
3	4760	4760	4760	4802	4802	4802	4760	4790	4760	4802	4802	4802
2	5906	5736	5574	6044	5926	5809	5989	5929	5827	6087	6022	5957
1	7292	7029	6758	7493	7306	7108	7454	7364	7188	7579	7507	7364
1/0	8924	8543	7973	9317	9033	8590	9209	9086	8707	9472	9372	9052
2/0	10755	10061	9389	11423	10877	10318	11244	11045	10500	11703	11528	11052
3/0	12843	11804	11021	13923	13048	12360	13656	13333	12613	14410	14118	13461
4/0	15082	13605	12542	16673	15351	14347	16391	15890	14813	17482	17019	16012
250	16483	14924	13643	18593	17120	15865	18310	17850	16465	19779	19352	18001
300	18176	16292	14768	20867	18975	17408	20617	20051	18318	22524	21938	20163
350	19703	17385	15678	22736	20526	18672	22646	21914	19821	24904	24126	21982
400	20565	18235	16365	24296	21786	19731	24253	23371	21042	26915	26044	23517
500	22185	19172	17492	26706	23277	21329	26980	25449	23125	30028	28712	25916
600	22965	20567	17962	28033	25203	22097	28752	27974	24896	32236	31258	27766
750	24136	21386	18888	28303	25430	22690	31050	30024	26932	32404	31338	28303
1000	25278	22539	19923	31490	28083	24887	33864	32688	29320	37197	35748	31959

"C" VALUES—Aluminum Conductors

| | Three Single Conductors | | | | | | Three Conductor Cable | | | | | |
| | Steel Conduit | | | Nonmagnetic Conduit | | | Steel Conduit | | | Nonmagnetic Conduit | | |
AWG or kcmil	600V	5KV	15KV	600V	5KV	15KV	600V	5KV	15KV	600V	5KV	15KV
14	236	—	—	236	—	—	236	—	—	236	—	—
12	375	—	—	375	—	—	375	—	—	375	—	—
10	598	—	—	598	—	—	598	—	—	598	—	—
8	951	950	951	951	950	951	951	951	951	951	951	951
6	1480	1476	1472	1481	1478	1476	1481	1480	1478	1482	1481	1479
4	2345	2332	2319	2350	2341	2333	2351	2347	2339	2353	2349	2344
3	2948	2948	2948	2958	2958	2958	2948	2956	2948	2958	2958	2958
2	3713	3669	3626	3729	3701	37672	3733	3719	3693	3739	3724	3709
1	4645	4574	4497	4678	4631	4580	4686	4663	4617	4699	4681	4646
1/0	5777	5669	5493	5838	5766	5645	5852	5820	5717	5875	5851	5771
2/0	7186	6968	6733	7301	7152	6986	7327	727	7109	7372	7328	7201
3/0	8826	8466	8163	9110	8851	8627	9077	8980	8750	9242	9164	8977
4/0	10740	10167	9700	11174	10749	10386	11184	11021	10642	11408	11277	10968
250	12122	11460	10848	12862	12343	11847	12796	12636	12115	13236	13105	12661
300	13909	13009	12192	14922	14182	13491	14916	14698	13973	15494	15299	14658
400	16670	15355	14188	18505	17321	16233	18461	18063	16921	19587	19243	18154
500	18755	16827	15657	21390	19503	18314	21394	20606	19314	22987	22381	20978
600	20093	18427	16484	23451	21718	19635	23633	23195	21348	25750	25243	23294
750	21766	19685	17686	25976	23701	20934	26431	25789	23750	29036	28262	25976
1000	23477	21235	19005	28778	26109	23482	29864	29049	26608	32938	31919	29135

CHART M (multiplier)

f	M	f	M
0.01	0.99	1.20	0.45
0.02	0.98	1.50	0.40
0.03	0.97	2.00	0.33
0.04	0.96	3.00	0.25
0.05	0.95	4.00	0.20
0.06	0.94	5.00	0.17
0.07	0.93	6.00	0.14
0.08	0.93	7.00	0.13
0.09	0.92	8.00	0.11
0.10	0.91	9.00	0.10
0.15	0.87	10.00	0.09
0.20	0.83	15.00	0.06
0.30	0.77	20.00	0.05
0.40	0.71	30.00	0.03
0.50	0.67	40.00	0.02
0.60	0.63	50.00	0.02
0.70	0.59	60.00	0.02
0.80	0.55	70.00	0.01
0.90	0.53	80.00	0.01
1.00	0.50	90.00	0.01
		100.00	0.01

$$M = \frac{1}{1+f}$$

"C" Values—Busways

Ampacity	Plug-In Busway		Feeder Busway		High Imped. Busway
	Copper	Aluminum	Copper	Aluminum	Copper
225	28700	23000	18700	12000	—
400	38900	34700	23900	21300	—
600	41000	38300	36500	31300	—
800	46100	57500	49300	44100	—
1000	69400	89300	62900	56200	15600
1200	94300	97100	76900	69900	16100
1350	119000	104200	90100	84000	17500
1600	129900	120500	101000	90900	19200
2000	142900	135100	134200	125000	20400
2500	143800	156300	180500	166700	21700
3000	144900	175400	204100	188700	23800
4000	—	—	277800	256400	—

Review Questions

1. Conductors may be installed in raceways in both solid or stranded configurations. What is the largest solid conductor that may be installed in a raceway?

2. What is the difference between an insulated conductor and a covered conductor?

3. Insulated conductor types are clearly identified by which letter when they are designed and approved for installations in wet locations?

4. When installing six #12 THWN conductors in a 3/4-inch rigid metal conduit, what is the percentage of ampacity adjustment required by the *NEC*®?

5. When installing three #4 THW conductors, three #10 THWN conductors, and three #12 THWN conductors, what ampacity adjustment is required and what is the allowable ampacity for these three sizes of conductors?

6. A two-wire branch circuit is routed in nonmetallic sheath cable, #12, two-wire with ground, 250 feet from the panel to serve a 120-volt appliance that draws 15.5 amps. What would the voltage drop be on this circuit?

7. What is generally the smallest conductor size for the grounded neutral and the ungrounded phase conductors to be run in parallel?

8. A small commercial building is supplied by two services, 480-volt/277-volt service used for power and lighting and a 208-volt/120-volt service used for appliance and branch-circuit loads. What color must the grounded neutral conductor be on these two systems?

9. Where circuits are run in parallel, what is the minimum size equipment-grounding conductor required to be run in each parallel raceway?

10. *NEC*® Section 250-51 requires the equipment-grounding conductor to be capable of carrying the available fault current that may be applied to a circuit. How many amperes will a #12 solid conductor carry for five seconds?

11. For how many seconds will a #2 copper conductor carry 10,000 amperes before it burns off?

Chapter 4

General Circuit Requirements

Conductors in an electrical system are divided into three categories:

- Service conductors (*NEC® Article 230*)
- Feeders (*Articles 215* and *225*)
- Branch-circuit conductors (*Articles 210* and *225*).

Service conductors, defined in *NEC® Article 100*, are those conductors that generally supply the premises. They extend from the street main or from the transformers to the meter and service equipment on the premises. (See Figures 4-1, 4-3.)

Feeders may consist of a single circuit of conductors connecting the service equipment to the branch-circuit panelboards or the main circuits or subfeeder circuits supplying distribution panelboards and branch-circuit panelboards within the building or structure supplied or may be supplying additional buildings or structures on the premises, respectively. Feeders are defined in *NEC® Article 100* as all circuit conductors between the service equipment or the source of a separately derived system in the final branch-circuit overcurrent device. Generally, a feeder will not supply a single overcurrent device, such as an outdoor heat pump or air conditioner condenser. The overcurrent device lo-

cated at the unit is termed Supplementary Overcurrent Device, and the circuit supplying it is a branch circuit, not a feeder.

A branch circuit is defined in *Article 100* as the circuit conductors between the final overcurrent device protecting the circuit and the outlet(s). See *Sections 240-9* and *240-10* for thermal relays, supplementary overcurrent protection, and other devices. This is to clarify that these supplementary overcurrent devices located within equipment such as thermal devices in light fixtures are not the final overcurrent device in the circuit and outlet. Branch circuit is further defined in *Article 100* as appliance branch circuits. An appliance branch circuit supplies energy to one or more outlets to which appliances are connected; such circuits have no permanently connected lighting fixtures that are not part of an appliance.

General purpose branch circuits, also defined in *NEC® Article 100*, supply a number of outlets for lighting and appliances. Individual branch circuits supply only one piece of utilization equipment, such as a branch circuit supplying a window air conditioner, a microwave oven, a dishwasher, and so on. A multiwire branch circuit is known by the industry as an Edison circuit. A multiwire branch circuit consists of two or more ungrounded conductors having a potential difference between them, and a grounded

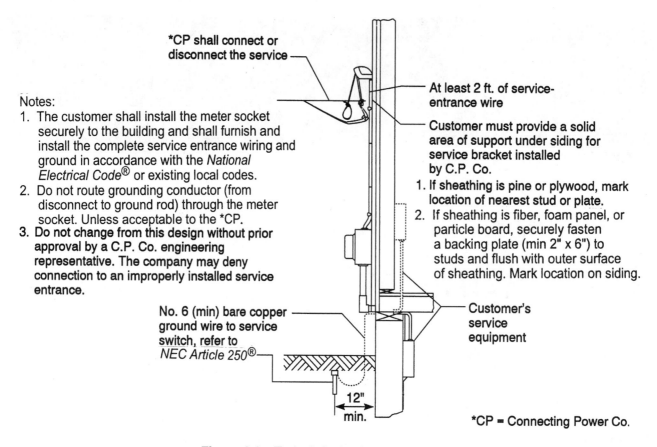

*CP shall connect or disconnect the service

Notes:
1. The customer shall install the meter socket securely to the building and shall furnish and install the complete service entrance wiring and ground in accordance with the *National Electrical Code®* or existing local codes.
2. Do not route grounding conductor (from disconnect to ground rod) through the meter socket. Unless acceptable to the *CP.
3. Do not change from this design without prior approval by a C.P. Co. engineering representative. The company may deny connection to an improperly installed service entrance.

No. 6 (min) bare copper ground wire to service switch, refer to *NEC Article 250®*

At least 2 ft. of service-entrance wire

Customer must provide a solid area of support under siding for service bracket installed by C.P. Co.

1. If sheathing is pine or plywood, mark location of nearest stud or plate.
2. If sheathing is fiber, foam panel, or particle board, securely fasten a backing plate (min 2" x 6") to studs and flush with outer surface of sheathing. Mark location on siding.

Customer's service equipment

12" min.

*CP = Connecting Power Co.

Figure 4-1   Typical single-phase service drop

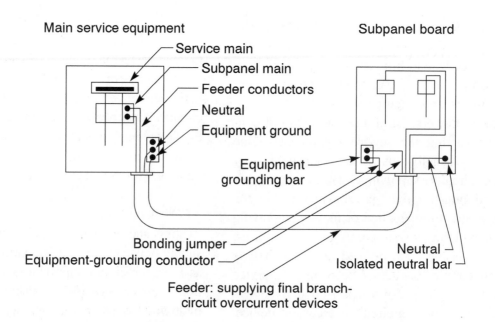

Main service equipment

Subpanel board

Service main
Subpanel main
Feeder conductors
Neutral
Equipment ground

Equipment grounding bar

Bonding jumper
Equipment-grounding conductor

Neutral
Isolated neutral bar

Feeder: supplying final branch-circuit overcurrent devices

Figure 4-2   Feeder supplying final branch-circuit overcurrent devices

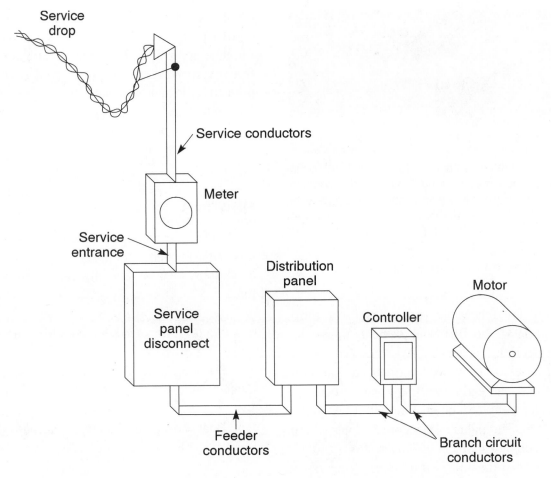

Figure 4-3 Specific conductor types as defined in *NEC® Article 100*

conductor having an equal potential difference between it and each ungrounded conductor of the circuit, and which is connected to the neutral (grounded) conductor of the system.

OVERCURRENT

Overcurrent protection devices, which we generally think of as fuses and circuit breakers, appear in the earliest editions of the *National Electrical Code®*. They are mentioned in several sections of the Underwriter's Association Rules and Requirements adopted August 31, 1897. However, they did not perform the same functions or have the degree of accuracy that they have today. Fuses as a form of overcurrent protection have been around a long time. Thomas A. Edison was granted his first patent on a fuse May 4, 1880. This was Patent No. 227,226. His patent referred to a safety conductor and did not use the word "fuse." Figure 4-4 shows one of Edison's handmade fuses from around 1887. His patent, in part, was worded as follows: "This small conductor has such a degree of conductivity as to readily allow the passage of the amount of current designed for its particular branch but no more. If, from any cause whatever, an abnormal amount of current, large enough to injure the translation devices or to cause a waste of energy, is diverted through a branch, the small safety wire becomes heated and melts away, breaking the overloaded branch circuit. It is desireable, however, that a few drops of hot molten metal resulting therefrom should not be allowed to fall upon carpets or furniture. Hence, I enclose the safety wire in a jacket or shell of non-conducting material."

The manufacture of fuse wire began in 1885. Manufacturers began to produce fuses in 1913 and 1914. Prior to this time most electrical fuses were manufactured by electricians. In 1921 the world's first clear-plug fuse was introduced, and in 1938

Figure 4-4 Quite a contribution to the electrical industry from an individual who had only three months of formal education in his lifetime, and was not one bit interested in financial matters. (Courtesy of Cooper Industries, Bussmann Division)

the time-delay renewable length fuse was introduced. In 1931 the first dual-element, time-delay fuse was introduced, and in 1934 the Type S-size rejecting-fuse stat fuses were introduced. The mid- to late 1940s and early 1950s saw the introduction and development of current-limiting fuses in the marketplace. The industry has come a long way in the high-tech development of modern fuses and circuit breakers, including fast-acting, time-delay, slow, combination fast-acting, time-delay, current-limiting, high-voltage, low-voltage, medium-voltage, and cable-limiting fuses.

The *National Electrical Code®* covers the general requirements for overcurrent protection in *Article 240.* Current in excess of the normal rating for conductors or equipment causes excessive heating and, unless properly and rapidly interrupted, may cause a fire. Overcurrent protection is provided to prevent this type of damage to conductors or equipment. Overcurrent protection for conductors usually does not provide adequate protection for equipment. Therefore, additional protective devices for most equipment are needed. The Code discusses the general requirements for overcurrent protection in *Article 240. In addition, specific overcurrent requirements are found in many articles of the Code. For example, overcurrent protection for motors is found in Article 430,* for appliances in *Article 422,* for air-conditioning equipment in *Article 440,* and for transformers in *Article 450.*

Proper overcurrent protection is an extremely important issue. Overcurrent protective devices are

the safety valve of an electrical system. Proper overcurrent protection keeps conductors, motors, transformers, appliances, and other electrical equipment from overheating because of overloads. Overcurrent protective devices also take the equipment off the line under short-circuit or ground-fault conditions.

There are two forces at work that cause damage to electrical systems under overcurrent load conditions.

HEATING

Heat is damaging to conductor insulation, motors, transformers, appliances, etc. For long-time conditions, heat is generally calculated by the formula

$$\text{Watts} = \text{Amperes Squared} \times \text{Resistance (I}^2\text{R)}$$

For short-time conditions, such as short circuits, the term used is

$$\text{Heat} = \text{Amperes Squared} \times \text{Seconds (I}^2\text{T)}$$

MAGNETIC

Magnetic forces can pull conductors out of their lugs and can force them against sharp edges, which can result in further damage. They can bend bus bars and rip them from their mountings. Magnetic forces vary as the square of the peak current. The effects of overcurrents on equipment is closely related to how much current will flow . . . and how long.

Protecting devices are actuated either thermally or magnetically. The most common types of overcurrent protective devices for conductors and some types of equipment are fuses and circuit breakers. For motors, thermal overload devices are often used. In larger installations, protective relays often are specified.

Standard ampere ratings for fuses and nonadjustable trip circuit breakers are from 15 to 6000; fuses have additional standard sizes of 1, 3, 6, 10, and 601, and many nonstandard sizes are also available. The specific standard sizes are listed in *NEC®* Section 240-6, necessarily for applying the requirements of the Code, which in many instances permit adjustment to the "next higher overcurrent device" or to the "next lower overcurrent device." In general, overcurrent devices must be located where they are readily accessible (as specified in *NEC® 240-24*) but not exposed to physical damage or to material that is readily combustible.

> *Voltage Rating:* Branch circuit breakers and the most commonly used power fuses are marked with their voltage rating, found on the label. Do not use these overcurrent devices at voltage levels greater than their rating. In general they can be used at lower voltages than their rated voltage. Most overcurrent devices are rated for AC current; for DC application check the markings on the device or manufacturer's data.

> **Example:** An electrical installation will be supplied from a 480-volt, 3-phase, Delta-connect system. What voltage rating should the fuses have? (a) 250 volts (b) 300 volts (c) 600 volts Answer: (c) 600 volts. *NEC® 240-60.*

Simple calculation of circuit loads and conductor ampacities may not always determine the proper overcurrent protection. In a circuit with high impedance, a serious delay may occur in the operation of an overcurrent protective device, increasing the hazard of fire and electrical shock.

Interrupting Rating

NEC® Section 110-10 requires that the overcurrent protective devices, the total impedance, and other characteristics of the circuit be so selected and coordinated that the circuit and its components are protected against damage that could result from either a phase-to-phase or phase-to-ground fault.

Overcurrent protective devices must have a sufficient interrupting rating to open the circuit under both simple overloads and maximum fault conditions. Without such an interrupting rating, an overcurrent protective device may fail physically at a time when its rapid operation is essential.

Series-Connected Ratings

Series ratings of overcurrent devices have become popular in recent years. This practice permits an overcurrent device with a lower interrupting rating than the system available fault current to be applied on the load side of another overcurrent device with a higher rating. One aspect of series ratings to remember is that on high faults within the system, both the main and branch-circuit device may open. This does not provide selective coordination of the overcurrent devices. If such selectivity is needed or desired, the use of a fully rated overcurrent device is required.

A popular application of a series ratings is the 22/10 rating. This system typically has a 22,000-ampere interrupting rating breaker as a main device and 10,000-ampere interrupting rating breakers connected as branch or feeder devices. When properly applied, this combination can be used on systems with up to 22,000 amperes of available fault current. Series ratings are available in many sizes and interrupting levels. Contact manufacturers to see if these ratings can be applied with their equipment.

Series ratings can be found for breaker-breaker and fuse-breaker combinations. However, it is critical to remember some important points when applying series ratings. Recognized combinations cannot be determined simply by looking at time-current curves. The combination is required by *NEC® Section 240-86(a)* to be marked on the end-use equipment (switchboards and panelboards, for example) with specific information about what devices are permitted in the series rating. This testing and listing is important because the interaction of the main and branch unit can be addressed during the listing procedure. Let-through charts and curves for overcurrent devices cannot be utilized on circuit breakers which begin to open in the first 1/2 cycle.

Refer to the *NEC®* for full representation of provisions affecting overcurrent protection as listed in *Section 240-2, Protection of Equipment* and *Section 240-3, Protection of Conductors.*

Modern electrical distribution systems can include various types of circuit-breaker overcurrent protection. Breakers are available that can be actuated thermally, magnetically, or through electronic sensing means. One advantage provided by circuit breakers is that they are normally capable of being reset after an overload or short circuit occurs. Breakers can also be provided with many accessories such as integral ground-fault sensing, shunt-trip mechanisms, alarm switches, and auxiliary contact assemblies.

CIRCUIT BREAKERS

Typical circuit breakers have a thermal-magnetic response mechanism. The thermal portion of the circuit breaker is actuated under overload conditions, and in many cases this mechanism is similar to the bimetal assembly in a thermostat. As current passes through the bimetal or a portion of the breaker assembly close to the bimetal, the heat generated by the current causes the bimetal to deflect and trip the breaker. Thermal assemblies are calibrated for the particular amperage breaker so that if the sustained overload continues, the breaker trips.

The magnetic portion of a breaker is designed to handle the short-circuit conditions an electrical circuit may experience. Magnetic assemblies are activated by passing the high-fault current through the breaker assembly and in turn the assembly provides a quick response time to clear the short circuit. When thermal and magnetic detection are combined, the

Figure 4-5b   Types IFL, IKL & ILL "I LIMITER" molded case current limiting circuit breakers (Courtesy of Square D Company)

result is a versatile overcurrent device that can provide reliable and resettable protection for a distribution system.

Recently, technological advancements have brought a new style of circuit breaker to the market, the electronic trip circuit breaker. It has replaced the thermal and magnetic assemblies with current transformers and electronic circuitry to provide an adjustable and versatile overcurrent device. Electronic trip breakers can be provided with field adjustments that allow tailoring to individual distribution systems. Response times of the breakers' functions can be adjusted to help coordinate the system and provide

Figure 4-5a Type MAL thermal-magnetic molded case circuit breaker (Courtesy of Square D Company)

Figure 4-5c   Full-function micrologic trip circuit breakers featuring: RMS sensing; full-function adjustability; defeatable instantaneous function zone electrive interlocking; trip indication; 100% rated; plus many additional features. (Courtesy of Square D Company)

excellent protection for increased loads. Some of these breakers can also be interconnected to provide increased selective coordination when clearing a fault condition by sending communication signals between breakers.

Circuit breakers also provide versatility owing to the availability of many accessories allowing the breaker to perform other tasks in addition to being an overcurrent device. Some of the included accessories are shunt trip units, which permit remote tripping of the circuit breaker by energizing a coil attached to the breaker frame, which in turn mechanically trips the breaker. Applications such as emergency stop or shutoff for equipment make this device useful.

Auxiliary contacts can be attached to the breaker to provide remote status indication of whether the breaker is open or closed. This permits remote monitoring of the breaker status and therefore also permits the breaker's connection to equipment whose operation requires knowledge of whether it is being open or closed.

Alarm switches provide remote monitoring of a breaker that has been tripped. Applications for this device include critical loads that, by shutting down for extended periods, create undesirable situations.

Undervoltage trip accessories trip the circuit breaker if the voltage of the system to which it is connected drops below a preset level. The breaker must then be reset after power is restored. Applications requiring that the equipment not be restarted automatically after a power outage can be supplied from a breaker with this frame. The breaker needs to be manually reset after power is restored.

Circuit breakers are covered in *Article 240* of the *NEC®*. UL489 is the standard for evaluation of these breakers.

FUSES

Types of Fuses

Modern branch circuit fuses are available in ampere ratings from 1/10 to 6000 amperes and in voltage ratings of 125, 250, 300, 480 and 600 volts. Figure 4-6 shows the internal parts of a common Fusetron® dual-element, time-delay fuse. Note in the cutaway illustration that the overload element opens

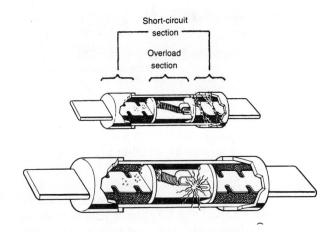

Figure 4-6 Cutaway view of a Fusetron® dual-element, time-delay fuse. 250 and 600 volts. 200,000 amperes interrupting rating. On overloads, the spring-loaded trigger assembly opens. On short-circuits or heavy ground faults, the fuse element(s) are generally made of copper. (Courtesy of Cooper Bussmann)

under overload conditions; the copper short-circuit element opens under short-circuit or ground-fault conditions. These fuses have an interrupting rating of 200,000 amperes and are available in 250-volt and 600-volt ratings. These fuses are Class RK5.

> **Example**: 1. A true dual-element fuse opens overloads in its (a) short-circuit section, (b) overload section. 2. It will open short-circuits in its (c) short-circuit section, (d) overload section. Answer: 1. (b) overload section. 2. (c) short-circuit section.

Some fuses, in addition to having the time-delay characteristic, also are classified as being current limiting. Figure 4-7 shows the internal parts of a Class RKI low-Peak dual-element, time-delay, current-limiting fuse. It is very similar to the Fusetron fuse in the overload range but much faster (more current-limiting) under short-circuit conditions. This is the result of a different design of the short-circuit section of the fuse. Generally, the short-circuit element is made of pure silver, although sometimes it is made of copper. These fuses can interrupt fault currents of not over 300,000 amperes rms symmetrical. Because they have an interrupting rating that is higher than the standard 200,000 amperes they are

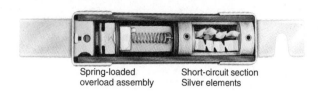

Spring-loaded
overload assembly

Short-circuit section
Silver elements

Figure 4-7 Cutaway view of a Low-Peak®dual-element, time-delay, current-limiting fuse. 250 and 600 volts. 300,000 amperes interrupting rating. On overloads, the spring-loaded trigger assembly opens. On short circuit or heavy ground faults, the fuse elements in the short-circuit section open extremely fast, thus the term "current limiting." The fuse elements are generally made of silver, although some types are made of copper. (Courtesy of Cooper Bussmann)

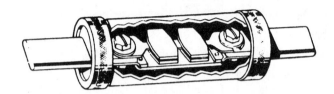

Figure 4-9 Class H cartridge fuse. Illustration shows a renewable-type fuse in which the Blown link may be replaced. 250 and 600 volts. Interrupting rating is 10,000 amperes. (Courtesy of Cooper Bussmann)

listed as special purpose fuses. They are available in 250-volt and 600-volt ratings. Some special purpose Class L and J fuses have interrupting rating of 300,000 amperes.

Some fuses have no intentional time-delay designed into them. They are intended to be fast over all values of overcurrent. Figure 4-8 illustrates the copper-on-silver short-circuit link inside a current-limiting fuse. The less-expensive ordinary fuses are called "one-time" fuses and might have zinc links. The fuse in Figure 4-6 has an interrupting rating of 200,000 amperes rms symmetrical. These are fast acting Class RK1 fuses.

The less costly fuses might have an interrupting rating of 10,000 amperes (Class H) or 50,000 amperes (Class K). Fast-acting fuses generally are used where there are no high momentary inrush currents,

such as occur when motors are started. Transformers also have high-inrush currents when energized.

> **Example:** Ordinary Class H fuses have no rejection feature. They have an interrupting rating of (a) 5,000, (b) 10,000, (c) 50,000 amperes. Answer: (b) 10,000 amperes.

Some fuses that can have their links replaced are called renewable fuses. They have a maximum interrupting rating of 10,000 amperes and are available in 250-volt and 600-volt ratings, 1 through 600 amperes. These fuses were common many years ago. But in today's modern electrical systems, where available fault currents are high, and where safety and proper overcurrent protection is a must, renewable fuses should not be installed. These are Class H fuses.

> **Example**: Renewable fuses have replaceable zinc links. These types of fuses are for use on circuits with not over (a) 5,000, (b) 10,000, (c) 25,000 amperes available. Answer: (b) 10,000 amperes.

Becoming more and more popular because of their small size are Low Peak®, Class J, time-delay, current-limiting fuses. This type of fuse is extremely fast acting under fault conditions but has time-delay

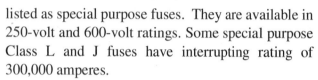

Figure 4-8 Cutaway view of a Limitron®current-limiting, fast-acting, single-element fuse. 250 and 600 volts. 200,000 amperes interrupting rating. This fuse has no intentional time-delay feature and is very fast under short-circuit or ground fault conditions. (Courtesy of Cooper Bussmann)

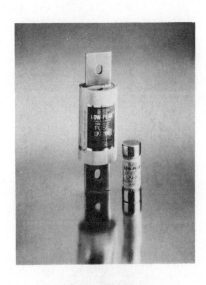

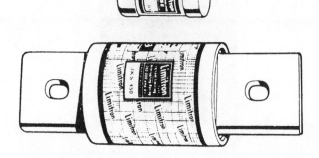

Figure 4-10 Class J, Low-Peak® current-limiting, time-delay fuse. 600 volts. 300,000 ampere interrupting rating. These fuses have excellent time-delay in the overload region, yet are extremely fast when short circuits or ground faults occur. They also have different dimensions than ordinary fuses. Therefore, they will fit only in switches suitable for use with Class J fuses. No other class of fuse will fit in Class J clips. (Courtesy of Bussmann Cooper Industries)

in the low-overload range. This feature provides the ultimate in protection for motor circuits and other circuits that are subjected to inrush currents. Class J fuses are much smaller than ordinary cartridge fuses. Special Purpose Low Peak® fuses have a 600-volt rating, a 300,000-ampere interrupting rating, and ampere ratings 1 through 600 amperes. Industry standard Class J fuses have a 200,000-ampere interrupting rating.

> **Example:** Class J fuses are available in time-delay types that are excellent for motor circuits. Properly sized they can provide Type 2

protection (no damage) for motor controllers. Class J fuses are (a) current-limiting, (b) not current-limiting. Answer: (a) current-limiting.

Another type of smaller-dimension power fuse is the Class T, current-limiting fuse. Suitable for use on motor circuits when sized large enough to handle the starting current, they find their application in OEM equipment where space is a consideration. Class T fuses are very fast acting. The interrupting rating of Class T fuses is 200,000 amperes in both 300-volt and 600-volt ratings. Class T fuses are available in 1 through 1,200 amperes.

All cartridge fuses that are current-limiting and that have a high interrupting rating have some sort of rejection feature, either by unique size or shape. They might have slots in their blades or annular rings in their ferrule (end cap). These are termed Class R fuses. Other fuses have holes in their blades to match corresponding mounting studs in the fuseholder. These are called Class T, J, or L fuses. Because of these rejection features, they cannot be replaced with ordinary Class H or Class K fuses. Thus, the installation of proper fuseholders ensures that only proper, high-interrupting-rating, current-limiting fuses will be used as replacements.

Class R fuses fit in both new Class R type fuse clips and in the older types of nonrejection fuse clips.

> **Example:** Fuses that are current-limiting and have a high interrupting rating have some sort of (a) rejection feature, (b) renewable link feature. Answer: (a) rejection feature.

Fuses rated over 600 amperes are called Class L fuses; they are rated 600 volts. Special Purpose Low Peak® Class L fuses have an interrupting rating of 300,000 amperes. Ordinary Class L fuses have an interrupting rating of 200,000 amperes. Class L fuses are mounted by means of nuts threaded onto studs in the switches. Generally, the switches that Class L fuses are installed in are of the "bolted pressure contact" type. Class L fuses are available in 601 through 6,000 amperes.

> **Example:** A time-delay, current-limiting fuse is marked 2,000 amperes, 600 volts. This fuse is a (a) Class RK1, (b) Class L, (c) Class H fuse. Answer: (b) Class L.

Figure 4-11 Class T, current-limiting, fast-acting. 200,000 amperes interrupting rating. 300 and 600 volts. Has very little time-delay. Generally used for protection of circuit breaker panels and for circuits that do not have high inrush loads, such as motors. When used on high-inrush types of loads, generally size at 300% so as to be able to override the momentary inrush current. These fuses have different dimensions than ordinary fuses. They will not fit in switches made for other classes of fuses. Nor will other classes of fuses fit into a Class T disconnect switch. (Courtesy of Bussmann Cooper Industries)

Plug fuses are available in the Edison base type and Type S fusestat. They come in time-delay types and normal speed. The time-delay type is preferred where the load serves one or more motors. The time-delay Edison base and Type S fuses can be found in 3/10 up to and including 30 amperes. When Type S fuse adapters are installed for a specific ampere rating, it is virtually impossible to insert a fuse of a higher ampere rating.

> **Example:** Code requirements for plug fuses are found in *Article 240, Part E, Sections 240-50* through *240-54*. Type S fuses require that an adapter be inserted into the Edison base fuseholder. For a given ampere rating, this adaptor has a unique feature that will not allow a higher-ampere-rated fuse to be installed. For an adaptor marked "SA15," which of the following sizes can be installed? (a) 15-ampere, (b) 20-ampere, (c) 30-ampere. Answer: (a) 15-ampere.

Figure 4-12 Low-Peak® current-limiting, time-delay, multi-element, 601-6000 ampere, 600-volt Class L fuse. 300,000 amperes interrupting rating. Links made of silver. Will hold 500% of rated current for minimum of four seconds. Good for use in high-inrush circuits (motors, transformers, and other inductive loads). The filler material is silica sand. Sometimes referred to as a "silver-sand" fuse. (Courtesy of Bussmann Cooper Industries)

BRANCH CIRCUITS

Branch circuits are classified according to the size of the overcurrent device and are limited to 15-, 20-, 30-, 40-, or 50-ampere sizes when supplying two or more outlets. Branch-circuit requirements for each rating are given in the *NEC® Table 210-24*. Circuits supplying a single outlet may be any size as needed.

For specific purpose branch circuits, see *NEC® Section 210-2*. The diagrams of different types of multiwire branch circuits are shown in Figures 4-14a and 4-14b. The calculation of branch circuits and feeders is covered in *Article 220*. The load demands

(A) Dual-element
type S fuse

(B) Adapter for
type S fuses

(C) Ordinary plug fuse, nontime-
delay, Edison-Base type

(D) Fusetron dual-element,
time-delay plug fuse,
Edison-base type

Figure 4-13 Type S fuses and adapter

and determination of conductor sizes for a wiring system should start with the branch circuits and work back through the feeders to the service conductors. The branch circuits should be sized to carry 125% of the computed load where the loads operate continuously for more than 3 hours, unless the branch-circuit panelboards are approved for continuous operation when supplying loads equal to 100% of the panelboard rating, or the branch-circuit conductors already have been derated to 80% or less because there are more than three current-carrying conductors in a raceway or cable. Derating a conductor to 80% is equivalent to providing 125% ampacity at the computed load.

Good engineering practice dictates that where loads may increase, the branch circuits should be loaded to not more than 50% of their rating to provide ampacity for future loads. Good practice also recommends one spare circuit for each five in use. Such spare circuits are not practical, however, in panelboards unless spare circuit conductors or empty conduits are run from flush-mounted panelboards to accessible junction boxes and capped off or otherwise sealed at the time the installation is first being made. Otherwise, conduit or cable cannot be run from the recessed panelboard without removing part of the wall finish. The minimum size wire permitted by the *NEC*® for branch circuits is No. 14 copper. No. 12 copper wire is a realistic and practical minimum branch-circuit conductor size in modern design practice. Particularly in commercial and industrial wiring, where the resistance of these minimum-size wires will not result in excessive voltage drop on

Multiwire Branch Circuits

Single-phase, 3-wire branch circuit

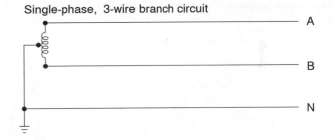

Requirements for a multiwire branch circuit:

1. Two or more ungrounded conductors.
2. Grounded neutral connected to neutral of system.
3. Potential difference between ungrounded conductors.
4. Equal potential difference between ungrounded conductors and neutral conductor.

Three-phase, 4-wire wye-connected branch circuit

Note: A 3-phase, 3-wire delta-connected circuit is not a multiwire circuit because the grounded (phase) conductor is not a neutral conductor. (See Figure 3-10)

Figure 4-14a (Courtesy of American Iron & Steel Institute)

Branch circuits supplying lampholders—Exceptions to 150 volt limit

Incandescent lamps

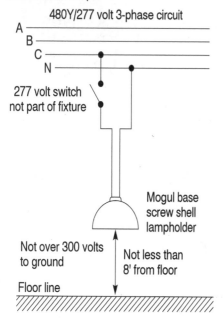

480Y/277 volt 3-phase circuit

A

B

C

N

277 volt switch
not part of fixture

Mogul base
screw shell
lampholder

Not over 300 volts
to ground

Not less than
8' from floor

Floor line

Electric discharge lamps

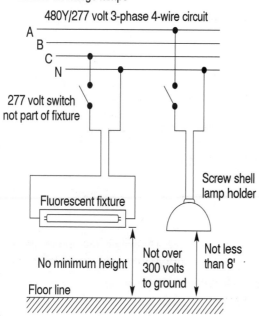

480Y/277 volt 3-phase 4-wire circuit

A

B

C

N

277 volt switch
not part of fixture

Screw shell
lamp holder

Fluorescent fixture

No minimum height

Not over
300 volts
to ground

Not less
than 8'

Floor line

Neither exception is applicable to residential occupancies except for electric discharge lamps in public areas of hotels.

Figure 4-14b   (Courtesy of American Iron & Steel Institute)

average length circuits. The basic formula for determining loads in amperes on a 2-wire lighting and appliance circuit is:

$$\text{Amperes} = \frac{\text{total connected load in watts}}{\text{line voltage}}$$

For a 3-wire circuit, the formula above may be used for each wire to neutral and may be considered separately, using line-to-neutral voltage. For an electric range on a 3-wire circuit, the voltage between the ungrounded phase conductors must be used in the formula. Load demand factors for domestic electric ranges and other cooking appliances are given in *NEC® Table 220-19*. The ampere load on the neutral of a range circuit may be taken as 70% of the load on the phase conductors, as permitted in *Section 220-22*, which states that for a feeder supplying household electric ranges, wall-mounted ovens, counter-mounted cooking units, and electric dryers, the maximum unbalanced load shall be considered as 70% of the load on the ungrounded conductors, as determined in accordance with the *Table 220-19* for ranges and *Table 220-18* for dryers.

Branch-circuit conductors supplying one motor are required to have ampacities equal to not less than 125% of the motor full load current rating, as per *NEC® Section 430-22*, which states generally that branch-circuit conductors supplying a single motor shall have an ampacity of not less than 125% of the motor full-load current rating. *Section 430-6* reminds us that the full-load motor current rating shall be the values given in *Tables 430-147, 430-148, 430-149, 430-150, 151(a), and 151(b)*, including the notes instead of the actual current rating marked on the motor nameplate. Separate motor overload protection shall be based on the motor nameplate current rating. Where the motor is marked in amperes but not in horsepower, the horsepower rating shall be assumed to be that corresponding to the values in amperes given in those tables or calculated accordingly if values are not given in the tables.

There are exceptions. For a branch circuit supplying several motors, the conductors supplying two or more must have an ampacity equal to the sum of the full-load current of all of the motors plus 25% of the largest rated motor in the group. Where one or more of the motors of the group are used on a short-time intermittent, periodic, or varying duty, the ampacity of the conductors shall be computed as follows: (1) determine the needed ampere rating for each motor used for other than continuous duty from *NEC® Table 430-22(a)* exception; (2) determine the needed ampere rating for continuous duty motor based on 100% motor full load current rating; (3) multiply the largest single motor ampere rating determined from (1) or (2) above by 1.25, add all other motor ratings from (1) and (2) above, and select the conductor ampacity for this total ampere rating.

There is one exception to that rule. Conductors supplying motors and other loads are covered in *NEC® Section 430-25*. These conductors are required where they supply a motor load and lighting or appliance load sufficient for the lighting/appliance load as computed in *Article 220* and other applicable sections, plus the motor load determined in accordance with *Section 430-24*, or for a single motor in accordance with *Section 430-22*. The exception to this rule is that the ampacity of the conductor supplying motor-operated, fixed, electric space-heating equipment shall conform with *Section 424-3(b)*. Where a circuit supplies more than one motor and combination load equipment, the conductors supplying these loads shall be not less than the minimum circuit ampacity marked on the equipment in accordance with *Section 430-7(d)*.

COLOR CODING

Good engineering practices dictate the marking or color coding of conductors to provide safety to those persons performing maintenance and routine inspection of the facility electrical system. *NEC® Section 200-6* requires the identification of grounded conductors. The grounded conductor of a branch circuit shall be white or natural gray. That is the general rule.

Where conductors of different systems are installed in the same raceway, box, or other type of enclosure, one system's grounded conductor, if required, shall have an outer covering of white or natural gray. Each other system's grounded conductor, if required, shall have a covering of white with an identifiable colored stripe (but not green) running along the insulation, or other and different means of identification. The NFPA Technical Committee CMP-2 responsible for *Articles 210, 220,* and *225* on branch-circuit requirements clarified that it was not their intent to permit gray for the additional system. White or gray is for a single system. White with a colored stripe is for an additional system. An example of this is where the primary nominal voltage power supply to the service was fed from a 480/277 wye system. A separately derived transformer is installed on the premises, reducing the voltage from 480/277 volts to 208/120 volts for general purpose outlets. If these two systems entered the same wireway, raceway, box, or enclosure, the 480/277 grounded conductor (neutral) would be white or natural gray. The 208/120 grounded (neutral) conductor would be white with a colored stripe, but not green. Additional requirements for identification of the grounded conductor can be found in *NEC® Section 310-12*.

A means for identifying the conductor with a higher voltage to ground, such as a high-leg, has specific requirements in *NEC® Section 215-8* for feeders, *Section 230-56* services, and *Section 384-3(e)* panelboards and switchboards. The conductor commonly known as the high-leg is found when the premises is served from a 4-wire, delta-connected secondary, where the midpoint of one phase is grounded to supply lighting and similar loads. The phase conductor with the higher voltage to ground shall be identified by an outer finish that is orange, or by tagging or other effecting means. Such identification shall be placed at each point where the connection is made if the neutral conductor is also present. This high-leg conductor shall be located at the middle phase in panelboards and switchboards. There are exceptions.

Multiwire branch circuits (see Figure 4-14b) are required to be identified and color coded in *NEC® Section 210-4(d)*. This section states that where more than one nominal voltage system exists in a building, each ungrounded system conductor shall be identified by phase and system. The means of identification shall be permanently posted at each branch-circuit panelboard. Each system phase conductor, wherever accessible, may be identified by separate color coding, marking tape, tagging, or other equally effective means. Although the *NEC®* does not require additional color coding for ungrounded conductor, identifying all conductors is recommended as good engineering practice for the purpose of safety and maintenance. Identification can be by tagging or color coding. Commonly used color coding for branch circuit wiring is given in the following table. The colors are not mandatory, but it is important to use the same coding throughout the system as far as possible to provide greater safety to personnel who come in contact with the electrical system throughout its life.

Industry Practices Not Required by NEC®
Common Color Coding for Ungrounded Conductors

No. of Conductors (including Neutral)	Recommended Colors
First system—3-wire	Black and Red
4-wire	Black, Red, and Blue
Second system—3-wire	Brown and Orange
4-wire	Brown, Orange, and Yellow

NEC® required identifications for the grounded conductor are as follows:

NEC® Required Identification for Grounded Conductors

Function	Identification
First system grounded conductor	White or natural gray
Second system	White with a stripe, not green
Grounding conductor	Green or green with one or more yellow stripes

Conductor with higher voltage to ground on 3-phase, 4-wire, delta connected

system (high-leg)	Orange or identified by tag or other means (middle phase)

Feeder Circuit Conductors

After the branch-circuit loads served by each panelboard have been determined, the size of the feeder for each panelboard must be calculated in accordance with NEC® Articles 215, 225, 220, and 310. The basic formula for determining the feeder size is:

$$\text{Amperes} = \frac{\text{Watts}}{K \times V \times PF}$$

Where there is phase-to-neutral voltage: K = one for 2-wire, single-phase, alternating-current or 2-wire, direct-current circuits; K = two for 3-wire, single-phase, alternating-current or 3-wire, direct-current circuits; and K = three for 4-wire, 3-phase wye-connected alternating-current circuits.

For phase-to-phase voltage: K = one for 3-wire, single-phase, alternating-current or 3-wire direct-current circuits; K = 1.732 for 3-wire or 4-wire, 3-phase, alternating-current circuits; V = voltage (phase-to-neutral or phase-to-phase as above). PF = power factor (1 for direct-current circuits). The approximate power factor for a feeder serving several motors may be calculated as follows: Multiply the horsepower of each motor by its power factor at 75% rated load; add the above products for each motor supplied; divide the sum by the total horsepower of all the motors connected to the circuit. Power factor is important when substantial motor loads are encountered. The lower the power factor, the more current is required for a given load, and the less efficiently the system will operate.

> **Example:** At 100% power factor on a 240-volt, 3-phase circuit, about 2.42 amperes per phase are required to do a 1-kilowatt; at 70% power factor about 3.45 amperes per phase are required for the same load.

The computed load of a feeder must not be less than the sum of all branch-circuit loads supplied by the feeder; after applying the demand factor permitted by NEC® Article 220, 220-3 and Tables 220-3(a) and 220-11, list the unit loads and demand factors that may be applied to the lighting part of the total load in determining the minimum feeder sizes. These minimum unit loads and feeder demand factors are also given in Figures 4-15 and 4-16.

Determination of feeder sizes should not be based solely on present needs. Adequate provision should be made for future load growth, because undersize feeders can effectively handicap system expansion. Design loads for feeders and panelboards and service conductors should not be more than 50% of the rated capacity, to allow for additional future loads. Voltage drop must be held within allowable limits. The size of the neutral may be less than the current-carrying conductors in some cases. The size,

Table 220-3(a). General Lighting Loads by Occupancies

Type of Occupancy	Unit Load per Square Foot (Volt-Amperes)
Armories and auditoriums	1
Banks	3½[b]
Barber shops and beauty parlors	3
Churches	1
Clubs	2
Court rooms	2
Dwelling units[a]	3
Garages — commercial (storage)	½
Hospitals	2
Hotels and motels, including apartment houses without provision for cooking by tenants[a]	2
Industrial commercial (loft) buildings	2
Lodge rooms	1½
Office buildings	3½[b]
Restaurants	2
Schools	3
Stores	3
Warehouses (storage)	¼
In any of the above occupancies except one-family dwellings and individual dwelling units of two-family and multifamily dwellings:	
Assembly halls and auditoriums	1
Halls, corridors, closets, stairways	½
Storage spaces	¼

Note: For SI units, 1 ft² = 0.093 m².

[a]See Section 220-3(b)(10).

[b]In addition, a unit load of 1 volt-ampere per square foot shall be included for general-purpose receptacle outlets where the actual number of general-purpose receptacle outlets is unknown.

Figure 4-15   Reprinted with permission from NFPA 70-1999.

Table 220-11. Lighting Load Demand Factors

Type of Occupancy	Portion of Lighting Load to Which Demand Factor Applies (Volt-Amperes)	Demand Factor (Percent)
Dwelling units	First 3000 or less at	100
	From 3001 to 120,000 at	35
	Remainder over 120,000 at	25
Hospitals*	First 50,000 or less at	40
	Remainder over 50,000 at	20
Hotels and motels, including apartment houses without provision for cooking by tenants*	First 20,000 or less at	50
	From 20,001 to 100,000 at	40
	Remainder over 100,000 at	30
Warehouses (storage)	First 12,500 or less at	100
	Remainder over 12,500 at	50
All others	Total volt-amperes	100

*The demand factors of this table shall not apply to the computed load of feeders or services supplying areas in hospitals, hotels, and motels where the entire lighting is likely to be used at one time, as in operating rooms, ballrooms, or dining rooms.

Figure 4-16   Reprinted with permission from NFPA 70-1999.

rating, or setting of the overcurrent protection devices (circuit breakers, fuses, relays, and the like) for feeder conductors is required, in general, to be determined by the conductor ampacities. The device is to be placed at the point where the feeder receives its supply (*NEC® Section 240-21*). There are restrictions to this rule for taps of 25 feet or fewer and taps 10 feet or fewer, as illustrated in Figure 4-17.

For motor loads where the overcurrent device may be selected according to the motor-starting currents, overloads imposed on conductors by normal motor-starting currents are of short duration and are therefore considered harmless to equipment and associated conductors. The use of time-delay overcurrent devices on all such loads permits the use of ratings more nearly equal to the conductor ampacity and still prevents needless tripping during motor-starting current surges. Where paralleled conductors are used as feeders, the derating factors mentioned earlier in this chapter must be applied to the allowable ampacity of the feeder conductors unless each set of parallel conductors is run in a separate raceway. If spare capacity cannot be designed into the feeder cables, installation of oversize conduit or raceway for selected feeder, or spare parallel conduit or raceway, will provide for future growth. Good engineering practices mandate this expansion capability.

The procedure for determining conductor and conduit sizes for a feeder is as follows:

Example: One 15-horsepower and two 10-horsepower motors are on separate branch circuits, and all branch circuits are to be in the same raceway. The motors are 240 volt, 3 phase, 60 Hz and operate at 85% power factor. Voltage drop is to be held to the maximum of 2%. Voltage drop in branch-circuit conductors is negligible. Feeder length is 200 feet. An increase of 50% in motor load is to be provided for. For branch-circuit conductors from *NEC® Table 430-150*, the full load current of a 15-horsepower motor is 42 amperes and each of the 10-horsepower motors is 28 amperes. Because there are nine conductors in one conduit, the conductor ampacities must be derated to 50%, as per section 310-15 in *Article 310*. The sizes of the motors and the conductor sizes for the branch-circuit serving the motors should be No. 4 for the 15-horsepower motors and No. 8 for the two 10-horsepower motors; all of the conductors are Type THW copper.

240-21. Location in Circuit. Overcurrent protection shall be provided in each ungrounded circuit conductor and shall be located at the point where the conductors receive their supply except as specified in (a) through (g) below. No conductor supplied under the provisions of (a) through (g) below shall supply another conductor under those provisions, except through an overcurrent protective device meeting the requirements of Section 240-3.

(a) Branch-Circuit Conductors. Branch-circuit tap conductors meeting the requirements specified in Section 210-19 shall be permitted to have overcurrent protection located as specified in that Section.

(b) Feeder Taps. Conductors shall be permitted to be tapped, without overcurrent protection at the tap, to a feeder as specified in (1) through (5) below.

(1) Taps Not Over 10 Ft (3.05 m) Long. Where the length of the tap conductors does not exceed 10 ft (3.05 m) and the tap conductors comply with all of the following:

(a.) The ampacity of the tap conductors is:

(1.)Not less than the combined computed loads on the circuits supplied by the tap conductors, and

(2.) Not less than the rating of the device supplied by the tap conductors, or not less than the rating of the overcurrent-protective device at the termination of the tap conductors.

(b.) The tap conductors do not extend beyond the switchboard, panelboard, disconnecting means, or control devices they supply.

(c.) Except at the point of connection to the feeder, the tap conductors are enclosed in a raceway, which shall extend from the tap to the enclosure of an enclosed switchboard, panelboard, or control devices, or to the back of an open switchboard.

(d.) For field installations where the tap conductors leave the enclosure or vault in which the tap is made, the rating of the overcurrent device on the line side of the tap conductors shall not exceed 10 times the ampacity of the tap conductor.

FPN: For overcurrent protection requirements for lighting and appliance branch-circuit panelboards, see Sections 384-16(a) and (e).

(2) Taps Not Over 25 Ft (7.62 m) Long. Where the length of the tap conductors does not exceed 25 ft (7.62 m) and the tap conductors comply with all of the following:

(a.) The ampacity of the tap conductors is not less than one-third of the rating of the overcurrent device protecting the feeder conductors.

(b.) The tap conductors terminate in a single circuit breaker or a single set of fuses that will limit the load to the ampacity of the tap conductors. This device shall be permitted to supply any number of additional overcurrent devices on its load side.

(c.) The tap conductors are suitably protected from physical damage or are enclosed in a raceway.

(3) Taps Supplying a Transformer [Primary Plus Secondary Not Over 25 Ft (7.62 m) Long]. Where the tap conductors supply a transformer and comply with all the following conditions:

(a.) The conductors supplying the primary of a transformer have an ampacity at least one-third of the rating of the overcurrent device protecting the feeder conductors.

(b.) The conductors supplied by the secondary of the transformer shall have an ampacity that, when multiplied by the ratio of the secondary-to-primary voltage, is at least one-third of the rating of the overcurrent device protecting the feeder conductors.

(c.) The total length of one primary plus one secondary conductor, excluding any portion of the primary conductor that is protected at its ampacity, is not over 25 ft (7.62 m).

(d.) The primary and secondary conductors are suitably protected from physical damage.

(e.) The secondary conductors terminate in a single circuit breaker or set of fuses that will limit the load current to not more than the conductor ampacity that is permitted by Section 310-15.

(4) Taps Over 25 Ft (7.62 m) Long. Where the feeder is in a high bay manufacturing building over 35 ft (10.67 m) high at walls, and the installation complies with all of the following conditions:

(a.) Conditions of maintenance and supervision ensure that only qualified persons will service the systems.

(b.) The tap conductors are not over 25 ft (7.62 m) long horizontally and not over 100 ft (30.5 m) total length.

(c.) The ampacity of the tap conductors is not less than one-third of the rating of the overcurrent device protecting the feeder conductors.

(d.) The tap conductors terminate at a single circuit breaker or a single set of fuses that will limit the load to the ampacity of the tap conductors. This single overcurrent device shall be permitted to supply any number of additional overcurrent devices on its load side.

(e.) The tap conductors are suitably protected from physical damage or are enclosed in a raceway.

(f.) The tap conductors are continuous from end-to-end and contain no splices.

(g.) The tap conductors are sized No. 6 copper or No. 4 aluminum or larger.

(h.) The tap conductors do not penetrate walls, floors, or ceilings.

(i.) The tap is made no less than 30 ft (9.14 m) from the floor.

(5) Outside Taps of Unlimited Length. Where the conductors are located outdoors, except at the point of termination, and comply with all of the following conditions:

(a.) The conductors are suitably protected from physical damage.

(b.) The conductors terminate at a single circuit breaker or a single set of fuses that will limit the load to the ampacity of the conductors. This single overcurrent device shall be permitted to supply any number of additional overcurrent devices on its load side.

(c.) The overcurrent device for the conductors is an integral part of a disconnecting means or shall be located immediately adjacent thereto.

(d.) The disconnecting means for the conductors are installed at a readily accessible location either outside of a building or structure, or inside nearest the point of entrance of the conductors.

(c) Transformer Secondary Conductors. Conductors shall be permitted to be connected to a transformer secondary, without overcurrent protection at the secondary, as specified in (1) through (4) below.

FPN: For overcurrent protection requirements for transformers, see Section 450-3.

(1) Protection by Primary Overcurrent Device. Single-phase (other than 2-wire) and multiphase (other than delta-delta, 3-wire) transformer secondary conductors are not considered to be protected by the primary overcurrent protective device. Conductors supplied by the secondary side of a single-phase transformer having a 2-wire (single-voltage) secondary, or a three-phase, delta-delta connected transformer having a 3-wire (single-voltage) secondary, shall be permitted to be protected by overcurrent protection provided on the primary (supply) side of the transformer, provided this protection is in accordance with Section 450-3 and does not exceed the value determined by multiplying the secondary conductor ampacity by the secondary to primary transformer voltage ratio.

(2) Transformer Secondary Conductors Not Over 10 Ft (3.05 m) Long. Where the length of secondary conductor does not exceed 10 ft (3.05 m) and complies with all of the following:

(a.) The ampacity of the secondary conductors is:

(1.) Not less than the combined computed loads on the circuits supplied by the secondary conductors, and (2.) Not less than the rating of the device supplied by the secondary conductors, or not less than the rating of the overcurrent-protective device at the termination of the secondary conductors.

(b.) The secondary conductors do not extend beyond the switchboard, panelboard, disconnecting means, or control devices they supply.

(c.) The secondary conductors are enclosed in a raceway,

Figure 4-17   Reprinted with permission from NFPA 70-1999.

which shall extend from the transformer to the enclosure of an enclosed switchboard, panelboard, or control devices, or to the back of an open switchboard.

FPN: For overcurrent protection requirements for lighting and appliance branch-circuit panelboards, see Sections 384-16(a) and (e).

(3) Secondary Conductors Not Over 25 Ft (7.62 m) Long. For industrial installations only, where the length of the secondary conductors do not exceed 25 ft (7.62 m) and complies with all of the following:

(a.) The ampacity of the secondary conductors is not less than the secondary current rating of the transformer, and the sum of the ratings of the overcurrent devices does not exceed the ampacity of the secondary conductors.

(b.) All overcurrent devices are grouped.

(c.) The secondary conductors are suitably protected from physical damage.

(4) Outside Secondary Conductors. Where the conductors are located outdoors, except at the point of termination, and comply with all of the following conditions:

(a.) The conductors are suitably protected from physical damage.

(b). The conductors terminate at a single circuit breaker or a single set of fuses that will limit the load to the ampacity of the conductors. This single overcurrent device shall be permitted to supply any number of additional overcurrent devices on its load side.

(c.) The overcurrent device for the conductors is an integral part of a disconnecting means or shall be located immediately adjacent thereto.

(d.) The disconnecting means for the conductors are installed at a readily accessible location either outside of a building or structure, or inside nearest the point of entrance of the conductors.

(5) Secondary Conductors from a Feeder Tapped Transformer. Transformer secondary conductors installed in accordance with Section 240-21(b)(3) shall be permitted to have overcurrent protection as specified in that Section.

(d) Service Conductors. Service-entrance conductors shall be permitted to be protected by overcurrent devices in accordance with Section 230-91.

(e) Busway Taps. Busways and busway taps shall be permitted to be protected against overcurrent in accordance with Sections 364-10 through 364-13.

(f) Motor Circuit Taps. Motor-feeder and branch-circuit conductors shall be permitted to be protected against overcurrent in accordance with Sections 430-28 and 430-53 respectively.

(g) Conductors from Generator Terminals. Conductors from generator terminals that meet the size requirement in Section 445-5 shall be permitted to be protected against overload by the generator overload protective device(s) required by Section 445-4.

Figure 4-17 (continued)

For sizing the raceway, *NEC® Table 1* in *Chapter 9* requires that the sum of the cross-sectional areas of the individual conductors be not more than 40% of the cross-sectional area of the conduit. From *Table 5*, the cross-sectional area for No. 4 THW is .0973 square inch. For No. 8 the cross-sectional area is .0556 square inch. Therefore:

$$3 \times .0973 = .2919 \text{ square inch}$$
$$6 \times .0556 = .3336 \text{ square inch}$$

The total for the nine conductors = .6255 square inch.

From *NEC® Table 4, Chapter 9*, a 1 1/4-inch IMC conduit is required, but a conduit not smaller than 2 inches should be used for good engineering design practices. For feeder conductors, feeder load as determined in *Section 430-24* and *Table 430-150* is:

Motor Horsepower	Percent Full Load	Full-Load Current	Amperes
15	125	42	52.5
10	100	28	28
10	100	28	28
Total load (rounded up) = 109 amperes.			108.5

Note: For a 50% load increase, a feeder ampacity of 163 is needed. *NEC® Table 310-16* lists size No. 2/0, 75-degree copper wire at 175 amperes, but this table makes no allowance for voltage drop. A No. 2 copper conductor carries 156 amperes for about 94 feet with a 1% voltage drop and 188 feet with about a 2% voltage drop; therefore a larger size than No. 2/0 is needed. A No. 3/0 copper conductor meets the desired voltage drop required. The feeder conduit from Appendix C Tables consists of three No. 3/0 conductors requiring a 2-inch conduit. Feeders supplying power to other buildings on the premises are covered in *Article 215* of the *NEC®*.

SERVICE-ENTRANCE CONDUCTORS

Conductor ampacity requirements become progressively larger working back from branch circuits through feeders to service conductors. The service conductors must have sufficient ampacity to take care of the entire electrical load proposed and anticipated of facility, plant, or building served. The most

Electrical service components

Overhead service

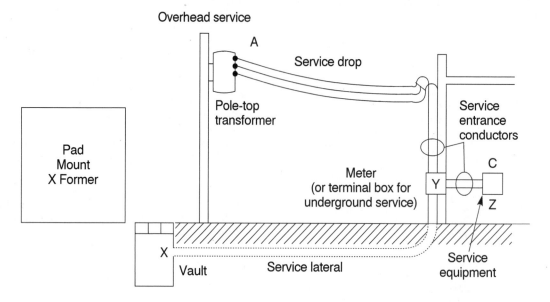

Figure 4-18 *NEC® Article 230-2* permits only one service per building. It is permissible to supply overhead or underground. Each service should end in a main disconnect (max. of 6) with overcurrent protection.

direct method of determining service-conductor size is by using computed loads in the formula:

$$\text{Amperes} = \frac{\text{total feeder load in watts}}{\text{Line voltage} \times \text{system power factor}}$$

Conductor sizes are selected from *NEC® Table 310-16*. Voltage drop in service and feeder conductors also should be considered in determining conductor size. The other method of determining service-conductor size is to recalculate all the various loads—lighting, appliance, motor, and special loads—for the entire building. Permissible demand factors may have to be applied to each type of load, provisions for space capacity made in each case, and allowance for voltage drop, to establish the total service-conductor load. Although bare aluminum neutrals are not permitted to make direct contact with the earth, bare copper neutrals are permitted unless soil conditions are unsuitable. Components of a service, for a single-occupancy building as defined in the *NEC®*, are illustrated in Figure 4-18.

Methods for serving multiple-occupancy buildings as permitted by *NEC® Section 230-2, 230-72,* and *230-90* are illustrated in Figures 4-19 and 4-20. In general, only one set of service drops or lateral conductor per building is permitted. Exceptions to

this rule for special-duty equipment, multiple occupancies, large-area buildings, and so on are listed in *Section 230-2*. Regardless of the service design selected, each individual set of service-entrance conductors must terminate in a readily accessible, manually operable means of disconnecting the building electrical system from the source of supply. In accordance with *Sections 230-70* and *230-72*, the disconnecting means for any set or subset of conductors may be a single switch or circuit breaker. There

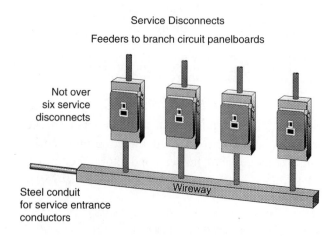

Figure 4-19 *NEC® Section 230-71(a)* permits up to six disconnects per service either in an enclosure or in separate enclosures.

a. No individual occupancies above second floor

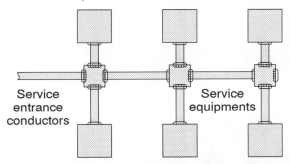

Service entrance conductors

Service equipments

Service-entrance conductors installed on exterior of building or embedded in not less than 2 inches of concrete or brick. *NEC® Section 230-6.*
Service-entrance conductors or subsets of service-entrance conductors may run to each occupancy. Each occupancy may have not more than six service disconnects and six sets of service overcurrent devices.

b. Individual occupancies above second floor

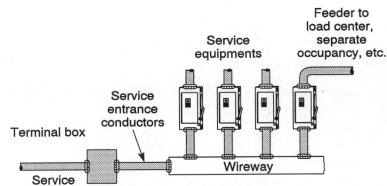

Feeder to load center, separate occupancy, etc.

Service equipments

Service entrance conductors

Terminal box

Service

Wireway

Service equipment grouped in common accessible space. *NEC® 230-72 EX.*
No more than six disconnects and six sets of overcurrent devices for entire building.
Each service equipment may supply feeders to load centers, including any number of meters, or separate occupancies.

Figure 4-20 Multiple occupancies per *NEC® Sections 230-2, 230-75, 230-90.*
(Courtesy of American Iron & Steel Institute)

may be a maximum of six switches or circuit breakers in a common enclosure, a group of separate enclosures, or in a switchboard at a readily accessible location and near the point of entrance of the service conductors as possible.

The service-disconnecting means, as per *NEC® Section 230-70*, must be suitable for the prevailing conditions. If installed in hazardous (classified) locations, they shall comply with the requirements of *Articles 500-516*. Each ungrounded service conductor must be protected with an overcurrent device with a rating or setting not higher than the allowable ampacity of the service conductor. The overcurrent device must protect all circuits and other service equipment. Exceptions to this rule, such as circuits for fire pumps or protective signaling systems, are listed in *Section 230-90*. No overcurrent device is permitted in the grounded service conductor except a circuit breaker that would simultaneously open all ungrounded conductors of the service. However, a means must be provided for disconnecting this service conductor from the interior wire. If this cannot be done by the service

switch or circuit breaker as required by *Section 230-75*, a terminal lug on the neutral bus of the entrance equipment may be provided.

Service-drop conductors are covered in *Part B* of *NEC® Article 230*. In general, the vertical clearance of service drops are based on weather conditions of 60ºF, no wind, with final unload sag in the wire, conductors, and cable, and they shall not be readily accessible, but where above the roofs shall have a clearance of not fewer than 8 feet from the roof surface, and the clearance is maintained for a distance of not fewer than 3 feet in all directions of the edge of the roof. There are exceptions to that rule; for example, the area above a roof surface subject to pedestrian or vehicle traffic shall have a vertical clearance from the roof surface in accordance with the clearance requirements in *Section 230-24 (b)*, and where the voltage does not exceed 300 volts and the roof has a slope of not fewer than 4 inches in 12 inches, a clearance of 3 feet shall be permitted. Where the voltage between conductors does not exceed 300 volts, a reduction in the clearance only

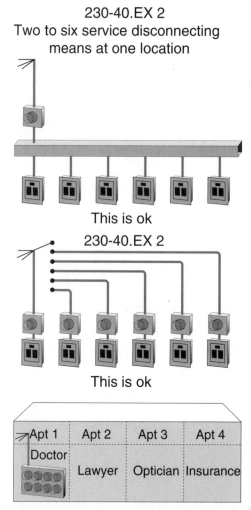

230-40.EX 2
Two to six service disconnecting
means at one location

This is ok

230-40.EX 2

This is ok

Apt 1	Apt 2	Apt 3	Apt 4
Doctor	Lawyer	Optician	Insurance

Each occupancy may have one set of service
entrance conductors and six disconnects*

Figure 4-21   Multiple-occupancy building

above the overhanging portion of the roof not fewer than 18 inches shall be permitted if not more than 4 feet of the service-drop conductors pass above the roof overhang and they terminate at a through-the-roof raceway or approved support.

Vertical-clearance-from-ground requirements dictate that where service-drop conductors of 600 volts or fewer, nominal, shall have a minimum clearance above final grade of 10 feet at the electric service entrance to buildings or at the drip-loop of a building electrical entrance, or above areas or sidewalks accessible to pedestrians only. This measurement is made from the final grade or other access surface only for service-drop cables supported on and cabled together with a grounded bare messenger

wire and are limited to 150 volts to ground: 12 feet for those areas over residential property and driveways and those commercial areas not subject to truck traffic and where the voltage is limited to 300 volts to ground; 15 feet over residential property, driveways, and those commercial areas not subject to truck traffic; 18 feet over public streets, alleys, roads, parking areas subject to truck traffic, driveways on other-than-residential property, and land traversed by vehicles (e.g., cultivated and grazing land, forests, and orchards). Clearances for conductors over 600 volts are given in the National Electric Safety Code (ANSI C-2) published by the Institute of Electrical and Electronics Engineers, Inc. See examples of acceptable clearances in Figure 4-22.

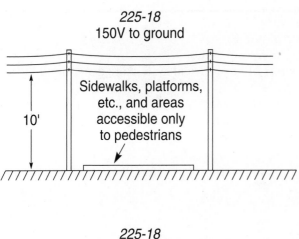

225-18
150V to ground

10'

Sidewalks, platforms, etc., and areas accessible only to pedestrians

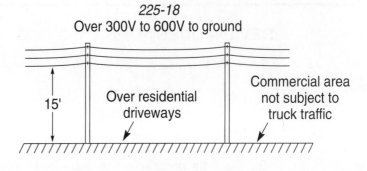

225-18
Over 300V to 600V to ground

15'

Over residential driveways

Commercial area not subject to truck traffic

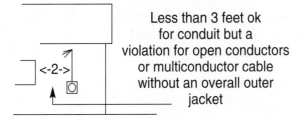

225-18
300V to ground

12'

Residential driveway

Commerical area not subject to truck traffic

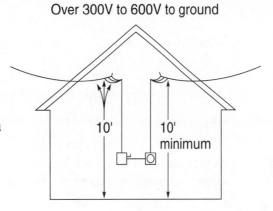

225-18
Over 300V to 600V to ground

10' | 10' minimum

230-9
Clearance from building openings

<-2->

Less than 3 feet ok for conduit but a violation for open conductors or multiconductor cable without an overall outer jacket

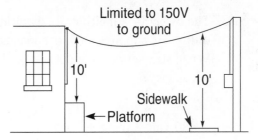

230-24(b)
Service drop cables with a grounded bare messenger

Limited to 150V to ground

10' | 10'

Platform

Sidewalk

230-9
Service conductors prohibited below material moving openings on farm or commericial buildings

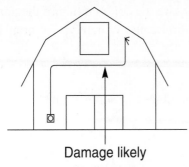

Damage likely

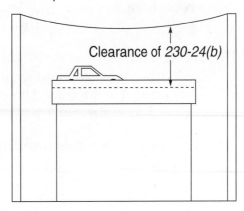

230-24(a), EX. 3
Rooftop clearance for vehicular traffic

Clearance of *230-24(b)*

Figure 4-22 Required clearances for service and feeder conductors

REVIEW QUESTIONS

1. Name the three categories into which conductors of an electrical system are generally divided.

2. Branch circuits are unique in that they originate from what?

3. The term *overcurrent* is defined in *NEC®* *Article 100*. Name four common overcurrent protective devices for electrical systems.

4. Name four advantages of utilizing circuit breakers as overcurrent devices.

5. Name three types of circuit breakers on the market today.

6. A set of No. 2 THWN conductors are routed in an underground duct 200 feet to a 75-horsepower motor. The motor is supplied by a 480-volt, 3-phase circuit. Are these conductors branch-circuit conductors, feeder conductors, or service conductors?

7. What size service conductors are required for a service drop feeding a 100-ampere service on a single-family dwelling?

8. The owner of a multifamily dwelling requests 3-phase, 208/120-volt power from the serving utility. He supplies each occupancy with a 120/208-volt, single-phase, 150-ampere service. What size service conductors and service-entrance conductors are required to feed each occupancy?

9. A 480-volt, 3-phase branch circuit supplies a piece of equipment with three individual motors. One motor is 15 horsepower, the second motor is 10 horsepower, and the third motor is 10 horsepower. What size branch-circuit conductors are required to feed these three motors? (The branch circuit is run to the equipment and the smaller motors are tapped from the main branch circuit.)

10. The *National Electrical Code®* does not generally require the color coding of the ungrounded conductors of a circuit. However, there are some requirements. Which type of branch circuits are required to be marked and how?

11. In an electrical system an overcurrent device must be placed in series with each _____ conductor.

12. A branch circuit supplies a single continuous-duty motor supplying a submersible pump

supplying water for a residence. Conductors must have an ampacity of at least _____ % of the full-load current rating.

13. How does one determine the ampacity of a motor when determining the size of conductors and overcurrent equipment design?

Chapter 5

General Raceway Installation Requirements

Before selecting a wiring method many factors must be considered: environment, physical protection requirements, temperature, and all other pertinent conditions that may effect the material selected. Once this has been done a wiring method may be selected from *Chapter 3* of the *NEC®*. Each installation generally requires several wiring methods to meet all the conditions, such as a method acceptable for "dry concealed locations," and a method that provides "physical protection." It may also be necessary to select a "flexible" wiring method for equipment connection and a material that can be used in "wet locations" and "underground." Some methods are acceptable in all locations. The final choice, however, must meet *Sections 110-3 (a) & (b)* and must be made by the designer. Wiring methods are all found in *Chapter 3*. Once this has been done, it is the responsibility of the installer to comply with the provisions in *Article 300* and the *NEC®* article that specifically covers each wiring method chosen when making the installation.

PROTECTION AGAINST PHYSICAL DAMAGE

The function of an electrical raceway is to facilitate the insertion and extraction of the conductors

Figure 5-1 Typical construction project's electrical steel raceways prefabricated and ready for installation in a concrete slab. (Courtesy of Allied Tube & Conduit)

Figure 5-2 A typical construction concrete pour. Demonstrating one example of the physical abuse to which raceways are subjected throughout the construction phase of all projects (Courtesy of Allied Tube & Conduit)

and to protect them from mechanical injury. All raceway types properly installed facilitate easy insertion and extraction of conductors. However, only rigid metal conduit, Intermediate Metal Conduit (IMC), and rigid nonmetallic conduit Schedule 80 provide protection of conductors against physical damage. Electrical Metallic Tubing (EMT) provides physical protection except where severe physical damage to the conductors is likely.

The designer and installer can no longer rely on all raceways to provide physical protection. Steel raceways, rigid metal conduit as covered in *NEC® Article 346*, and Intermediate Metal Conduit (IMC) as covered in *Article 345* provide the greatest physical protection and are permitted in all applications of the Code. Schedule 80 PVC as covered in *Article 347* does provide excellent physical protection. However, a Fine Print Note (FPN) in *Section 347-2* warns that in cold temperatures PVC may become brittle and crack or break under impact. Therefore, Schedule 80 may be used for physical protection, but consideration must be given to the environment in which it is installed and the physical protection needed under those conditions. Schedule 40, nonmetallic conduit cannot be used for physical protection. See *NEC® 347-3 (c)*.

Electrical Metallic Tubing (EMT) as covered in *NEC® Article 348* provides great physical protection for all but severe physical damage locations.

Electrical Nonmetallic Tubing (ENT) as covered in *NEC® Article 331* is a flexible, pliable, nonmetallic raceway system that does not provide physical protection for conductors and is not permitted where subject to physical damage. In most applications this product must be concealed within walls or floors with specific characteristics or otherwise protected.

From the time a raceway or cable is shipped from the factory until it is installed, it is frequently subject to physical damage. It is bundled, hoisted, conveyed, dropped, stored, tooled, measured, bent, and drilled. It lies in the midst of construction activity involving heavy trucks and loaded wheelbarrows that can strike and run over it. It may be struck with a chisel, screwdriver, or hammer while being mounted and anchored. It can be subject to damage while tied to steel reinforcing rods, while cables are pulled in, or during the pouring of concrete, or by the vibrators used to settle concrete around the reinforcing steel. It can be struck where it emerges from the wall, ceiling, or floor.

Raceways are known to frequently outlast repeated reconstruction and remodeling of buildings they serve. All these things must be considered when selecting the raceway for a specific installation. Great care is needed in making these decisions, as one of the important factors in selecting a raceway is the long-term economy of the installation. In the case of electrical modernization, old wires can often be pulled out and new wires pulled in without structural remodeling. It is up to the designer to select a raceway type to provide this protection, and service.

NEC® Section 300-4 requires conductors to be adequately protected where subject to physical damage in both exposed and concealed locations, except for Intermediate Metal Conduit (IMC), rigid metal conduit, rigid nonmetallic conduit, and Electrical Metallic Tubing (EMT). Raceways and cables shall be protected where run through bored holes or notches in wood, through metal framing members, and where run parallel to the framing members when a distance of 1 1/4 inches cannot be maintained. The raceway or cable type shall be protected from penetration by screws or nails with a steel plate sleeve or equivalent of at least 1/16 inch thick. It is the responsibility of the designer to

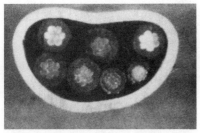
Cross-section of 1" of IMC after impact test.

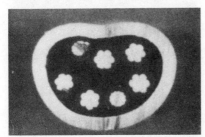
Cross-section of 1" GRC after impact test.

Test result summary: The after-impact internal area exceeded the 40% allowable fill in all tests, for both IMC and rigid conduit.

Deformation did not cause rupture of the insulation or severing of the conductors in either IMC or rigid conduit, and in all cases the conductors were easily withdrawn.

Figure 5-3 Examples of actual test samples after severe impact (Courtesy of Allied Tube & Conduit)

select a raceway type that will provide protection to the conductors in the environment in which it is installed. The degree of physical damage dictates the raceway type to be selected. For installations over 600 volts nominal, above-ground installations, the conductors shall be protected by installing them in rigid metal conduit, intermediate metal conduit, electrical metallic tubing, or rigid nonmetallic conduit. Conductors may also be run in cable trays, bus ways, or cable bus in some installations. The installation requirements for over-600-volt installations are also found in *Article 300. Section 300-37 (a)* governs above-ground installations; *Section 300-50(b)* provides the physical protection requirements for underground installations over 600 volts where they emerge from the ground.

Although raceways were originally designed to protect conductors from mechanical injury, as well as to provide versatility for future changes by facilitating the insertion and extraction of conductors, in the past 30 years special-use raceways have been

developed, accepted, and included in the *NEC®*. The specific physical protection capabilities of each raceway are addressed in the individual sections of Chapter 6 of this book.

UNDERGROUND REQUIREMENTS

NEC® Section 300-5 covers the requirements for underground under 600 volt installations. *Section 300-5* gives the minimum cover requirements to install cable or conduit. All underground installations shall be grounded and bonded in accordance with *Article 250*. Where approved for direct burial, cable installed under a building shall be in a raceway where it is extended beyond the outside wall of the building. Direct buried cables and cables emerging from the ground shall be protected in a raceway extending from the minimum cover distance required in *Section 300-5* below grade to a point at least 8 feet above finish grade. Conductors entering a building shall be protected to the point of entrance. Where a raceway or enclosure is subject to physical damage, the conductors must be installed in rigid metal conduit, intermediate metal conduit, or Schedule 80 rigid nonmetallic conduit or its equivalent. Raceways or cables shall not be back-filled with a material that will damage, prevent adequate compaction, or contribute to corrosion of raceways or cables.

When necessary to prevent physical damage to a raceway or cable select material (such as sand), suitable running boards, sleeves, or other approved means shall be used. Raceways through which moisture may contact energized live parts, such as conductors entering the building from outside into panelboards or switchboards that may contact live parts, shall be sealed or plugged at either or both ends. If hazardous gases or vapors are present, this may also require the sealing of underground raceways where they enter the building. A bushing or terminal fitting shall be placed on the end of a conduit or raceway that terminates underground where cables of a type approved for direct burial, such as Type "UF," emerge as a direct-burial wiring method. A seal incorporating the physical protection characteristics of a bushing is permitted in lieu of the bushing. *NEC® Section 300-50* has similar requirements for underground installations of over

Table 300-5. Minimum Cover Requirements, 0 to 600 Volts, Nominal, Burial in Inches (Cover is defined as the shortest distance in inches measured between a point on the top surface of any direct-buried conductor, cable, conduit, or other raceway and the top surface of finished grade, concrete, or similar cover.)

Location of Wiring Method or Circuit	Column 1 Direct Burial Cables or Conductors	Column 2 Rigid Metal Conduit or Intermediate Metal Conduit	Column 3 Nonmetallic Raceways Listed for Direct Burial Without Concrete Encasement or Other Approved Raceways	Column 4 Residential Branch Circuits Rated 120 Volts or Less with GFCI Protection and Maximum Overcurrent Protection of 20 Amperes	Column 5 Circuits for Control of Irrigation and Landscape Lighting Limited to Not More than 30 Volts and Installed with Type UF or in Other Identified Cable or Raceway
All locations not specified below	24	6	18	12	6
In trench below 2-in. thick concrete or equivalent	18	6	12	6	6
Under a building	0 (in raceway only)	0	0	0 (in raceway only)	0 (in raceway only)
Under minimum of 4-in. thick concrete exterior slab with no vehicular traffic and the slab extending not less than 6 in. beyond the underground installation	18	4	4	6 (direct burial) 4 (in raceway)	6 (direct burial) 4 (in raceway)
Under streets, highways, roads, alleys, driveways, and parking lots	24	24	24	24	24
One- and two-family dwelling driveways and outdoor parking areas, and used only for dwelling-related purposes	18	18	18	12	18
In or under airport runways, including adjacent areas where trespassing prohibited	18	18	18	18	18

Notes:
1. For SI units, 1 in. = 25.4 mm.
2. Raceways approved for burial only where concrete encased shall require concrete envelope not less than 2 in. thick.
3. Lesser depths shall be permitted where cables and conductors rise for terminations or splices or where access is otherwise required.
4. Where one of the wiring method types listed in Columns 1–3 is used for one of the circuit types in Columns 4 and 5, the shallower depth of burial shall be permitted.
5. Where solid rock prevents compliance with the cover depths specified in this table, the wiring shall be installed in metal or nonmetallic raceway permitted for direct burial. The raceways shall be covered by a minimum of 2 in. of concrete extending down to rock.

Figure 5-4 Burial depth requirements as per *NEC*® for 600 volts or less installations (Reprinted with permission from NFPA 70-1999.)

600 volts nominal are found in Section *300-50(a)-(e)* and *Table 300-50* which provide specific requirements for those installations.

EXPOSED TO DIFFERENT TEMPERATURES

Raceways exposed to different temperature requirements are covered in *NEC*® *Section 300-7*, which requires that where portions of interior raceways are exposed to widely different temperatures, such as in refrigeration or cold-storage plants, circulation of air from the warmer to the colder section through a raceway shall be prevented by sealing the open ends of the wiring method where the temperature change occurs. This reduces condensation from building up in the equipment being served by that raceway.

Table 300-50. Minimum Cover Requirements (*Cover* is defined as the shortest distance in inches measured between a point on the top surface of any direct-buried conductor, cable, conduit, or other raceway and the top surface of finished grade, concrete, or similar cover.)

Circuit Voltage	Direct-Buried Cables	Rigid Nonmetallic Conduit Approved for Direct Burial*	Rigid Metal Conduit and Intermediate Metal Conduit
Over 600 V through 22 kV	30	18	6
Over 22 kV through 40 kV	36	24	6
Over 40 kV	42	30	6

Note: For SI units, 1 in. = 25.4 mm.

*Listed by a qualified testing agency as suitable for direct burial without encasement. All other nonmetallic systems shall require 2 in. (50.8 mm) of concrete or equivalent above conduit in addition to above depth.

Exception No. 1: Areas subject to vehicular traffic, such as thoroughfares or commercial parking areas, shall have a minimum cover of 24 inches (610 mm).

Exception No. 2: The minimum cover requirements for other than rigid metal conduit and intermediate metal conduit shall be permitted to be reduced 6 inches (152 mm) for each 2 inches (50.8 mm) of concrete or equivalent protection placed in the trench over the underground installation.

Exception No. 3: The minimum cover requirements shall not apply to conduits or other raceways that are located under a building or exterior concrete slab not less than 4 inches (102 mm) in thickness and extending not less than 6 inches (152 mm) beyond the underground installation. A warning ribbon or other effective means suitable for the conditions shall be placed above the underground installation.

Exception No. 4: Lesser depths shall be permitted where cables and conductors rise for terminations or splices or where access is otherwise required.

Exception No. 5: In airport runways, including adjacent defined areas where trespass is prohibited, cable shall be permitted to be buried not less than 18 inches (457 mm) deep and without raceways, concrete enclosement, or equivalent.

Exception No. 6: Raceways installed in solid rock shall be permitted to be buried at lesser depth where covered by 2 inches (50.8 mm) of concrete, which shall be permitted to extend to the rock surface.

Figure 5-5 Burial depth requirements as per *NEC*® for over 600 volt installations. *NOTE:* Utility or Local jurisdiction requirements may exceed these minimum requirements. (Reprinted with permission from NFPA 70-1999.)

Although the *NEC*® only requires this precautionary measure for interior installations, it is also recommended where wiring methods pass from outside to environmentally controlled areas. Good installation practices dictate entry into equipment in a manner so that the condensation will not drip on exposed live parts (see Figure 5-6). It is not uncommon for improper installations to short-out, causing expensive down time. Special fittings designed for sealing conduit openings are not required for this application. Field methods for sealing may be accomplished using "duct-seal" or other removable putty that will not deteriorate conductor insulation. Expansion joints are required where necessary to compensate for thermal expansion and contraction. The specific requirements for nonmetallic raceways that have a higher coefficient of expansion are covered below.

EXPANSION PROPERTIES OF RACEWAYS

As do all construction materials, raceway systems expand and contract with variations in temper-

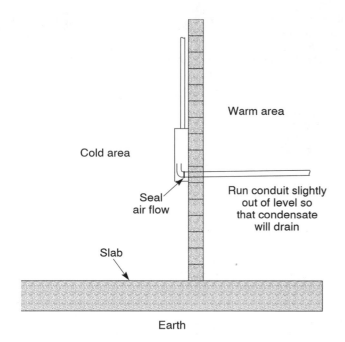

Warm area

Cold area

Seal
air flow

Run conduit slightly
out of level so
that condensate
will drain

Slab

Earth

Figure 5-6 Seals are required to comply with *NEC*®
300-7 where the raceway is subjected to more than one
temperature such as shown.

concerned with thermal expansion and contraction in steel and aluminum installations because of the low coefficient of linear expansion. 1993 *NEC*® *Section 347-9* states that an expansion coupling is needed wherever the change in length due to temperature variation exceeds .25 inch. *Section 300-7(b)* states: "Raceways shall be provided with expansion joints where necessary to compensate for thermal expansion and contraction." Expansion must be a consideration in both metallic and nonmetallic raceways. Owing to the low coefficient of expansion in steel raceways (only about 5% of that of PVC nonmetallic rigid conduit), expansion joints are rarely necessary to comply with *Section 300-7(b)*. However, expansion joints are often necessary in all raceway types, both ferrous and nonferrous, metallic and nonmetallic, for architectural considerations. Where electrical raceway systems cross building expansion joints, which are commonly found in large buildings, the electrical raceway system should include an expansion joint to permit the raceway to move with the building.

Other considerations for expansion may be encountered where buildings are designed as earthquakeproof, such as are commonly found in West Coast areas. Where these flexible architectural joints are encountered, the electrical raceway system should also include expansion joints to compensate for any movement encountered during an earthquake.

Nonmetallic raceways, however, because of the

ature. The coefficient of linear expansion for various conduits is (a) PVC conduit 3.38×10^{-5} in./in./°F, (b) aluminum conduit 1.2×10^{-5} in./in./°F, and (c) steel conduit 0.6×10^{-5} in./in./°F. These figures all were exposed to a high degree of temperature change.

The designer and installer generally need not be

Table 347-9(A). Expansion Characteristics of PVC Rigid Nonmetallic Conduit Coefficient of Thermal Expansion = 3.38×10^{-5} in./in./°F

Temperature Change (°F)	Length Change of PVC Conduit (in./100 ft)	Temperature Change (°F)	Length Change of PVC Conduit (in./100 ft)
5	0.2	105	4.2
10	0.4	110	4.5
15	0.6	115	4.7
20	0.8	120	4.9
25	1.0	125	5.1
30	1.2	130	5.3
35	1.4	135	5.5
40	1.6	140	5.7
45	1.8	145	5.9
50	2.0	150	6.1
55	2.2	155	6.3
60	2.4	160	6.5
65	2.6	165	6.7
70	2.8	170	6.9
75	3.0	175	7.1
80	3.2	180	7.3
85	3.4	185	7.5
90	3.6	190	7.7
95	3.8	195	7.9
100	4.1	200	8.1

Table 347-9(B). Expansion Characteristics of Fiberglass Reinforced Conduit (Rigid Nonmetallic Conduit) Coefficient of Thermal Expansion = 1.5×10^{-5} in./in./°F

Temperature Change (°F)	Length Change of Fiberglass Conduit (in./100 ft)	Temperature Change (°F)	Length Change of Fiberglass Conduit (in./100 ft)
5	0.1	105	1.9
10	0.2	110	2.0
15	0.3	115	2.1
20	0.4	120	2.2
25	0.5	125	2.3
30	0.5	130	2.3
35	0.6	135	2.4
40	0.7	140	2.5
45	0.8	145	2.6
50	0.9	150	2.7
55	1.0	155	2.8
60	1.1	160	2.9
65	1.2	165	3.0
70	1.3	170	3.1
75	1.4	175	3.2
80	1.4	180	3.2
85	1.5	185	3.3
90	1.6	190	3.4
95	1.7	195	3.5
100	1.8	200	3.6

Figure 5-7 To utilize these tables for metal raceways: multiply the length, change values in the table by .01 for aluminum raceways and .05 for steel raceways. (Reprinted with permission from NFPA 70-1999.)

Example

380 ft. of conduit is to be installed on the outside of a building exposed to the sun in a single straight run. It is expected that the conduit will vary in temperature from 0°F in the winter to 140°F in the summer (this includes the 30°F for radiant heating from the sun). The installation is to be made at a conduit temperature of 90°F. From the table, a 140°F temperature change will cause a 5.7 in. length change in 100 ft. of conduit. The total change for this example is 5.7″ × 3.8 = 21.67″ which should be rounded to 22″. The number of expansion couplings will be 22 ÷ coupling range (6″ for E945, 2″ for E955). If the E945 coupling is used, the number will be 22 ÷ 6 = 3.67 which should be rounded to 4. The coupling should be placed at 95 ft. intervals (380 ÷ 4). The proper piston setting at the time of installation is calculated as explained above.

$$0 = \left[\frac{140 - 90}{140} \right] 6.0 = 2.1 \text{ in.}$$

Insert the piston into the barrel to the maximum depth. Place a mark on the piston at the end of the barrel. To properly set the piston, pull the piston out of the barrel to correspond to the 2.1 in. calculated above.

See the drawing below.

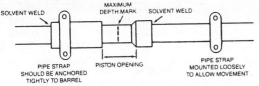

Figure 5-8 Rigid Nonmetallic Conduit. For steel raceway 21.7 x .05 = 1.08 inches total expansion, an expansion fitting would not generally be required; for aluminum raceway 21.7 x .1 = 2.17 inches an expansion fitting may be required. (Courtesy of Carlon, A Lamson & Sessions Company)

high coefficient of expansion, have special requirements in *NEC® Article 347. Section 347-9* states that expansion joints for rigid nonmetallic conduit shall be provided to compensate for thermal expansion and contraction except where less than .25 inches of expansion is encountered. The code in *Section 347-9* clarifies that all installations of rigid nonmetallic conduit encountering an expansion of over .25 inches are required to have an expansion joint installed. A good rule of thumb in utilizing *Table 347-9A* for other than PVC rigid nonmetallic conduit is to apply the table as follows. A multiplier of .05 results in the projected expansion of ferrous (steel) metal conduit, or about 1/20th of PVC. A multiplier of .1 results in the projected expansion of (aluminum) nonferrous rigid metal conduit, or about 1/10th of PVC. To clarify, see the following example. For fiberglass conduit see *Table 347-9E*.

A manufacturer of nonmetallic conduit recommends that 30° F be added to the estimated temperature range when PVC conduit is installed in direct sunlight to allow for radiant heating. This is owing to the gray color of the raceway, which absorbs that radiant heating.

An expansion coupling consists of two sections of conduit, one telescoping inside another. When expansion couplings are installed, alignment of the piston and the barrel is important. Mount the expansion joint level or plumb for best performance. For a vertical run the expansion coupling must be installed close to the top of the run, with the barrel jointing down so that rainwater does not run into the opening. The lower end of the conduit run must be secured at the bottom so that any length change due to temperature variation will result in an upward movement.

If this same 380-foot run of conduit was installed using steel conduit, the expansion would be 21.67 × .05, or approximately 1.1 inches. If run in aluminum conduit, expansion would be 21.67 × .1, or approximately 2.2 inches.

Summary: (1) Anticipate expansion and contraction of PVC conduit in all above-ground exposed installations. (2) Use an expansion coupling when the length change due to temperature variation will exceed .25 inch. (3) PVC conduit expands 4.1 inches for each 100 feet run and 100° of temperature change. (4) Steel conduit expands approximately .2 inch per each 100 foot run and 100° temperature change. (5) Aluminum conduit would expand approximately .4 inch for each 100 feet run and 100° F temperature change. (6) Align expansion coupling with conduit run to prevent binding (plumb or level). (7) Follow the instructions to set the piston opening. (8) Rigidly fix the outer barrel of the expansion coupling so it cannot move. (9) Mount the conduit connected to the piston loose enough to allow the conduit to move as the temperature changes.

> **NOTE:** A manufacturer recommends that for all above-ground installations of PVC raceway where temperature change in excess of 25° F (14° C) is anticipated, expansion joints shall be installed.

SECURING AND SUPPORT REQUIREMENTS

Although the specific requirements for securing and supporting each wiring method type are covered in the article governing that wiring method, general securing and supporting requirements are covered in *NEC® Article 300*. All support methods are made up from hardware items that may not be "listed" by an approved testing laboratory. The integrity of support methods must be field evaluated. Careful consideration should be given to the methods used because the integrity of the wiring method may be jeopardized by inadequate support methods.

Section 300-11 requires that raceways, cable assemblies, boxes, cabinets, and fittings shall be securely fastened in place. Ceiling support wires that do not provide secure support are not permitted as the sole support. This *NEC®* section permits branch-circuit raceways to be supported on support wires where the equipment is supported by or located below the suspended ceiling, but are not permitted to be used for the support in fire-rated ceilings. Remember, this covers only branch circuits. *Sections 300-23, 725-6, 760-7, 770-7, 800-5,* and *820-5* require access to electrical equipment behind panels, including suspended ceilings that are designed to allow access, and further states that access shall not be denied by an accumulation of wires and cables that prevents access.

Caution: the designer should coordinate this support system with the architect designing the ceiling-grid system. Building inspectors may not permit the use of these support wires and may require independent support of raceway systems. Raceways are not to be used to support other raceways, cables, or nonelectrical equipment. There are some exceptions in the *NEC®*. Good support methods are important in maintaining the integrity of the raceway system. Generally, each wiring component must be independently supported. For example, the outlet or junction box must be supported in accordance with *Article 370*. The wiring method must be supported in accordance with the article governing it, such as Electrical Metallic Tubing (EMT), *Article 348*. Lighting fixtures are supported in accordance with *Article 410*, etc.

Determine The Piston Opening

The expansion joint must be installed to allow both expansion and contraction of the conduit run. The correct piston opening for any installation condition should use the following formula:

$$0 = \left[\frac{T\ max - T\ installed}{\Delta T}\right]E$$

where: 0 = piston opening (in.)

$T\ max$ = maximum anticipated temperature of conduit (°F)

$T\ ins$ = temperature of conduit at time of installation (°F)

ΔT = total change in temperature of conduit (°F)

E = expansion allowance built into each expansion coupling (in.)

STANDARD EXPANSION COUPLINGS

E945 series expansion couplings are designed to compensate for length changes due to temperature variations in exposed conduit runs. See explanation on previous page.

NOTE: Do not use expansion coupling when encased in concrete. Conduit is immobilized by the concrete and will conform to the expansion rate of the concrete.

Expansion Couplings

Part No.	Size	Ctn. Qty.	Lay Lengths		Available Length Expansion-Contraction
			Stop to Stop Total Closed	Stop to Stop Total Open	
E945D	1/2	50	12 1/4	18 5/8	6
E945E	3/4	50	12 1/4	18 5/8	6
E945F	1	45	12 3/4	19 1/8	6
E945G	1 1/4	30	12 3/4	19 1/8	6
E945H	1 1/2	25	12 3/4	19 1/8	6
E945J	2	15	13 1/2	19 7/8	6
E945K	2 1/2	10	14	20 3/8	6
E945L	3	10	16 1/2	23	6
E945M	3 1/2	5	16 1/2	23	6
E945N	4	5	17 1/2	24	6
E945P	5	3	18 1/2	24 1/2	6
E945R	6	2	20 1/2	26 1/2	6

SHORT EXPANSION COUPLINGS

(Expands to a maximum of 2″)

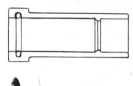

Part No.	Size	Ctn. Qty.
E955D	1/2	40
E955E	3/4	40
E955F	1	25
E955G	1 1/4	15
E955H	1 1/2	10
E955J	2	6

Figure 5-9 (Courtesy of Carlon, a Lamson & Sessions Company)

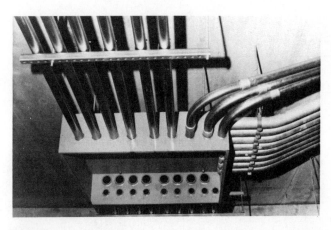

Figure 5-10 Trapeze supports as well as individual raceway support in accordance with *NEC®* (Courtesy of Allied Tube & Conduit)

MECHANICAL AND ELECTRICAL CONTINUITY

Metal or nonmetallic raceways, cables, armors, and cable sheaths shall be continuous between cabinets, boxes, fittings, and other enclosures or outlets.

> **NOTE:** Wrench-tight fittings and couplings and the proper supporting of the wiring method insures compliance with these requirements.

There is an exception to this requirement in *NEC® Section 300-12*. The electrical continuity of conductors in raceways shall be continuous, and there shall be no splice or tap within the raceway itself. Again, there are some exceptions to that rule for auxiliary gutters, wireways, boxes, fittings, surface metal raceways, and busways.

INDUCTION HEATING

When an alternating current is flowing, an electromagnetic field exists around each energized conductor. It varies in strength as the current in the conductor varies from zero to maximum during each half-cycle. If the conductors in a circuit are in separate steel raceways, the changing electromagnetic field induces an electromotive force in each raceway. Current will then flow in each raceway. The magnitude of the current flow is determined by the impedance of the raceway as a current path. Current flow through this impedance may generate sufficient heat to raise the temperature of the raceway high enough to damage the conductor insulation. The current may also cause an additional voltage drop by inducing a back electromotive force in the conductor itself. One solution is to have a conductor of each phase and the neutral of the circuit and equipment-grounding conductor where required in each raceway. The sum of the currents in one direction equals the sum of the currents in the opposite direction at any instant, and the changing electromagnetic fields balance out by canceling each other out. Inductive heating therefore is minimized. (See Figure 5-12.)

NEC® Section 300-20 requires electrical circuits enclosed in metal raceways to have each phase conductor, neutral, and equipment-grounding conductor, where used, installed in the raceway. The same requirement applies to parallel-conductor circuits carried over two or more parallel metal raceways. Where used, the neutral conductor and the equipment-grounding conductor must be run in each raceway. Inductive heating may occur where the conductors of the circuit pass through the individual openings in the wall of the raceway, metal pole box, or cabinet. When feeders are terminated at a panelboard, the effect can be minimized by cutting slots in the metal between conductor openings. Another method is to cover a rectangular opening cut in the metal cabinet with an insulated block containing individual openings for separate conductors. One or the other of these methods is required by *NEC® Section 300-20*. There is an exception to this requirement where current is negligible, such as in secondary conductors of X-ray and electric discharge sign circuits. Because aluminum is not a magnetic metal, there is no heating due to hysteresis; however, induced current is present. It is not of sufficient magnitude to require the grouping of conductors or special treatment in passing the conductors though aluminum wall sections.

NEC® Section 300-3 (b) also requires all conductors of the same circuit, and, where used, the

300-11(a)
Securing and Supporting

Raceway

Branch circuit wiring supplying equipment
within, supported by or below suspended
ceiling may be supported by ceiling support
wires of non-fire-rated ceilings

Figure 5-11 Assorted hardware designed for supporting electrical wiring methods. (Courtesy of B-Line Systems Company, and Raco Inc.)

Nonmetallic Clamps

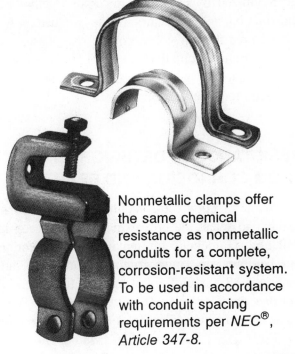

Nonmetallic clamps offer the same chemical resistance as nonmetallic conduits for a complete, corrosion-resistant system. To be used in accordance with conduit spacing requirements per *NEC®*, *Article 347-8.*

Snap Strap Conduit Wall Hangers

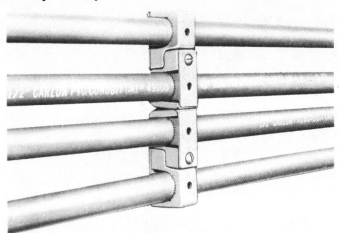

Clamp solves bowing problems resulting from the expansion and contraction of conduit caused by varying temperature changes.

Snap Strap Clamps

To be used in accordance with conduit spacing requirements per *NEC®*, Article 347-8.

One Hole

Two Hole

Examples of Outlet Box Supports

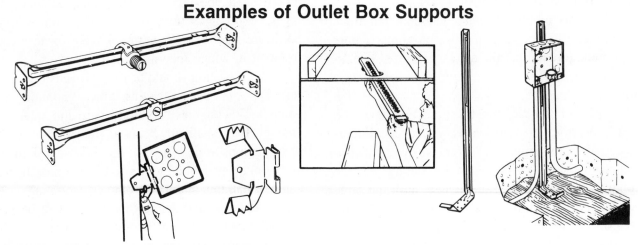

Figure 5-11 (continued) Assorted hardware designed for supporting electrical wiring methods. (Courtesy of Carlon, a Lamson & Sessions Company, and Raco Inc.)

Induced Currents in Steel Raceways

NEC Sec. 300-20
Unbalanced load, 3-phase, 4-wire system

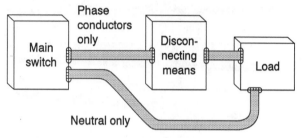

Entire loop subject to inductive heating
(violates *NEC Section 300-20*).

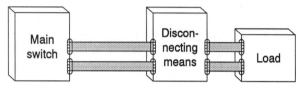

Neutral and phase conductors in each raceway.

Figure 5-12 (Courtesy of American Iron & Steel Institute)

neutral and all equipment-grounding conductors to be contained within the same raceway, cable, tray, trench, cable or cord, with some exceptions. *Section 300-5(a)* has similar requirements for direct buried circuitry in a raceway or cable; they shall be installed in close proximity in the same trench to direct buried cables. *Exception 2 of Section 300-5(i)* permits isolated phase installations in nonmetallic raceways in close proximity where the conductors are paralleled, as permitted in *Section 310-4*, and where the conditions of *Section 300-20* are met.

This is not a recommended practice; however, *NEC® Section 310-4* makes very clear the importance of paralleling conductors to prevent imbalanced loading or induction by stating that the conductors must be the same length, have the same conductor material, be the same size in circular mil area, have the same insulation type, and be terminated in the same manner. Fine Print Note (FPN) to

Section 310-4 alerts us that the differences in inductive reactance and the unequal division of current can be minimized by choice of materials, methods of construction, and orientation of the conductors. It is not required that the conductors of one-phase, neutral, or grounded-circuit conductors be the same as those of another phase-neutral or grounded-circuit conductor to achieve balance.

RESISTANCE TO DETERIORATION FROM CORROSION AND REAGENTS

NEC® Section 110-3 requires the designer and installer to select the proper materials for each installation. Guidelines for making these selections can be found in Underwriters Laboratories standards, the UL Green Book, known as the *Electrical Constructions Materials Directory, or the* UL White Book, also known as *General Information Directory for Electrical Construction, Hazardous Location, and Electrical Heating and Air Conditioning Equipment.* In addition, *Section 110-11* reminds us that deteriorating agents contribute to corrosion. Unless identified for use in the operating environment, no conductors or equipment shall be located in that environment. *Section 110-11* states: "Some cleaning and lubricating compounds can cause severe deterioration of many plastic materials used for insulating and structural applications in equipment."* *Section 300-6* reminds us that metal raceways, armored cable, boxes, cable sheathing, cabinets, elbows, couplings, fittings, supports, and support hardware shall be of material suitable for the environment in which they are installed.

Outer coating should be of approved corrosion-resistant material, such as zinc, cadmium, or enamel. Where protected from corrosion solely by enamel, they shall not be used out-of-doors or in wet locations as described in the following. Ferrous and nonferrous metal raceways, cable armor, boxes, cable sheathing, cabinets, elbows, couplings, fittings, supports, and support hardware shall be permitted to be installed in concrete or in direct contact with the earth and areas subject to severe corrosive

*(Reprinted with permission from NFPA 70-1999.

influences when made of material judged suitable for the condition and when provided with corrosion protection approved for the condition. In portions of dairies, laundries, canneries, and such locations where the walls are frequently washed, or where there are surfaces of absorbent materials such as damp paper or wood, the entire metallic wiring system shall be mounted so that there is at least 1/4 inch of air space. Nonmetallic raceways, boxes, and fittings shall be permitted to be installed without air space. In general, in areas where acids and alkali chemicals are stored or may be present such corrosive conditions exist, particularly when wet or damp. Severe corrosive conditions may also be present in portions of meat-packing plants, tanneries, glue houses, some stables, and installations immediately adjacent to seashore and swimming pool areas. Corrosive conditions are also present where chemical deicers are used and in storage cellars or rooms for hides, casings, fertilizer, salt, and bulk chemicals.

NEC® Article 547 gives some additional alerts when the installation falls within the scope of agriculture buildings and there are concerns about corrosive atmospheres. These include poultry and animal facilities that may be damp from washing or sanitizing, where cleaning agents and similar conditions exist; livestock confinement areas; environmentally controlled poultry houses; and similar enclosed areas of an agriculture nature. With these sections there is ample warning to the designer to make a correct and proper selection of the material to be specified.

Since 1905, rigid metal conduit has been manufactured with corrosion protection, originally a coating of black enamel, later by means of electrogalvanizing and hot-dip galvanizing. In 1961, the first PVC-coated rigid steel conduit was offered. Today numerous products on the market provide for specialty coatings with specific concerns relating to corrosion and deteriorating factors from the presence of severe environment and chemicals. These coatings are applied both inside and outside the listed Galvanized Rigid or IMC conduit. Underwriters Laboratories, based on their tests and evaluations, state in the *General Information for Electrical Construction, Hazardous Location, and Electric Heating and Air Conditioning Equipment 1998* (White Book)

that intermediate metal conduit (IMC), as covered by *NEC®* Article 345, and rigid metal conduit, as per Article 346, installed in concrete do not require supplementary corrosion protection.

Rigid metal and IMC installed in contact with soil does not generally require supplementary corrosion protection. EMT as covered in *Article 348* generally does not require supplementary protection in concrete on grade or above, but generally does if in concrete below grade or in direct contact with the soil. Aluminum raceways in concrete or direct contact with soil always requires supplementary protection. In the absence of specific local experience, soils producing severe corrosive effects are generally characterized by a low resistivity (fewer than 2000 ohm-centimeters). Wherever ferrous metal conduit or tubing runs directly from concrete encasement into soil, severe corrosive effects are likely to occur on metal in contact with the soil.

Supplementary nonmetallic coatings as part of the conduit have not been investigated for resistance to corrosion; manufacturers should be contacted for this data. Supplementary nonmetallic coatings of greater than .010-inch thickness applied over metallic protective coatings are investigated with respect to flame propagation and detrimental effects to the basic corrosion protection provided by the protective coatings. Many manufacturers today, in addition to the galvanized coating on metal raceways, also provide an additional layer of chromate or other organic coating to prevent formation of "white rust," the powdery substance that otherwise forms as a corrosive process of zinc. This provides some additional protection, which is not evaluated by UL. However, it certainly adds to the appearance as well as the protection on those raceway systems.

Many questions exist as to the suitability of the galvanized protective coating applied to rigid, IMC and EMT conduit. The zinc coating need not be unscratched to remain effective. If a scratch exposes bare steel, moisture in the air forms an electrolyte on the surface of the raceway, and through galvanic action, zinc is transported to the exposed steel and deposited or plated out as an electrolytic band. This healing action of zinc affords protection even where wrench marks or scratches are as much as 0.1 inch wide. These metallic raceway types have provided

Rigid Steel Conduit, Intermediate Steel Conduit (IMC), and Electrical Metallic Tubing (EMT) are *Corrosion Resistant* for Wide Applicability

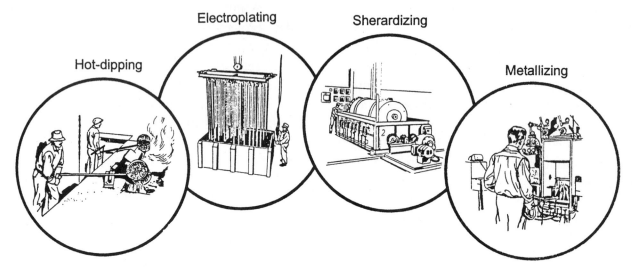

Both the exterior and the interior of tubular steel raceways are coated to protect from corrosion. Coatings are usually of zinc.

Zinc coating, or galvanizing, effectively seals the steel surfaces from corrosive moisture and other corrosive agents. In addition, the zinc coating may slowly sacrifice itself—and thus save the steel surfaces from corrosion—in those cases where the conduit may have been damaged and the steel exposed by a blow.

There are four methods of galvanizing steel conduit:

1. Hot-dipping
2. Electroplating
3. Sherardizing
4. Metallizing

In the hot-dipping process, the clean tube is immersed in a molten zinc for a suitable time and then withdrawn. The tube is wiped with a superheated steam or air to remove any excess of zinc on either the exterior or interior, and to control the thickness of zinc. The zinc alloys with the iron at the surface.

In the electroplating process, the clean tube is placed in a solution of zinc salts and made a cathode of an electrical circuit so that the zinc is plated from the solution onto the outside of the tubing. Positive contact with the solution is obtained through zinc anodes, which replace the zinc content of the solution as it is used up.

In the sherardizing process, the conduit is first coated with zinc by heating a predetermined amount of commercially pure zinc powder and the steel pipe in a slowly revolving, sealed retort. A second method is also employed of electroplating a controlled thickness of zinc coating to the steel pipe. Both methods of zinc coating are subjected to accurately controlled furnace heat.

In the metallizing process, finely divided particles of zinc are metal-sprayed in a heated, semimolten condition onto the surface of a specially prepared tube to form an adherent coating.

Although hot-dipping and electroplating are the most common, any of the four methods of coating gives the tube a uniform coating of zinc of excellent corrosion resistance. (See A.S.A. for thickness of coating and test procedures.)

Figure 5-13

excellent serviceability in countless installations over the past 100 years. Corrosion is generally not a problem in most soil types. However, the final decision must be made by the designer and approved by the authority having jurisdiction.

Many think that rigid nonmetallic conduit (*NEC® 347*) is impervious to all corrosive atmospheres. Not true! Rigid nonmetallic conduits, both Schedule 40 and Schedule 80, are generally acceptable for many environments; however, when making an installation in environments containing chemicals, check the manufacturer's recommendations. If there are any questions for specific suitability in a given environment, prototype samples should be tested under actual conditions. Underwriters Laboratories' *General Information From Electrical Construction Materials Directory* (White Book) states that listed PVC conduit is inherently resistant to atmospheres containing common industrial corrosive agents and also withstands vapors or mist from caustic, pickling acids, plating bath, and hydrofluoric and chromic acids. PVC conduit, elbows, and bends including couplings that have been investigated for direct exposure to other reagents may be identified by printing "reagent resistant" on the surface of the product.

Such special uses are described as follows: PVC conduit, elbows, and bends where exposed to the following reagents at 60° C or less, acetic, nitric (25C only) acids, acids in concentrations not exceeding 1/2 normal; hydrochloric acid in concentrations not exceeding 30%; sulfuric acid in concentrations not exceeding 10% normal; sulfuric acid in concentrations not exceeding 80% (25C only); concentrated or dilute ammonium hydroxide; sodium hydroxide solutions in concentrations not exceeding 50%; saturated or dilute sodium chloride solution; cottonseed oil, or ASTM #3 petroleum oil.

NOTE: See Appendix, Figure A-12, for manufacturer's corrosion resistance data.

An additional alternative was offered in 1961 when the first PVC-coated rigid steel was pioneered. Today several different types of coated conduit are designed for different corrosive environments. These manufacturers offer a complete wiring method in spe-

TABLE OF METALS IN GALVANIC SERIES
Corroded End (anodic or less noble)

MAGNESIUM
ZINC
ALUMINUM
CADMIUM
IRON OR STEEL
TIN
LEAD
NICKEL
BRASS
BRONZES
NICKEL-COPPER
ALLOYS COPPER

Any one of these metals and alloys will theoretically corrode while offering protection to any other that is lower in the series, so long as both are electrically connected.

As a practical matter, however, zinc is by far the most effective in this respect.

Because zinc is more active in galvanic couples than iron and steel, it provides these metals with electrolytic protection against rust. This protection is so effective that even though there be a small exposed area on the base metal, the attack of the elements will be directed to the zinc, and protection will continue as long as sufficient zinc remains.

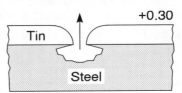

This is what happens at a small exposed area in a coating of a metal having a lesser tendency to go into solution than the base metal. The steel rusts away, protecting the tin.

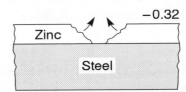

This is what happens at a small exposed area in a coating of a metal having a greater tendency to go into solution than the base metal. The zinc is consumed while protecting the steel from any attack.

Position in the galvanic series is not an infallible guide to galvanic protection because of other influences. Zinc is the one metal that will protect iron and steel under conditions of exposure to be expected.

cial environments, including coated conduit, fittings and accessories, and a complete package line. They also furnish experienced sales and engineering personnel capable of recommending the proper system for your application and provide a complete bill of material including all conduit fittings, junction boxes, conduit bodies, and hardware for a complete system. Installation is made simple with standard rigid conduit tooling, and damage to the surface is easily repaired by the installer at the time the damage occurs.

The manufacturers of these products begin with standard listed galvanized rigid conduit in conformance with NC C80.1 and Underwriters Laboratories UL6 in accordance with *NEC® Article 346* or listed intermediate metal conduit in compliance with UL 1242 and NC C80.6 in accordance with *NEC® Article 345* listed products. That material has a protective coating from 15 to 40 mils. Protection is available on the outside, inside, or both in various mil thicknesses, according to the specific needs, and is suitable in all hazardous, heavy industrial, manufacturing, and corrosive areas, bridges, highways, utility plants and substations, and the like. These products afford excellent protection from corrosion and reagents and deterioration from reagents, as well as excellent physical protection, and are ideal where both these conditions exist.

> **NOTE:** See interior and exterior chemical-resistance charts in Appendix, Figure A-13.

Today's designers and installers are offered an excellent selection of materials to meet all the needs of today's state-of-the-art installations.

PENETRATION OF FIRE-RESISTANT ASSEMBLIES

There are three major building codes. In the year 2000 these three codes will be replaced by the International Building Code (IBC). These codes include many provisions the purpose of which is to provide construction that not only withstands the effects of fire in a portion of a building, but also reduces the likelihood of fire spreading within the building. For extremely large or high structures building codes usually require fire-resistant con-

Figure 5-14 Examples of PVC-coated rigid steel conduit (Courtesy of Robroy Industries & Ocal Inc.)

struction and noncombustible construction materials. Building codes prescribe that in fire-resistant buildings, the columns, beams, girders, walls, and floor constructions must have a resistance to fire measured by a standard fire test procedure (ASTM E119, *Standard Method of Fire Tests of Building Construction and Materials*). The specific requirements of the assembly design under the classification system are found in Underwriters Laboratories *Fire Resistive Directory 1999* (Orange Book) and the *Building Materials Directory* (Manila Book). Other accredited labs also conduct tests to establish fire ratings. The three model building codes govern fire-resistance requirements and stipulate the time-rated period for various types of occupancies (such as 1-hour, 2-hour). These requirements are found in the BOCA National Building Code, ICBO Uniform Building Code, and SBCCI Standard Building Code.

Many fire-resistant assemblies use cellular floors, decks, and suspended ceilings. Openings for recessed lighting fixtures, electrical outlets, and air vents may be included in these fire-resistant assemblies. The ceiling usually is, but may not be, tested as part of the assembly. The details of the construction are important if the required fire resistance is to be achieved. Fire-resistant assemblies tested by UL are "classified." This is the same approval as "listed" but connotes testing for a very specific application in a specific manner. It is essential that a listed fire-resistant assembly conform to the tested specimen in all particulars after field installation. The materials and dimensions must be at least equal to those of the test specimen and constructed as per the classified design. Any holes or apertures in the fire protection must be limited in number, area, and distribution in the same manner as in the test assembly. Engineering analysis also must show that fire resistance of an assembly is not critically affected by a proposed arrangement of openings for lighting fixtures or ducts.

The Code permits openings for outlets, air-conditioning ducts, and similar equipment where such openings do not exceed the percentage of the ceiling area established by fire-test data. Because such openings allow only a small amount of heated gases generated by a fire to enter a concealed space, their hazard is considered negligible. Openings of this type are often described as membrane penetrations. Obviously, no such penetration is permitted to extend through an entire wall or floor and ceiling assembly. The design and construction of fire-rated assemblies are quite specific, even including specific nails and screws. Minor changes in design, such as thickness or the method of fastening of construction materials, may have a significant effect on the assembly's behavior during a fire. Wires supporting a suspended grid ceiling, for example, are critical; for this reason the *NEC®* does not allow such wires to be used for support of electrical wiring when it is a rated ceiling (see *Section 300-11(a)*).

If a wall or ceiling is penetrated for the passage of raceways or for work space around them, the fire resistance of the wall or ceiling is effectively destroyed unless the opening around the penetrating item is properly sealed. This danger is recognized in *NEC® Section 300-21*, which states: "Electrical installations in hollow spaces, vertical shafts, and ventilation or air-handling ducts shall be so made that the possible spread of fire or products of combustion will not be substantially increased. Openings around electrical penetrations through fire-resistance rated walls, partitions, floors, or ceilings shall be firestopped using approved methods to maintain the fire-resistance rating."

Because most plumbing, air-handling, electrical, communications, and other building services must pass through fire-resistant assemblies, coordination of building design, type of materials, and installation methods for these systems is essential to minimize the spread of fire. Changes in design or type of materials used in a fire-rated assembly should not be introduced without preapproval of the authority having jurisdiction.

Plans for proposed buildings may not indicate all eventual penetrations. The absence of detailed provisions for electrical or communications systems in the floor plan does not mean that such systems will not be installed later. If the electrical and communications circuits are to be added after construction is completed, it may be necessary either to construct chases (shafts) in fire-rated floors, walls, and ceilings for cables or raceways to pass to other areas, or openings may be made in the walls, floors, or ceilings. Such penetrations through an entire assembly may void the intended fire protection of the construction and allow the spread of any of the products of combustion (smoke, hot or toxic gases, or flame). Creation of passageways through which smoke, gases, and heat may pass only increases the hazards to building occupants. It is the responsibility of the authority having jurisdiction (AHJ) to determine whether to approve a particular firestop method or material. This is a significant responsibility for the AHJ, who usually must look to others for information to verify that a particular method or material has been adequately evaluated.

The building codes provide two recognized methods for sealing openings around penetrating items. These are through-penetration, fire-stop systems and annular space-filler materials. Through-penetration systems are available for both combustible and noncombustible penetrating items. Annular space fillers may be used only with noncombustible penetrants. Why the difference?

> **Warning!** The building codes generally consider all materials, except for steel and copper, as combustible.

When combustibles (e.g., nonmetallic sheathed cables, insulated conductors in cable tray, aluminum-sheathed cables, and PVC conduit) are exposed to heat, they burn or melt away and materials surrounding them fall away, thus leaving a hole through which the fire continues to travel either vertically or horizontally to adjoining floors or compartments. In the 1970s a disastrous cable fire spread through the

Brown's Ferry nuclear plant in just this manner. Because of that event new methods of sealing were developed, among them materials that swell up when exposed to heat and fill the hole as the nonmetallics burn away. This swelling process is called intumescing. Other through-penetration systems are the endothermic type. They release chemically bound molecules of water when exposed to heat. Both types must be tested in accordance with ASTM E814. UL Standard 1479 is the same as ASTM E814, ("Fire Tests of Through-Penetration Firestops").

Through-penetration systems classified by UL are found in the "Orange Book" (*Fire Resistance Directory*). This directory is very explicit about the assembly design, size, and type of the penetrating item(s), the size of the opening around the item, and the thickness and construction of the protection materials. It is very important that the listing criteria be followed precisely, as the systems are designed to ensure enough volume to fill the opening when the penetrating item burns or melts away or to control the temperature on the unexposed side. UL specifies that individual components are not rated, nor are they intended to be mixed with other systems. The classified rating applies only to the complete system. Some think through-penetration firestop systems prevent the passage of smoke and toxic gases. Although this may be true in some instances, the E814 test does not provide test criteria for determining the passage of smoke and toxic gases.

Through-penetration fire-stop systems are absolutely necessary for combustible penetrations. For noncombustible penetrations (e.g., EMT, IMC, and rigid conduit), either through-penetration systems or annular space fillers may be used. The systems for metallic penetrations are also found in the UL Orange Book. To use this book for both combustibles and non-combustibles, you will need to know the specifics of the assembly being penetrated, such as the fire-rating, type of studs, thickness of gypsum or concrete, and the size and type of penetrating item(s).

The Orange Book indicates an "F" rating and a "T" rating. The F rating is based on the time period for which the system prohibits flame passage to the unexposed side (as evidenced by flame occurrence) in conjunction with an acceptable hose-stream test performance.

Unless the penetration is contained in a wall cavity, you may also need to know the T rating of the specific system. The T rating is the test time expired before the sealant and penetrating item reach a temperature of 325°F above ambient on the non-fire-exposed side of the assembly in addition to containing the flame and passing the hose-stream test. Hose-stream testing is the application of a specified water stream from a fire hose to evaluate the structural integrity of the assembly after fire exposure. It is not relative to fire-fighting activities during an actual fire. You should consult the applicable building code for exact language regarding T ratings, as this is still in the process of change and varies in different codes.

Annular space fillers, which are permitted to be used only with noncombustible penetrants, do not require T ratings. However, the building codes do require that, to qualify as annular space fillers, materials must be able to withstand ASTM E119 time-temperature conditions under positive pressure for the full period during which an assembly has been rated without igniting cotton waste placed in contact with it.

Common construction materials have long been used with great success as annular space fillers. These are products such as cement, mortar, grout, handyman caulk, mineral wool, even joint compound. As an aid to the industry, the NEMA, Steel Rigid Conduit and Tubing Section sponsored an investigation at UL to document the performance of these longtime annular space fillers.

These generic-type tests easily affirmed the performance of these materials. Based on this testing, the three national model building codes have added language that automatically accepts cement, mortar, and grout for sealing openings in masonry assemblies. The applicable code should be reviewed for specifics because there are some variables in the precise language.

A summary (or full) report on this annular space testing is available from NEMA, Section 5RN, 1300 North 17th Street, Suite 1847, Rosslyn, VA 22209. Ask for UL Special Services Investigation Fire NC 546 Project 90NK11650.

This report documents that everyday joint compound can maintain the fire-resistance rating of a gypsum wallboard assembly for two hours—even after damage to the fire-exposed side due to a furnace explosion. It is a good document for your library to use in securing AHJ approval when planning to seal with annular space filler.

Electricians and designers must become educated about the sealing of openings around electrical wiring to comply with *NEC® Article 300-21* and with the building codes.

COMBUSTIBILITY OF RACEWAY MATERIALS AND CONDUCTOR INSULATION

In the past 12 to 15 years a great deal of attention has been given to hazards of fire relative to combustible electrical systems (along with combustible construction, furnishings, and finishing materials). Much of this attention was created by the expansion by the *NEC®* of nonmetallic wiring methods inside by buildings. The degree of hazard remains controversial, as does the risk involved. Certainly individuals draw different conclusions from the same set of facts; additionally, risk tolerance varies from person to person, otherwise we would not have deep-sea divers and fighters of oil-well fires. Other questions have also arisen in this controversy, such as how many lives is it permissible to lose?—we can't feasibly get to zero. The age-old response then becomes, Your family or mine? In this section I present some of the issues involved. The user, designer, and jurisdiction then have to determine the acceptable level of risk for the particular situation in their own view.

The *NEC®* provides minimum code requirements and allows choices in wiring methods. However, in a few instances it has directly addressed the matter of fire concerns. One specific area is in plenums and other environmental air spaces. The whole construction code arena recognizes the dangers inherent in allowing the products of combustion (smoke, heat, and toxic gases) to spread throughout a building by way of the air handling system. Materials allowed in such spaces are carefully controlled (although a series of events in recent years is making this borderline).

Article 300-22 of the *NEC®* controls the electrical system in environmental air spaces. The wiring methods in ducts or plenums used for environmental air [*300-22(b)*] must be EMT, IMC, rigid metal conduit, flexible metallic tubing, MI cable, or Type MC cable employing a smooth or corrugated impervious metal sheath without an overall nonmetallic covering. To connect physically adjustable equipment and devices permitted in these areas, lengths of flexible metal conduit or liquidtight flexible metal conduit not over 4 feet long are permitted.

A plenum is defined in the *NEC®* as a "compartment or chamber to which one or more air ducts are connected and which forms part of the air distribution system."* The BOCA National Building Code contains the definition: "An enclosed portion of the building structure which forms part of an air distribution system and is designed to allow the movement of air."

Thus, it is important to recognize that a ceiling space through which the building air moves may be considered a plenum. If there is ever any doubt relative to the *NEC®* definition, the building code governs.

In environmental air spaces other than ducts or plenums [*300-22(c)*], other wiring methods are added to those named previously. These include totally enclosed nonventilated insulated busways having no provisions for plug-in connections, Type AC cable, or other factory-assembled, multiconductor control or power cable that is specifically listed for the use. Some of these are called "plenum cable," a misnomer because they are not for use in actual plenums. Other means of enclosing the conductors, beyond those conduits named in *300-22(b)*, are surface metal raceway or wireway with metal covers, or solid-bottom metal cable tray with solid-metal covers. [For minor exceptions refer to *300-22(c)*.]

What are some of the concerns regarding combustible wiring methods, whether or not they are in environmental air spaces? The primary nonmetallics used in buildings for raceways and conductor insulation are PVC and nylon. When PVC reaches a temperature in the range of 450°

---

*Reprinted with permission from NFPA 70-1999.

to 500°F, it starts to break down chemically (thermally decompose), although some formulations may not ignite until almost 1000°F. This decomposition releases a very corrosive and irritating hydrogen chloride gas (HCl) and at this point may not even produce visible smoke. This HCl is very irritating to the eyes, nose, and lungs. As the decomposition progresses, thick black smoke is also produced; actual burning (flaming) is not required. The smoke obscuring the path of escape, burning eyes, and choking sensation make evacuation very difficult. Even if one gets out, sufficient exposure to HCl as a product of combustion (attached to soot particles) can produce severe respiratory problems that are frequently irreversible and at times lead to delayed death. (In toxicity tests many animal deaths from PVC combustion do not occur until several days later.)

PVC also produces large amounts of carbon monoxide. Although nonplasticized PVC, such as that used in raceways, is ignition resistant, it burns once the chlorine and other fire retardants are driven off. Plasticized PVC, used in conductor insulation, burns more quickly. Another gas of concern is benzene, which is both flammable and carcinogenic. Burning nylon produces cyanide.

Testing at the University of Pittsburgh concluded that 18 to 30 inches of PVC conduit burned in an average room is sufficient to create a dangerous situation and kill humans in 10–15 minutes if decomposed between 300°C (572°F) and 500°C (932°F). It should be noted that the fire created lethal levels of CO that were present before the ENT gave off lethal products of combustion.

An additional effect of HCl is its corrosive action on electrical contacts and other electrical and electronic equipment. Deposits left by burning PVC in equipment rooms, even after thorough scrubbing, have been known to cause corrosion weeks and months later. Computer equipment exposed to PVC combustion products during testing has been knocked out of service for weeks. The corrosive effects on electrical contacts of HCl from overheated conductors could affect operation of safety and signaling equipment. Testing on the nonmetallic side of the issue shows that HCl from PVC adheres to walls, floors, and room objects, and the position is taken by some people that

much of the HCl is therefore removed from the atmosphere—how much under the great variety of fire conditions is the unanswered question. Certain Code articles also require that nonmetallic conduit be placed behind a thermal barrier to inhibit its exposure to heat for a projected fifteen minutes. In such instances it must be ensured that inadvertent openings are not made and left unsealed, or that thermal barrier ceiling tiles are not removed.

Noncombustible raceways have long been the recognized and proven method of inhibiting fire spread by way of the electrical system. Even if the conductor insulation inside burns, the spread will be limited, the smoke will be more contained, and the raceway itself will not produce smoke or hazardous gases.

It is incumbent on users to evaluate the specific set of conditions for an electrical system, including the quantity and type of occupants that might be anticipated in a building and its design, and thus determine the level of risk they wish to take with combustible materials. State supreme courts have found that Code compliance alone does not necessarily preclude fire-injury liability. This does not apply only to the electrical system, but is a consideration for all design.

WIRING IN DUCTS, PLENUMS, AND OTHER AIR HANDLING SPACES

The provisions for installation of electrical wiring and equipment in ducts, plenums, and other air handling spaces are covered generally in *NEC® Article 300, Section 300-22*. However, building codes and local fire codes take precedence and are usually more restrictive. *Article 100* defines "plenum" as "A compartment or chamber to which one or more air ducts are connected and which forms part of the air distribution system." It is advised that the designer or installer check with the chief building official for guidance before encroaching on these areas. No wiring systems of any type must be installed in ducts used to transport dust, loose stock, or flammable vapors. No wire of any type shall be installed in any duct or shaft used for vapor removal or for ventilation of commercial cooking equipment.

Wiring methods approved for ducts or plenums used for environmental air are Type MI cable, Type MC cable employing a smooth or corrugated impervious metal sheath, electrical metallic tubing (EMT), flexible metal conduit, intermediate metal conduit, or rigid metal conduit. These are permitted to be installed in ducts or plenums specifically fabricated to transport environmental air.

Flexible metal conduit and liquid-type flexible metal conduit not exceeding 4 feet is permitted where adjustable equipment and devices are located in these chambers. The connectors used for flexible metal conduit shall close openings in the connection. Equipment and devices shall be permitted within such ducts and plenum chambers only where necessary for the direct action or sensing of the contained air. Where such equipment or devices are installed and illumination is necessary to facilitate maintenance and repair, only enclosed gasket-type fixtures are permitted. In other spaces for environmental air used for purposes other than ducts and plenums as described previously, such as a space over a hung ceiling used for environmental air handling purposes, wiring methods approved in these areas are totally enclosed, nonventilated, insulated busways having no provision for plug-in connections; wiring methods consisting of MI cable; Type MC cable without an overall nonmetallic covering; Type AC cable; or other factory assembly multiconductor control or power cable that is specifically listed for this use.

Other types of cables and conductors shall be installed in electrical metallic tubing (EMT), flexible metallic tubing, intermediate metal conduit, rigid metal conduit, flexible metal conduit, or, where accessible, surface metal raceway or wireway with metal covers or solid-bottom metal cable tray with solid-metal covers. Electrical equipment with a metallic or nonmetallic enclosure listed for use and having adequate fire-resistance, low smoke-producing characteristics, and associated wiring materials suitable for the ambient temperature shall be permitted in such areas unless specifically prohibited in other sections of the *National Electrical Code®*. There are limited exceptions to these rules. *Section 300-22(d)* permits data processing and raised-floor areas, which are covered in *Article 645*, provided that all the conditions in *Section 645-2* are met. The requirements are noted in *Section 645-2*. Disconnecting means must be installed to dis-

connect all electronic equipment in the room and all power to dedicated HVAC heating, ventilating, and air-conditioning equipment serving the room and cause all required fire and smoke dampers to close. The disconnects must be grouped and identified and shall be controlled from locations readily accessible at all principal exit doors.

In addition, a separate heating, ventilating, air-conditioning (HVAC) system is provided that is dedicated to the electronic computer and data processing equipment's use and is separated from other areas of occupancy. However, other HVAC equipment may serve these areas provided that fire/smoke dampers are installed at the point of penetration of the room boundary. Such dampers must operate on the activation of smoke detectors and the disconnecting means required. Only listed electronic computer and data processing equipment is to be installed in the room. The room is to be occupied only by personnel needed for maintenance and functional operation of that room, and the room is to be separated from other occupancies within that building or floor by fire-resistant rated walls, floors, and ceilings, with protected openings. The building construction shall comply with the applicable building codes.

Good engineering design and close coordination with the authority having jurisdiction responsible for these occupancies, such as the chief building official, fire marshal, or chief electrical inspector, should all be consulted for their interpretation in classifying the room before applying *Article 645* to an electrical system. Remember, *NEC® Article 645* permits more lenient wiring methods; therefore greater caution must be taken to confine a fire to that area and to ensure that it does not spread to the rest of the building. Additional steps must be taken to ensure easy egress for those working in that room in an emergency.

RACEWAY ALLOWABLE CONDUCTOR FILL

Allowable conductor fill requirements are found in *NEC® Chapter 9, Tables* and *Notes to the Tables*. Additional tables for raceway fill limitations where all conductors are the same size can be found in Appendix C. In all types of conductors except lead covered, three conductors or more in a raceway are to be limited to 40% of the cross section of the conduit or tubing. Other allowable percentages of fill are listed in *Table 1* of *Chapter 9*. Equipment-grounding or bonding conductors, when installed, must be included in the calculations. The actual dimensions shall be used in making the calculation whether these conductors are insulated or bare. Conduit nipples not exceeding 24 inches between enclosures shall be permitted to be filled to 60% of the total cross-sectional area. Conductors not included in *Chapter 9* shall use the actual dimensions in making the calculations where conductors' dimensions are found in *Table 8* of *Chapter 9*. Multiconductor cable of two or more conductors shall be treated as a single conductor when calculating the percentage of conduit fill. For cables that have an elliptical cross section, the calculations shall be based on the major diameter of the ellipse as the circle diameter. The trade size of the raceway in *Table 4, Chapter 9* shall be used as the internal diameter in inches. (This information may not be found in the 1999 *NEC®*; in such case, consult the manufacturer for wire fill instructions.)

> **Example:** What size IMC conduit is needed to enclose (9) nine No. 12 THW conductors? Answer: 3/4 inch, ref. *NEC® Appendix C-4*

> **Example:** What size rigid PVC Schedule 80 is needed to enclose (3) three 500 kcmil THWN, (1) one 4/0 THWN insulated conductor, and (1) one 1/0 bare conductor? Answer: 3 1/2 inch, ref. *NEC® Chapter 9, Tables 1, 4, 5,* and *8*

Although the raceway is sized according to the allowable conductor fill in *Chapter 9* of the *NEC®*, other considerations are necessary in sizing the conductors, which must be considered before the raceway system itself is sized, for example, the ampacity adjustment factors in *Table 310-15(b)(2)(a)* for more than three conductors per raceway, the ambient temperature considerations, which require ampacity correction where the ambient temperatures exceed levels listed at the bottom of each ampacity table. Good engineering practice, however, recommends that when all these things are considered, the raceway system should be oversized by 50% to handle future expansion and growth to the electrical facility.

CONDUIT AND TUBING BENDING

The guidelines for maximum number of bends permitted in raceways are covered in *Chapter 3* of the *National Electrical Code®*. Each wiring method has specific guidelines within its own article. However, experienced installers rarely put the maximum number of bends allowed by the *NEC®* in a raceway system. Experienced installers realize that ease of installing the fish tape and pulling the conductors is an important factor. They weigh the number of conductors, length of run, size of the raceway, and location of the raceway, all based on their prior experience in determining the distance between pull points. There can be no substitute for experience.

NEC® 300-17 states that the number and size of conductors in the raceway shall be not more than will

Figure 5-15a
Mechanical Benders for EMT, IMC, Rigid Steel, and Aluminum Conduit
Durable ratchet mechanism for bending. Bending degree indicator and charts for fast, easy, accurate bending. Bypass the ratchet for fast, direct bending on smaller sizes. Each shoe has its own permanently mounted swing-away hook to eliminate loose parts. All shoe and bending accessories can be mounted on the frame unit and secured with a chain and lock. Heavy-duty undercarriage with wheels. (Courtesy of Greenlee Textron Inc.)

Figure 5-15b
Site-Rite II Malleable Iron Hand Benders
Patented sight indicator to eyeball the key bending angles easily and fast. Nonslip, contoured pedal for added leverage. Strong, stable square hook design. Cast-in markings for stub-ups, saddle bends, back-to-back bends, and head-up or -down bending. Patented design on shape of bender. (Courtesy of Greenlee Textron Inc.)

permit dissipation of heat, ready installation, and easy withdrawal without damage to the conductors or their installation. *Section 300-18* tells us that raceways shall be installed as a complete system between outlet junction or splicing points prior to installation of the conductors. Therefore, the specific requirements in the article covering the wiring method that limits the equivalent of four quarter bends (360° total) between pull points (e.g., conduit bodies and boxes) is not realistic in many installations. The location of the bends and the length of the run sets these limits based on experience. These limitations comprise all bends, including box offset, 90s, kicks, and saddles to be an accumulated total of not more than 360°.

Conduit bending has changed much over the past 30 years in our industry. There was a time when all bends were made in the field with segment benders on rigid conduit. By this I mean that the older

segment hydraulic benders made each bend at 3° to 7°, requiring many bends to develop a full 90° in a conduit. These benders would kink the conduit if the installer tried to bend them more than that without moving the bending shoe. EMT, though, has been bent by the old hand-hook benders for many years. Today many benders on the market no longer require mathematical calculations and segment bending. They require only setting the dial and pushing the button. Many manufacturers today produce high state-of-the-art benders that make calculations for the installer. Examples of these are shown in Figures 5-16 and 5-17a following. Generally, in all tubular raceways, both metallic and nonmetallic, the bends shall be made so that the raceway is not damaged, the internal diameter is not effectively reduced, and the radius of the curve of the inner edge of the bend shall not be less than what is shown in Figure 5-19, *NEC® Table 346-10*. There are exceptions to that rule. Non-metallic raceways, as covered in *Article 347*, may be bent only with equipment identified for the purpose.

Bending Electrical Metallic Tubing

Electrical metallic tubing (EMT), covered in *NEC® Article 348*, was developed by the Republic Steel Company in 1929. Shortly thereafter, Jack Benfield patented and marketed the Benfield Conduit Bender. Many hand-held benders using the hook-type method have come on the market since

Figure 5-16 Flip-top hydraulic bender, EMT, IMC & rigid metal conduit (Courtesy of Greenlee Textron Inc.)

and employ similar design. The method for bending EMT small sizes through 1 1/4 inch has changed little since the inception of EMT. Larger sizes of EMT must be bent by hydraulic benders, much the same as rigid or IMC conduit benders (see Figure 5-18). The Benfield Bender and similar types require very little calculation, only that you mark the tubing and place the arrow at the mark as per the instructions on the bender itself, and you will get remarkable results. It is important when using the hand-held benders that you apply heavy, consistent pressure with your foot on the bender and pull the handle back steadily while applying that pressure. The result will be a clean bend without kinks or wrinkles. With a little practice and common sense, anyone can soon learn to install EMT quickly, ef-

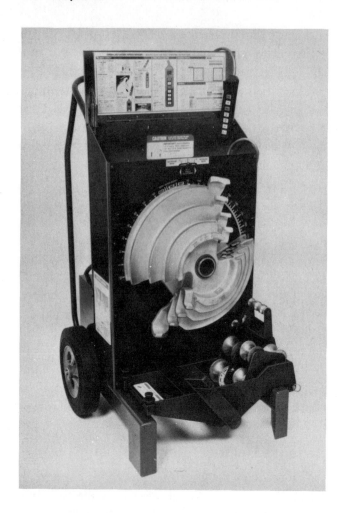

Figure 5-17a Speed Bender—Complete EMT, IMC, and rigid bender with memory for repeat bending and override feature for minor adjustment (Courtesy of Greenlee Textron Inc.)

Figure 5-17B Electric PVC heater bender 1/2" through 6" (Courtesy of Greenlee Textron Inc.)

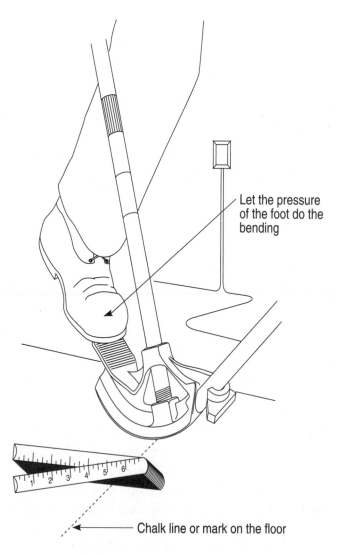

Let the pressure of the foot do the bending

Chalk line or mark on the floor

Figure 5-18 Most hand benders are marked with easily identifiable grooves for accurate offsetts and back-to-back bending after a little practice.

ficiently, and in a workmanlike manner. (See Figure 5-15.)

Bending Rigid Metal Conduit and Intermediate Metal Conduit (IMC)

IMC is covered in *NEC® Article 345*. IMC was developed by Allied Tube & Conduit, Inc. in 1975. *Article 345-10* directs the installer to *Article 346* for the methods and minimum radius permitted for IMC, which is exactly the same as those permitted for rigid conduit. IMC is a harder steel; therefore, only full shoe-type benders are recommended. However, once the installer is set up to use the product, it can be installed as easily and as quickly as rigid metal conduit. Threading is also accomplished with ease after making only slight field adjustments to standard I.P.S. pipe dies.

346-10. Bends—How Made

Bends of rigid metal conduit shall be so made that the conduit will not be damaged and that the internal diameter of the conduit will not be effectively reduced. The radius of the curve of the inner edge of any field bend shall not be less than indicated in Table 346-10.

Table 346-10. Radius of Conduit Bends

Size of Conduit (in.)	Conductors Without Lead Sheath (in.)
½	4
¾	5
1	6
1¼	8
1½	10
2	12
2½	15
3	18
3½	21
4	24
5	30
6	36

Note: For SI units, 1 in. = 25.4 mm (radius).

Figure 5-19   This *NEC®* table is applicable to *NEC® Articles 331, 345, 346, 347* and *348* (Reprinted with permission from NFPA 70-1999.)

Exception: For field bends for conductors without lead sheath and made with a single operation (one shot) bending machine designed for the purpose, the minimum radius shall not be less than that indicated in Table 346-10, Exception.

Table 346-10. Exception, Radius of Conduit Bends

Size of Conduit (in.)	Radius to Center of Conduit (in.)
½	4
¾	4½
1	5¾
1¼	7¼
1½	8¼
2	9½
2½	10½
3	13
3½	15
4	16
5	24
6	30

Note: For SI units, 1 in. = 25.4 mm (radius).

Figure 5-20   This *NEC®* table is applicable to *NEC® Articles 331, 345, 346, 347,* and *348* (Reprinted with permission from NFPA 70-1999.)

How to Thread and Cut IMC

The same equipment used to thread GRC can be used to thread IMC.

To field thread IMC, use a standard 3/4" taper N.P.T. die head. Loosen the screws or locking collar holding the cutting dies in the head so that the dies can freely move away from the thread. Using a factory-threaded piece, turn the die head completely on the finished thread without cutting any metal. Have the individual dies fitting loosely on the threads. Now tighten the screws or locking collar so that the dies are held tightly in the holder. The die is now ready for use.

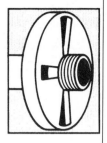

In a die head with an adjusting level, set the head to cut a slightly oversized thread. This will ordinarily be one thread short of being flush with the face of a thread gage when the gage is hand tight.

This is within the tolerance limits, which allow the thread to be one thread short or long of being flush with the gage face.

It is good practice to thread one thread short because this will prevent butting of conduit in a coupling and will allow the coupling to cover all of the threads on the conduit when wrench tight.

After adjusting the dies as outlined above, proceed as follows:

1. Cut off conduit with either a saw or disk cutter. Be careful to get a straight cut.
2. Dress off the outside burr with a file.
3. Ream the inside burr with a pipe reamer. The reamer may be turned by hand or held stationary while the conduit is being power driven. The chamfer on the inside should be about 3/64" wide and even all around the inside edge.
4. Start the die head on the conduit with some forward hand pressure in a straight line.
5. Stop the cutting as soon as the die has taken hold and apply thread-cutting oil freely to the dies and the areas to be threaded.
6. Thread one thread short of the end of chaser.
7. Back off die head and clean chips off thread.

You may ask why the same equipment used to thread GRC can be used to thread IMC. Rigid metal conduit, regardless of whether it is on the high side of the dimensional tolerance or the low side, is threaded with a 3/4"-per-foot tapered NPT chaser. The same chaser is used for IMC. The resulting thread is no different than the thread on rigid metal conduit.

The thinner wall does not create threading problems because, when threading rigid conduit, you are simply cutting away the excess wall material. When threading IMC, only the portion of the wall necessary to put the required thread on the conduit is cut away.

Because IMC trade sizes vary from 1/2 inch to 4 inches, several types of bending equipment are recommended to achieve proper, accurate bending of IMC. Most-often-used trade sizes are 1/2 and 3/4 inch. They are bent predominantly with hand benders. Several bending manufacturers have developed hand benders expressly for IMC and are recommended. However, IMC can be bent with 1/2-inch and 3/4-inch rigid-type hand benders. The often-asked question, "Can IMC be bent with a hickey?" is extremely difficult to answer. Hickey design puts total dependence for support of the walls onto the conduit itself. Therefore, bending success is directly proportional to the skill of the hickey user. This is true to a certain extent for any hickey use. IMC needs support along the lineal dimension of the conduit; this the hickey does not provide. Although not recommended, bending IMC with a hickey should not be totally ruled out. A good source of information for methods to use with a hickey can be found in manufacturer's literature.

One-inch IMC can be bent by hand with a

How to Bend IMC

Because IMC trade sizes vary from 1/2" to 4", Allied recommends several types of bending equipment (hand and power) to achieve proper, accurate bending of IMC.

The most-often-used trade sizes are the 1/2" and 3/4" and they are bent predominantly with hand benders. Several bender manufacturers have developed hand benders expressly for IMC. We recommend their use. However, IMC can be bent with 1/2" and 3/4" rigid-type hand benders.

The often-asked question, "Can IMC be bent with a hickey?" is extremely difficult to answer. The design of a hickey puts total dependence for support of the walls onto the conduit itself. Therefore, bending success is directly proportionate to the skill of the person using the hickey. (This is true to a certain extent in any use of a hickey.) IMC needs support along the lineal dimension of the conduit. This the hickey does not provide. Although Allied does not recommend bending IMC with a hickey, a contractor should not totally rule out its usage. Your Klein supplier is a good source of information for methods to be used with a hickey.

One-inch IMC can be bent by hand with a bender expressly designated to do that job. However, due to the strength of material, the job is not an easy one. The amount of pressure needed to make a successful bend must be a combination of foot and leg strength on the shoe pedal and a good hard tug on the handle of the bender. There is no way to say 1" IMC is easy to bend by hand. In fact, Allied recommends that 1" IMC be bent on a "Chicago-type" bender as produced by Lidseen of North Carolina and various other manufacturers as shown in the bending guide. A fast and adequate job on sizes 1/2" through 1 1/2" can be accomplished by using hand benders and a "Chicago-type" bender. If a job requires sizes larger than 1 1/2", power bending is necessary, using equipment specifically developed for IMC.

Power benders are available for all ten sizes of IMC. A word of caution, however: Some existing equipment requires modification with a conversion kit for bending IMC. Others—such as Greenlee 555 SBC, 882 CB, 881, and 881 CT Ensley 666; and Enerpac-Eegor B-448—do not require modification.

The latest in IMC power bending equipment is the Greenlee 881 for sizes 2 1/2" thru 4", 882 CB for sizes 1 1/4" thru 2", and the combination of an Enerpac-Eego and Mini-Eegor for sizes 1" thru 4". These power benders are certainly an investment to investigate as they also bend EMT and GRC.

bender expressly designed to do that job; however, owing to the strength of the material, the job is not easy. The pressure to make a successful bend must be a combination of foot and leg strength on the shoe pedal and a good hard tug on the handle of the bender. One-inch IMC is not easy to bend by hand. In fact, it is recommended that it be bent with a Chicago-type bender. These benders are produced by several manufacturers, including Greenlee and Lidseen. A fast, adequate job on sizes 1/2 through 1 1/2 inch can be accomplished by hand benders and the Chicago-type bender.

If the job requires larger than 1 1/2 inch, then power bending is necessary, using equipment specifically developed for IMC. Power benders are available for all ten sizes of IMC. A word of caution, however—some existing equipment requires modification with a conversion kit for bending IMC. Others, such as Greenlee 555-SBC, 882-CB, 881 and 881-CT, Insley 666, and Enerpac-EEGOR B448, do not require modification. The latest IMC power bending equipment is the Greenlee 881 for sizes 2 1/2 through 4 inches, 882-CB for sizes 1 1/4 through 2 inches, and a combination of an Enerpac-EEGOR and a mini-EEGOR for sizes 1 inch through 4 inches. These power benders are certainly an investment to consider, as they also bend EMT and GRC.

Bending Rigid Nonmetallic Conduit

NEC® Section 347-13 requires that all field bends shall be made so that conduit is not damaged and the internal diameter is not effectively reduced. All bends shall be made with equipment identified

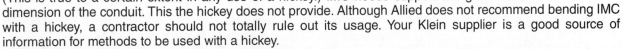

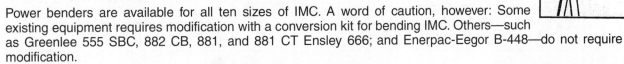

Which HAND BENDERS to use.
IMC Bending Equipment Guide:

Available Equipment	1/2	For bending IMC Trade Sizes: 3/4	1	1-1/4	1-1/2	2	2-1/2	3	3-1/2	4

HAND BENDERS

Equipment	1/2	3/4	1	1-1/4	1-1/2	2	2-1/2	3	3-1/2	4
Klein-Benfield	51216	51217	51206							
Appleton	IM-50R	IM-75R	IM-100R							
Ideal	74-002	74-003	74-006							
Gardner	940	941	942							
Greenlee	841	842								
Enerpac	B-7	B-10								
Ensley	E714	E715								

MECHANICAL BENDERS

Equipment	1/2	3/4	1	1-1/4	1-1/2	2	2-1/2	3	3-1/2	4
Lidseen of N.C. (Chgo. Type) 5100	●	●	●							
Lidseen of N.C. (Chgo. Type) 5200				●	●					
Ensley (K.C.) E777	●	●	●	●						
Greenlee 1800	●	●	●							
Greenlee 1801				●	●					
Greenlee 1818	●	●	●	●						
Chicago Misc.	●	●	●							
Enerpac (Sidewinder) BW11	●	●	●							
Enerpac (Sidewinder) BW12				●	●					
Enerpac (Sidewinder) BW15	●	●	●	●	●					
Enerpac (Sidewinder) BW27	●	●	●	●	●					
Enerpac (Sidewinder) BW30	●	●	●	●	●					

This guide compiled from catalog material furnished by the manufacturers listed. For exact specifications and capabilities of each manufacturer's products, reference should be made to their published material. No warranty or guarantee is made by Allied Tube & Conduit Corporation.

Figure 5-21 IMC bending equipment guide (Courtesy of Allied Tube & Conduit)

Which POWER BENDERS to use.
IMC Bending Equipment Guide:

Available Equipment	1/2	For bending IMC Trade Sizes:								
POWER BENDERS Electric & Hydraulic		3/4	1	1-1/4	1-1/2	2	2-1/2	3	3-1/2	4
Greenlee 555 and 555 SBC	●	●	●	●	●	●				
Ensley E-666CP and E-666M (For 1-½" thru 2" add attachment set E-6688)	●	●	●	●	●	●				
Ensley E-551				●	●	●				
Ensley E-120 and E-121 One Shot—24" Radius (For 2-½" add 41-156 shoe; for 3" add 41-157 Shoe)							●	●		
Ensley E-124 One Shot—24" Radius							●	●		
Greenlee 881 and 881 CT							●	●	●	●
Greenlee 882CB					●	●	●			
Greenlee 882 plus attachment RIGI					●	●	●			
Greenlee 884 plus attachment (Use Attachment Set 881-A)							●	●	●	●
Greenlee 885 plus attachment (Use Attachment Set 881-A)							●	●	●	●
Enerpac Cyclone B 2000 Series	●	●	●	●	●	●				
Enerpac-Mini Eegor B-200 Series			●	●	●	●				
Enerpac-Eegor B-448 For 2" add BZ-20 2" Shoe; BZ-22 2" Follow Bar; BZ-24 2" U-Strap						●	●	●	●	●

This guide compiled from catalog material furnished by the manufacturers listed. For exact specifications and capabilities of each manufacturer's products, reference should be made to their published material. No warranty or guarantee is made by Allied Tube & Conduit Corporation.

Figure 5-21   *(continued)* IMC bending equipment guide (Courtesy of Allied Tube & Conduit)

for the purpose and the radius of the curve of the centerline of such bends shall not be less than shown in *Table 346-10*. Although the *Code* in *Section 346-10* requires measurement of the radius at the inner edge, testing laboratories and manufacturers make measurements to the centerline. There are several manufacturers of bending equipment designed to bend nonmetallic conduit. The infrared types are fast and efficient, lightweight, and simple to use. Although innovative electricians have found many ways to heat and bend nonmetallic conduit in the past, none of them are recommended; they are in violation of the Code.

The proper way to bend nonmetallic conduit is to insert the conduit into the bender and heat the bend evenly over its entire length. Small sizes will heat quickly and become very malleable—sizes 1/2 through 1 1/2 inches can be shaped into almost any configuration. When bent to the installer's measurements the conduit may be sponged with cool water until it hardens. The conduit may then be installed. Larger sizes, 2 inches and larger, require several minutes to heat and require internal support to prevent crimping or deforming during the bending process. Bending plugs are available; they are to be installed by inserting into the conduit section prior to heating. The air inside the conduit expands when heated, providing internal pressure that provides the needed internal support so the conduit can be bent without deforming. The conduit must be cooled by air or by sponging with water before removing the plugs.

REVIEW QUESTIONS

1. When an installer selects a wiring method, it must meet at least what two disciplines?

2. All the specific wiring method types are found in *NEC® Chapter 3*. Wiring methods are specified throughout the *NEC®* in individual sections for specific uses. Where are the general wiring method conditions found in the *National Electrical Code®*?

3. Name three specific wiring methods that provide physical protection for conductors.

4. What is the burial depth for a residential branch circuit rated at 120 volts and protected by a ground-fault circuit interrupter (GFCI) rated at 20 amperes, supplying a post lantern along the side of the driveway?

5. Is it permissible to install that cable supplying the post lantern directly under the slab-on-grade residence to the existing service panel located ten feet inside the building?

6. The service supplying a residence is fed from the street underground in rigid nonmetallic conduit to the building, extending up into the meter on the outside of the building. What is the burial depth required for the rigid nonmetallic conduit enclosing the service lateral conductors?

7. What precaution must be taken where condensation may occur, such as raceways extending from the meter enclosure outside of a building inside to an environmentally controlled room terminating in a panelboard that contains live parts?

8. Can a 1-inch EMT tubing containing three #4 conductors feeding a small lighting and branch-circuit panelboard be supported on suspended ceiling support wires?

9. An electrical metallic tubing (EMT) is routed through the suspended ceiling of an office building. The installer installed a green equipment-grounding conductor within the tubing. Does the EMT have to be wrench tight and properly supported, even though the equipment-grounding conductor is enclosed and run with the conductors in accordance with *NEC*® *300-3*?

10. Is rigid metal conduit permitted to be run exposed within a commercial cooking vent hood to supply a light fixture?

11. A rigid conduit is routed from a main switchboard to the rear of an industrial building, terminating in a wireway containing several branch circuits to feed equipment in that area. The 4-inch conduit contains three #4 THWN types, four #2 THWN types, eight #6 THWN types, and three #1 THWN types. What size conduit is required?

Chapter 6

Raceway Systems

HISTORY OF RACEWAY

Only about a hundred twenty years ago our great electrical industry was first growing like wildfire. This was a result of such great inventors and great minds as Thomas Edison, holder of hundreds of patents promoting DC voltage; Nikola Tesla, inventor of the 3-phase motor and a promoter of AC voltage; William Merrell, founder of Underwriters Laboratories, along with George Westinghouse and others. Thomas Edison's lighting system was announced in December 1879 and the first commercial installation was put in operation in May 1880. Within just 5 years there were over 500 plants in operation, serving over 100,000 lamps. In 1886 George Westinghouse introduced an alternating current system. As a result of this rapid growth in the use of electricity there were many fires, electrocutions, and electrical accidents. With arcing and sparking and resultant fires occurring almost daily, and with keen competition among delivery methods, a group met in 1881 to discuss safer electricity use. The National Association of Fire Engineers met in Richmond, Virginia, and from this meeting came a proposal that served as a basis for the first national electrical code.

In the earliest beginnings of the electrical era, before the turn of the century, designers began to devise various means of distributing the wonderful servant made practical by the work of these great inventors—electricity. Ordinary electrical conductors strung without protection soon proved dangerous and wasteful. Some means by which they could

A Chronology of Steel Electric Conduit

1879	Edison demonstrated his multiple incandescent lamp system
1880	First commercial installation of the Edison System.
1882	Pearl Street Station of the New York Edison Company put into operation.
1886	Westinghouse alternating-current system exhibited.
1888	Zinc tube conduit system in New London placed by Greenfield.
1889	Greenfield-Johnson paper tube conduit produced.
1891	Brass (and iron)-armored paper tube conduit announced. Paper-lined gas pipe proposed.
1894	Paper lining applied to commercial gas pipe for conduit use.
1895	Wood-lined, fiber-lined, and composition-lined iron pipe conduits appeared.
1896	Wiring in both insulated and uninsulated iron pipe permitted by Conference Code.
1897	First National Electrical Code published.
1898	Enameled iron pipe appeared in both standard and light weights. Garland's cleaning method.
1899	Flexible steel conduit produced.
1902	Electro-galvanized conduit produced.
1903	Lined conduit practically abandoned.
1908	Sherardized galvanized conduit appeared.
1912	Hot-dipped galvanized conduit offered.
1913	Lined conduit rules disappear from National Electrical Code.
1928	"Electrical metallic tubing" permitted by National Electrical Code for exposed wiring.
1975	Intermediate metal conduit permitted by National Electrical Code for exposed locations requiring physical protection.

Figure 6-1

be enclosed and protected permanently against mechanical and electrical damage was clearly needed.

Harry Greenfield introduced the first known conduit system in 1888: zinc tubes with copper elbows. In 1889 Gus Johnson and Harry Greenfield introduced insulated paper tube conduit. The wire could be pulled after the conduit was in place and the building construction complete. The merit of the new conduit system was quickly realized and many miles were installed over the next two years. Other similar types of conduit were introduced—woven fabrics, fibers, and flexible glass—but these were not accepted and Greenfield and Johnson enjoyed a near-monopoly for the next decade on interior conduit. They produced variations, attempting to develop a product with greater mechanical strength. Brass armored tube with paper liners was the most successful of these. In the early 1890s other companies began to produce gas pipe conduit with wood, fiber, or clay base liners. Installations using black painted steel pipe, which originally ran through buildings to provide gas for lighting, was being used with some success and was relatively inexpensive. The first approved unlined pipe appeared in 1897 and was enameled inside and out.

From that evolved today's galvanized rigid steel conduit. Later the lighter-weight steel electrical tubing for such wiring became more widespread. The earlier conduits were of zinc with copper elbows; then spirally-wrapped paper tubing with brass joining sleeves was introduced. As the shortcomings of these conduits were revealed, brass-enclosed paper tubing with brass elbows and couplings was introduced. But this, too, was found to be unsatisfactory, chiefly because of inadequate physical protection and mechanical strength.

The concept of a conduit into which electrical conductors could be pulled, and pulled out, remained worthwhile; the sole question to be answered was what type of material and processing would provide the best and most economical conduit?

Iron-armored conduit was introduced in 1894. It consisted of ten-foot lengths of standard-weight, wrought-iron gas pipe, with paper tube lining and threaded couplings and nipples. This conduit was a great advance but still far from satisfactory. The linings reduced the usable interior area to an unacceptable degree. Paper or fiber linings would crumble, and wooden linings splinter, making field bending impossible. The linings were not moistureproof. Finally, lined conduits were expensive.

The introduction of good rubber-insulated electrical conductors made possible the development of unlined steel conduit, which made its appearance in 1897. Here at last was a conduit that provided full mechanical and electrical protection for conductors. The system could be readily expanded to provide maximum design flexibility, and included a host of other desirable features, all in one easily installed, convenient raceway system. Although the quick acceptance of this new steel conduit by the electrical industry was followed by rapid improvements in the product, it was not without its problems. The gas fitters immediately claimed the right to install their wrought-iron gas pipe and there ensued great disputes between electrical craftsmen and gas fitter craftsmen. This was finally resolved, and electricians were allowed to install conduit.

About this same time groups began to cut wrought-iron gas pipe into ten-foot lengths and bootleg it over the counters to the electricians as conduit. The problem was that electrical conduit was rough on the exterior and smooth on the interior, whereas wrought-iron gas pipe was just the opposite—it had a smooth exterior with a rough interior. The smoothness of conduit interior was important so as not to tear the conductors' rubber insulation as they were being pulled in. Fortunately, with rapid advances in the industry and improvements in the product, these problems were soon far behind. The first exterior coating to prevent oxidation was enamel. Then in 1902 the first electrogalvanized rigid steel conduit was introduced. In 1908 sherardized steel conduit appeared, followed in 1912 by hot-dip galvanizing.

Figure 6-2 Raceways provide physical protection and excellent reusability; many metal raceway systems installed over 50 years ago are still being used today. (Courtesy of Allied Tube & Conduit)

A little later, the metallizing process was introduced. In 1931 the Empire State Building was completed with over 1,200,000 feet of rigid metal conduit that is still in service today.

ADVANTAGES OF RACEWAY SYSTEMS

Since 1905 the two primary functions of electrical raceways have been to facilitate the insertion and extraction of conductors and to protect the conductors from mechanical injury.

Today's raceway systems combine these functions with a high degree of safety and greater distribution and utilization flexibility. It is the responsibility of the designer and installer to select the correct raceway for each application, and all things must be considered. *NEC® Section 110-3* requires the judging of equipment, and considerations such as the following must be evaluated: suitability of the installation and its conformity with the Code; suitability of equipment; suitability of

the product for the specific purpose, environment, and application; suitability of the equipment as evidenced by listing or labeling; mechanical strength; durability—including parts designed to enclose and protect the other equipment; wire-bending and connection space; electrical insulation; heating effects under normal and abnormal use; arcing effects; classification by type, size, voltage, current capacity, and specific use; and other factors that safeguard persons likely to come in contact with the equipment.

NEC® 110-3 (b) states that listed and labeled equipment shall be used or installed in accordance with any instructions included in the listing or labeling. Therefore, improper selection of a raceway system or improper installation of that system is a *National Electrical Code®* violation, which means the product may not provide the intended service or safety. Selection of a wiring method should include consideration of the type and design of the building and the number of prospective occupants, and their ability to escape in the event of fire. Combustibility aspects of raceways or other wiring methods is

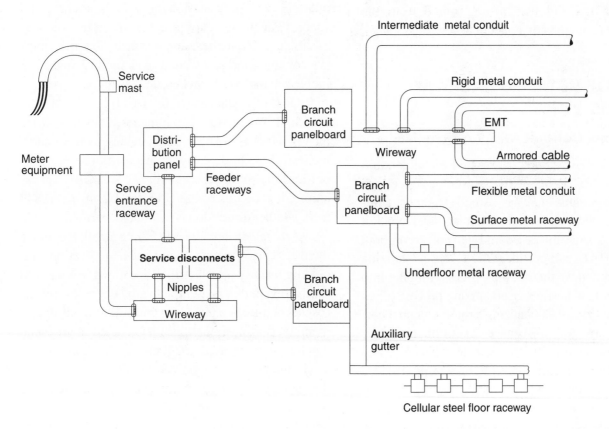

Figure 6-3 For raceways see *NEC® Chapter 3. NOTE:* Locations of overcurrent devices not shown.

Figure 6-4   Rigid steel conduit, per *NEC® Article 346* (Courtesy of Allied Tube & Conduit)

important because the wiring system is installed throughout the building. Also, some installations have large quantities of overhead wiring. Often many branch circuits and feeders are routed from the equipment room over or adjacent to corridors.

TYPES OF METALLIC RACEWAY SYSTEMS

Rigid Metal Conduit *NEC® Article 346*

NEC® Article 346 covers rigid metal conduit, also called GRC, RMC, ARC, and stainless steel. All rigid metal conduit must be "listed." The most common is the heaviest-weight, tubular steel conduit (GRC). It is smooth-walled and galvanized on both the inside (I.D.) and outside (O.D.). The galvanizing may be applied by the hot-dip process (dipping into molten zinc) or the electrogalvanizing process (electrodeposited zinc). Galvanizing protects against corrosion. Corrosion resistance is influenced by environmental conditions such as moisture and chemical exposure. Also covered under this article is nonferrous metal conduit. Both products are evaluated by UL 6. The nonferrous products most commonly produced are aluminum rigid metal conduit and silicone bronze conduit, which are evaluated according to ANSI C80.5.

UL 6 and ANSI C80.1 are the standards for GRC and RMC. The international standard is IEC 981. These IEC soft metric conversions are proposed as the metric designations when the United States converts to the metric system. An example is "1/2-inch GRC," which becomes "16GRC." (See Figure 10 Appendix.) These conduit types are currently listed and produced to UL and ANSI standards. Both have been adopted by the federal government. The federal specification was formerly WW-C-581, Class 1/Type A. The dimensions and weights are shown in A-10 of the Appendix.

Galvanized rigid steel conduit is a threaded product, with threads conforming to ANSI B2.1 and *NEC® Section 346-8*. The UL listing requires a listed piece of conduit to have a coupling attached to one end. This coupling is electrogalvanized to avoid zinc buildup on the interior threads that might occur with hot-dipping. All rigid conduit is shipped with the coupling on one end and the open threads on the opposite end protected by color-coded plastic caps to provide easier field identification. The red caps indicate quarter sizes such as 3/4 inch, 1 1/4 inches, and so on. The black caps are for 1/2 sizes such as 1/2 inch, 1 1/2 inches, and so forth. The blue caps are for the even sizes such as 1 inch, 2 inches, 3 inches, 4 inches, 5 inches, and 6 inches.

GRC is produced in trade sizes 1/2 inch, 3/4 inch, 1 inch, 1 1/4 inches, 1 1/2 inches, 2 inches, 2 1/2 inches, 3 inches, 3 1/2 inches, 4 inches, 5 inches and 6 inches. These trade designations are not actual O.D. or I.D. dimensions. The I.D. for 1/2-inch GRC, for example, is 0.632 inch; the O.D. for 1/2 inch is 0.840 inch. Neither dimension is the .500 one might expect.

Article 346, and the *NEC®* in general, recognize the use of rigid metal conduit in *all* applications, including hazardous (classified) locations. Galvanized rigid steel conduit is exempted only where severe corrosive conditions exist. Even then, additional supplementary coatings, such as PVC-coated rigid metal conduit, can be used for those applications. Rigid steel conduit is the only conduit listed virtually throughout the *NEC®* without exceptions. It is the oldest as well as the most widely used raceway product in production today. GRC conduit is still found in buildings built around the turn of the century and still provides good, safe service. Many buildings remod-

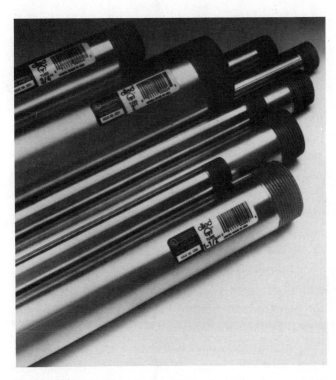

Figure 6-5 IMC (intermediate steel conduit), per *NEC® Article 345* (Courtesy of Allied Tube & Conduit)

eled today utilize their existing rigid steel conduit systems for all or a portion of their new wiring.

Galvanized steel rigid conduit evolved from black gas pipe, which was already in place and a logical means of protecting those "newfangled" electrical conductors Tom Edison was touting! GRC is considered the premier electrical wiring method. It provides superior physical protection for conductors, is not affected by extreme temperatures (there are no code or standards limitations), and can be used indoors, outdoors, and underground, concealed or exposed.

Rigid conduit is a code-recognized equipment-grounding conductor in accordance with *NEC® Section 250-118(2)*. See Appendix A-4.

Intermediate Metal Conduit (IMC)

Intermediate metal conduit, commonly known as IMC, is covered in *NEC® Article 345;* all IMC is galvanized steel and required to be listed. It was developed in the early 1970s by Allied Tube & Conduit. It was intended originally as an entirely separate system but has evolved with threads and dimensions that make it interchangeable with GRC. In fact, the same couplings are used on both IMC and GRC. *Articles 345* and *346* are virtually identical, and the *NEC®* recognizes both products for exactly the same applications, without limitations.

The standards for listing and production of IMC are UL 1242 and ANSI C80.6. UL 6 has been adopted as a federal specification. IMC was formerly covered by WW-C-581 as Class 2/Type A (ANSI C80.6 has not yet completed the adoption process).

IMC is galvanized on the outside diameter (O.D.) and has a UL-approved, corrosion-resistant, organic or inorganic coating on the interior (I.D.). Although the IMC wall is less than the wall thickness of GRC, its method of manufacture and steel chemistry provide physical strength equivalent to, or often greater than, that of GRC, thus providing equivalent physical protection. Listed IMC is also a conduit threaded to ANSI B2.1, with an attached coupling. The color codings for the protective end caps are 1/4-inch sizes green, 1/2-inch sizes yellow, and even sizes orange. This provides easy visual differentiation between IMC and GRC. Further identification is provided by indenting the letters IMC on the pipe at regular intervals. The I.D. (internal diameter) of IMC is slightly larger than that of GRC, with 1/2 inch having a nominal .660 I.D.; the O.D. nominal for 1/2 inch is 0.815 inch, making pulling the same number of conductors in IMC easier. The dimensions for nominal trade sizes of IMC can be found in Figure A-10 of the Appendix.

Wall thickness comparison is 1/2 inch IMC 0.070 inch and 1/2-inch GRC 0.104 inch. Compatible threads are possible because less steel is cut away in threading IMC. As with rigid conduit, each length is designed to be 10 feet including the coupling.

IMC has become a very popular replacement for GRC because it provides the equivalent physical protection, has a larger I.D., and may be installed more easily because of its lighter weight. It is a totally different product than GRC and requires adjustment to common-standard threading machines before threading and special bending equipment or adapters on standard bending tools. However, once contractors are set up to install IMC, they usually find it much quicker and easier to install. See Appendix 16-20.

Intermediate metal conduit is a code recognized equipment grounding conductor in accordance with *NEC® Section 250-118(3)*. See Appendix A-4 for circuits above or below 600 volts nominal.

Figure 6-6 EMT (electrical metallic tubing) as per *NEC®* *Article 348* (Courtesy of Allied Tube & Conduit)

Electrical Metallic Tubing *(Article 348)*

Electrical metal conduit, commonly known as EMT, is the lightest-weight tubular metal raceway manufactured. EMT is most commonly produced as galvanized streel; however, it may also be produced in aluminum. Both types are required to be "listed." It is an unthreaded plain-end product joined together by setscrew, indentation, or compression-type connectors and couplings. The coupling is not provided as a part of the listed EMT. EMT was developed in 1928 but did not gain popularity until the start of World War II, when the entire nation was trying to conserve steel. Soon after the use of EMT began and tradesmen learned how to bend and install the product, it gained the massive popularity it enjoys today. It is widely used throughout the industry for branch-circuit and feeder raceways. It is very versatile because the unthreaded design can be altered, reused, and redirected with ease. The conductors can be inserted and extracted easily because of the very smooth interior,

making conductor installation fast and simple. The internal diameter (I.D.) is sufficiently large to provide ease in pulling the maximum number of conductors permitted by the *NEC®*. EMT has one of the larger I.D.s among the many raceway types being manufactured today. Despite the lightweight steel construction, it still provides substantial physical protection and may be used in most exposed locations, except where subject to severe physical damage.

The listing and production standards for EMT are UL 797 and ANSI C80.3. These standards have also been adopted by the federal government, replacing WW-C-563.

Electrical metallic tubing has a galvanized O.D. and a UL-approved, corrosion-resistant organic or inorganic coating on the I.D. For identification purposes each ten-foot length is indented "EMT."

Again, the dimensions are trade-size designations only, not actual 1/2 inch. I.D. is .620 inch and O.D. is a nominal 0.706 inch. Other weights and dimensions for electrical metallic tubing can be found in Figure A-10 of Appendix.

The *NEC®* authorizes EMT for above and below 600 volts, and for both exposed and concealed work. It may be used in most applications except where subject to (1) severe physical damage, or (2)

Figure 6-7 Integral coupling rigid and IMC meets *NEC®* *Section 250-118,* which recognizes these products as a grounding conductor installed in accordance with *Article 346* for rigid, and *Article 345* for IMC (Courtesy of Allied Tube & Conduit and Triangle Wire and Cable Company)

Galvanized Electrical Metallic Tubing (EMT)

Industry Standards:
UL 514B - Fittings for conduit and outlet boxes
UL 797 - Electrical metallic tubing
Federal Spec. WW-C-536-A - Conduit, metal, rigid
       electrical, thin-wall steel type
ANSI standard C80-3 - Electrical, metallic tubing
       zinc coated

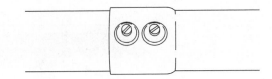

Galvanized Electrical Metallic Tubing (EMT) Compression Uni-Couple (Not produced at this time)

Industry Standards:
UL 514B - Fittings for conduit and outlet boxes
UL 797 - Electrical metallic tubing
Federal Spec. WW-C-536-A - Conduit, metal, rigid
       electrical, thin-wall steel type
ANSI standard C80-3 - Electrical, metallic tubing
       zinc coated

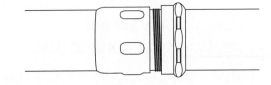

Figure 6-8 Integral coupling EMT meets *NEC® Section 250-118(4)*, which recognizes these products as a grounding conductor installed in accordance with *Article 348* (Courtesy of Allied Tube & Conduit and Triangle Wire and Cable Company)

cinder concrete, cinder fill, or permanent moisture, unless protected on all sides by a layer of noncinder concrete at least 2 inches thick. It is also limited in some hazardous locations. EMT is permitted in wet locations when used with proper fittings. Both rain-tight and concrete-tight fittings are available to be used in wet locations and in poured concrete. Where installed in locations subject to extreme temperature change, such as rooftops, expansion fittings must be used in accordance with *Section 300-7(b)*. The support requirements for electrical metallic tubing can be found in *NEC® 348-13*. Bending shall be in accordance with *348-12* and shall not exceed four quarter bends or 360° total between pull points. In most instances EMT shall be supported at least every 10 feet and within 3 feet of outlet boxes, junction boxes, and conduit bodies. In ceiling joists, bar joists, and other areas where structural members do not readily permit support within 3 feet, a distance of 5 feet from each outlet box, junction box, or conduit body shall be permitted. See Appendix 16-20.

Electrical metallic tubing is a code recognized equipment grounding conductor in accordance with *NEC® Section 250-118(4)*. See Appendix A-4.

Special Feature Tubular Steel Conduits Integral Coupling Steel Raceways

IMC and GRC may be purchased with an integral coupling such as Kwik-Couple, which allows joining of two lengths of Rigid or IMC together by turning the coupling rather than the conduit. This coupling performs similarly to a union ("Erickson") in that it can be tightened without rotating the conduit. This type of conduit can provide substantial savings where routing entails difficulty in joining the conduit lengths together or where several unions are required because of the architectural design, obstructions, or in existing installations. Many installers integrate these special-feature raceways with the traditional raceways to accomplish an easier and better installation. Where field bends or architectural appurtenances have made joining difficult or impossible without a union (Erickson) in the past, these new state-of-the-art products make installations easy and fast.

EMT is available with a belled end fitting with installed setscrews or integral compression coupling, which eliminates the need for a separate coupling. Substantial savings can be achieved in installed cost.

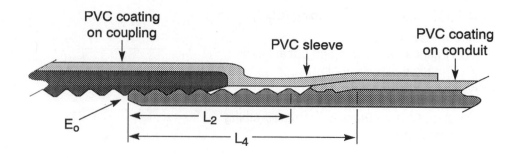

Figure 6-9   PVC-coated conduit ideal for high corrosive areas and areas subject to physical damage. (Courtesy of Rob Roy Industries)

Coated Rigid Conduit and Intermediate Metal Conduit (IMC)

Coated metallic raceways are manufactured in accordance with NEMA standard RN-1 for polyvinyl chloride (PVC), externally coated, galvanized, rigid steel conduit and intermediate metal conduit. It must be installed in accordance with the article that covers the metallic conduit material used, such as galvanized rigid metal conduit, which must be installed as per *NEC® Article 346*. Intermediate metal conduit must be installed in accordance with *Article 345*. A copy of NEMA Standard RN-1 is available through the National Electrical Manufacturers Asso-

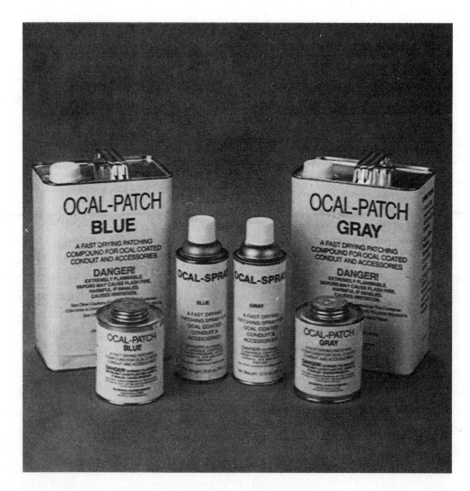

Figure 6-10   If PVC-coated metal conduit is damaged during coated raceways installation, repairing the damaged coating is quick and easy with touch-up compounds. (Courtesy of Ocal Inc.)

ciation (NEMA), Section 5RN, 1300 North 17th Street, Suite 1847, Rosslyn, VA 22209.

This product adopts in whole or in part ANSI C80.1 and C80.6. It also adopts ASTM D149/87 test methods for dielectric breakdown voltage and dielectric strain of solid electrical insulating materials at commercial power frequencies: D638-87 for tensile properties of plastics, D1790-83 for brittleness temperature of plastic film by impact, D2240-86 for rubber property durometer hardness, G6-83 for abrasion resistance of 5-ply coatings, G10-83 for specific bendability of pipeline coatings, G23-81 practice for operating light, and water exposure apparatus (carbon-arc type) for exposure of nonmetallic materials (Underwriters Laboratories UL6 and UL1242).

PVC-coated rigid conduit was first introduced in 1961 by Rob Roy Industries. Today there are three major manufacturers of PVC-coated steel conduit: Rob Roy Industries, OCAL Inc., and Perma-Cote Industries. These major manufacturers have recognized the need for a wiring method that provides complete corrosion protection and excellent installation qualities. PVC-coated conduit combines the strength of metal with the corrosion resistance of a bonded plastic coating for a permanent, troublefree installation. It is an attractive product, and identifies your electrical conduit system.

In service it functions economically through complete and lasting corrosion protection. The adhesive qualities of the PVC coating effectively prevent corrosive fume seepage; there is no undercreep or corrosion travel. Overlapping pressure-sealing sleeves on couplings and conduits create tight, pressure-sealed joints. You can just install it and forget it. There is no need for maintenance. The manufacturer works closely with the designer and the installer to provide a complete material list. The coating process begins with listed rigid metal or intermediate metal conduit. Lighting fixtures can also be coated. Conduit bodies of all types and sizes are coated with overlapping pressure seals. Support channel, all hardware, straps, and supports are coated and corrosionproofed the same as the conduit. Motor-starting switches, unions, and the like can all be coated to provide a complete corrosion-resistant system. In addition, the manufacturer provides a coating touch-up kit that consists of spray-on or paint-on protection that effectively repairs all damage to the original integrity. However, proper installation, damage free, is possible and essential.

This product has been widely accepted for industrial facilities, both for above-ground exposed and for direct burial in all environments. There is no equal for this state-of-the-art raceway system. Coated conduit offers the combination of excellent physical protection by using UL-listed, galvanized rigid steel conduit or galvanized intermediate steel conduit plus unequaled corrosion protection from a 40-mil-thick PVC exterior coating. For severe environments, where corrosive vapors may also attack the conduit from the inside, a two-mil-thick urethane interior coating is also supplied with these products. These and other coatings can also be applied to rigid aluminum conduit to fit the requirements of each unique installation.

> **NOTE:** See Figure A-13 in the Appendix for chemical resistance charts.

Flexible Metal Tubing

NEC® Article 349 describes a raceway not commonly found on general construction projects. This raceway is circular in cross section, flexible, metallic, and liquidtight without a nonmetallic jacket. Flexible metal tubing is required to be "listed." Its installation must comply with *Article 300, Article 349*, and *Article 110-21*. It may be installed only in dry locations that are accessible and protected from physical damage or concealed above, such as suspended ceilings. It is limited to a 1000-volt maximum for branch circuits only. It is not permitted in hazardous locations, storage battery rooms, hoistways, underground, or embedded in concrete and is not permitted in lengths over 6 feet. It may be manufactured in trade size 3/8 inch, 1/2 inch, and 3/4 inch only. The number of conductors permitted in 1/2 inch and 3/4 inch shall not exceed the percentage of fill specified in *NEC® Table 1, Chapter 9*. 3/8 inch shall not exceed that as permitted in *Table 350-12*. It shall comply with *NEC® 250-118(8), Exception 1*. The number of bends are specifically listed and limited to those radii as per *Tables 349-20 (a)* and *(b)*. See *Section 250-118(8)* for for limited permitted uses as an equipment grounding conductor.

350-10. Size.

(a) Minimum. Flexible metal conduit less than ½-in. electrical trade size shall not be used unless permitted in (1) through (5) below for ⅜-in. electrical trade size.

(1) For enclosing the leads of motors as permitted in Section 430-145(b)
(2) In lengths not in excess of 6 ft (1.83 m)
    (a) For utilization equipment, or
    (b) As part of a listed assembly, or
    (c) For tap connections to lighting fixtures as permitted in Section 410-67(c)
(3) For manufactured wiring systems as permitted in Section 604-6(a)
(4) In hoistways, as permitted in Section 620-21(a)(1)
(5) As part of a listed assembly to connect wired fixture sections as permitted in Section 410-77(c)

(b) Maximum. Flexible metal conduit larger than 4-in. electrical trade size shall not be used.

FPN: Metric trade numerical designations for flexible metal conduit are ⅜ = 12, ½ = 16, ¾ = 21, 1 = 27, 1¼ = 35, 1½ = 41, 2 = 53, 2½ = 63, 3 = 78, 3½ = 91, and 4 = 103.

350-12. Number of Conductors. The number of conductors permitted in a flexible metal conduit shall not exceed the percentage of fill specified in Table 1, Chapter 9, or as permitted in Table 350-12 for ⅜-in. flexible metal conduit.

Table 350-12. Maximum Number of Insulated Conductors in ⅜-in. Flexible Metal Conduit*

Size (AWG)	Types RFH-2, SF-2		Types TF, XHHW, AF, TW		Types TFN, THHN, THWN		Types FEP, FEPB, PF, PGF	
	Fittings Inside Conduit	Fittings Outside Conduit	Fittings Inside Conduit	Fittings Outside Conduit	Fittings Inside Conduit	Fittings Outside Conduit	Fittings Inside Conduit	Fittings Outside Conduit
18	2	3	3	5	5	8	5	8
16	1	2	3	4	4	6	4	6
14	1	2	2	3	3	4	3	4
12	—	—	1	2	2	3	2	3
10	—	—	1	1	1	1	1	2

*In addition, one covered or bare equipment grounding conductor of the same size shall be permitted.

Figure 6-11 Special requirements when installing 3/8-inch flexible conduit. (Reprinted with permission from NFPA 70-1999.)

Flexible Metal Conduit

Flexible metal conduit shall be "listed" and must comply with the applicable provisions of NEC® Article 350 or Article 300. The basic standard used to investigate products in this category is UL1 (Underwriters Laboratories Standard for Flexible Metal Electrical Conduit). "Greenfield" was invented in 1902 by Harry Greenfield and Gus Johnson and listed by Sprague Electric Co. as "Greenfield flexible steel conduit." Today the term "Greenfield" is commonly used for all flexible metal conduit. The sizes permitted include flexible aluminum and steel conduit in trade sizes 3/8 inch to 4 inches and reduced wall-type RW trade sizes 3/8 inch to 3 inches. However, Section 350-10 generally does not permit flexible metal conduit less than 1/2-inch electrical trade size.

There are exceptions for enclosing the leads of motors. As permitted in Article 430, 3/8-inch flexible metal conduit may be used in lengths not in excess of 6 feet as part of a listed assembly or for tap connections to lighting fixtures as required in Article 410 or for utilization equipment. 3/8-inch trade size is also permitted for manufactured wiring systems, as covered in Article 604. 3/8-inch trade size shall be permitted as part of a listed assembly to connect wire sections as permitted in NEC® Section 410-77 (c). The maximum number of insulated conductors in 3/8" flexible metal conduit shall be limited as per Table 350-12 (see Figure 6-11).

For 1/2-inch through 4-inch trade size, the number of conductors shall not exceed the percentage of fill as specified in NEC® Table 1, Chapter 9. See Section 250-118(5) and (6) for limited permitted use as an equipment grounding conductor.

Liquidtight Flexible Metal Conduit

Liquidtight flexible metal conduit is manufactured in trade sizes 3/8 inch through 4 inches for conductors in circuits of 600 volts nominal or less in accordance with *NEC® Article 351 Part A* and *UL Std. 360.* Liquidtight flexible metal conduit that is suitable for direct burial and poured in concrete is marked "Direct Burial." It is a raceway of circular cross section, having an outer liquidtight, nonmetallic, sunlight-resistant jacket over an inner flexible metal core, with associated couplings, connectors, and fittings, and is approved for the installation of electrical conductors. Installations of liquidtight flexible metal conduit must comply with *Articles 300* and *351* and specific sections, where applicable, of *Articles 250, 501, 502, 503,* and *505.* The marking requirements for this product must comply with *Section 110-21.*

It may be used where listed and marked for direct burial in earth and exposed and concealed work where the conditions of operation and maintenance require flexibility or protection from liquids, vapors, or solids. It shall not be used where subject to physical damage, or where any combination of ambient or conductor temperatures produce an operating temperature in excess of that for which the material is approved.

Where it is installed as a fixed raceway, it shall be secured at intervals not exceeding 4 1/2 feet and within 12 inches on each side of every outlet, box, junction box, cabinet, or fitting, except where permitted to be fished or not exceeding 3 feet in length for flexibility and not exceeding 6 feet in length for fixture tap conductors. It is permitted as a grounding conductor where both the conduit and the fittings are approved for grounding. Where a bonding jumper is

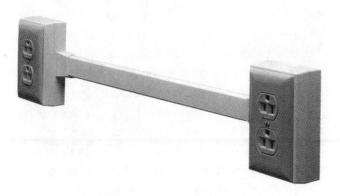

Figure 6-12a Surface raceways systems are ideal for wiring work stations or adding convenience outlets in existing areas. (Courtesy of Carlon, A Lamson & Sessions Company)

required it shall be installed in accordance with *NEC® Section 250-102.* Except where used for flexibility, an equipment-grounding conductor must be installed. It shall be permitted as a grounding means in 1 1/4-inch trade sizes and smaller if the total length of the liquidtight flexible metal conduit and any ground-return path is 6 feet or fewer, the conduit is terminated in fittings listed for grounding, and the circuit conductors contained therein are protected by an overcurrent device rated at 20 amperes or fewer for 3/8-inch and 1/2-inch trade sizes, and 60 amperes or fewer for 3/4-inch through 1 1/4-inches trade sizes. See *Section 250-118 (7)* for specific permitted uses. Where used as a fixed-wiring method, there shall not be more than the equivalent of four quarter bends, 360° total, between pole points (e.g., conduit bodies and boxes).

Metallic and Nonmetallic Surface Raceways

Metallic and nonmetallic wireways are required to be "listed" and are intended for installation in accordance with *NEC® Article 352 Parts A* and *C* and UL 5.

Surface Metal Raceways. Surface metal raceways come in various sizes and types of materials. This product is generally referred to as plug mold or plug strip by field installers. However, it may include heavy industrial listed surface metal raceway that can withstand physical damage and is approved for industrial locations. This *NEC®* section also includes products suitable for use in dwelling, kitchen counter areas, in laboratories for test benches, and in retrofitting old dwellings where it is difficult or impossible to conceal the wiring methods.

Many of these products can be painted to match the decor and provide a quick, decorative, efficient way to comply with modern-day requirements for convenience outlets located throughout the facility. This surface wiring method is also used to provide protection for conductors routed from the ceiling to desk locations in secretarial open spaces or where partition walls do not extend to the ceiling. These products must be installed in accordance with the manufacturer's instructions and the manufacturer's listing, because many special use types are available

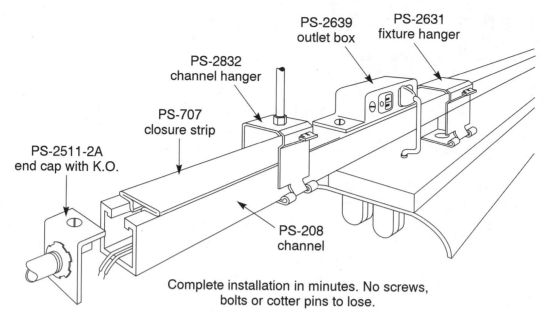

PS-2639
outlet box

PS-2631
fixture hanger

PS-2832
channel hanger

PS-707
closure strip

PS-2511-2A
end cap with K.O.

PS-208
channel

Complete installation in minutes. No screws,
bolts or cotter pins to lose.

Figure 6-12b   Strut may serve as surface metal raceway for which it is "listed" or as a support system, or both. It is manufactured in enameled steel, galvanized, stainless steel, and aluminum. (Courtesy of Allied Tube & Conduit)

on the market today. They are listed by UL in various categories for grounding where installed in accordance with the manufacturer's instructions. Surface metal raceways are permitted to extend through drywall, dry partitions, and dry floors. Where a combination of surface raceways are used for signaling, lighting, and power circuits, the surface metal raceway shall be compartmentalized to separate the systems and shall be identified by sharp contrasting colors in the interior finish. The same position of the compartments must be maintained throughout the premises for which this type of surface metal raceway is installed.

Splices and taps are permitted, but they shall not be permitted to fill the raceway to more than 75% of its area at the point the splice or tap occurs. Where splices and taps are employed in metal raceways without removable covers, they shall be made only in junction boxes and in an approved manner. Surface metal raceways must be of such construction as to distinguish them from other raceways. Their elbows, couplings, and similar fittings shall be designed so that the sections can be electrically and mechanically coupled together and installed without subjecting the wires to abrasion. Where covers and accessories of nonmetallic materials are used on surface metal raceways, they shall be identified for such use. Surface

metal raceway shall not contain conductors larger than the specifications for which the raceway designed. Surface metal raceway must comply with the applicable provisions in *Article 300*; it is permitted only in dry locations and shall not be used where subject to physical damage, unless approved for that purpose. It shall not be used where the voltage is 300 volts or more unless it has a metal thickness of .040 inch nominal. It is not permitted where subject to corrosive vapors, in hoistways, or in hazardous locations, except where permitted by *NEC® Exception, Section 300-4(b)* and generally shall not be concealed.

The heavy wall surface metal raceway as covered in *Part C* of *Article 352* is frequently used for the support of fluorescent fixtures and the protection of conductors supplying the fixtures is called "strut." This type of surface raceway is also used extensively for the fabrication of racks and trapeze-type raceway hangers. It also provides channel raceway that can be installed in areas subject to physical damage. The advantage of this wiring method is that the conductors can be placed in the channel where necessary prior to placing the cover on the channel.

Surface Nonmetallic Raceways. The requirements for surface nonmetallic raceways shall be of such a construction as to distinguish them from other raceways. Their fittings, couplings, and elbows shall

NEMA 3R RAINTIGHT GUTTER

UL LISTED

Utility company sealing provision

NEMA 1 LAY-IN WIREWAY

UL LISTED

CONNECTOR

END CAP

90° ELBOW

45° ELBOW

TEE

CROSS

Figure 6-13   Typical listed wireway or auxiliary gutter (Courtesy of Unity Manufacturing)

be designed so that sections can be mechanically coupled together and installed without subjecting the wires to abrasion. Where combination surface metal raceways are used for signaling, lighting, and power circuits, the different systems shall be run in separate compartments, shall be identified by printed legend or by sharply contrasting colors of the interior finish, and the same relative position of the compartments shall be maintained throughout the premises. Surface nonmetallic raceway and fittings shall be of a suitable nonmetallic material that is resistant to moisture and chemical atmospheres. It must be flame retardant and resistant to impact, crushing, distortion from heat under conditions likely to be encountered in service, and low temperature effects.

It is permitted only in dry locations and shall not be used where concealed, where subject to severe physical damage, or where the voltage is 300 volts or more between conductors, unless listed for higher voltage. It cannot be used in hoistways, hazardous locations except as permitted in *NEC® Section 502-4(b)*, where ambient temperatures exceed those for which the nonmetallic raceway is listed, or where conductors whose insulation temperature limitations exceed those for which the nonmetallic raceway is listed. Nonmetallic raceway shall also comply with the applicable provisions of *Article 300*. No conductor larger than that for which the raceway is designed shall be installed in the surface nonmetallic raceways.

Metallic and Nonmetallic Wireways—Auxiliary Gutters

Wireways, auxiliary gutters, and their associated fittings are intended for installation in accordance with *NEC® Article 362* for wireways, *Article 374* for auxiliary gutters, and UL 870, which evaluates both products. However, in reality they are the same product. In many instances they are junction boxes and must comply with *Article 370*. An auxiliary gutter and a wireway look exactly alike in physical stature and manufacture. Application and use dictate whether it is classified as a wireway or an auxiliary gutter. Field installers generally refer to them both as gutter spaces. However, it is very important that the installer distinguish between the two articles, because they are used very differently.

NEC® Article 362 permits this material, which

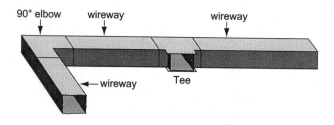

Of sheet metal, one side hinged or removable.
Wires loose in wireway.
May be used only in exposed work.
Not more than 30 current-carrying conductors in any cross-section (exceptions—*Sec. 362-5. NEC®*). Splices and taps may be in wireway, if accessible. Securely supported at intervals not exceeding five feet unless approved for supports at greater intervals.

Figure 6-14 Wireways as in *NEC® Article 362*

has been evaluated in accordance with UL 870, to be utilized as a raceway. However, *Article 374*, which covers the requirements for auxiliary gutters and is also evaluated in accordance with UL 870, does not permit this material to be installed as a raceway but only as a supplement to the wiring spaces at meter centers, distribution centers, switchboards, and similar points of wiring systems. It may enclose conductors or bus bars but shall not enclose switches, overcurrent devices, appliances, or other similar equipment. An auxiliary gutter is permitted to extend up to 30 feet beyond the equipment that it supplements. A wireway, however, has vastly different rules. In this text we will discuss them separately.

Caution: Use them separately and install them correctly. There are vastly different uses, and the restrictions in the *NEC®* contain very different rules.

Metal Wireways

Wireways are sheet-metal troughs with hinged or removable covers for housing and protecting electrical wires and cables. Wireways are intended to be installed in accordance with *NEC® Article 362* and UL Std 870. The conductors are to be laid after the wireway has been installed as a complete system. Wireways are permitted to be used only for exposed work, and when installed in wet locations shall be of a

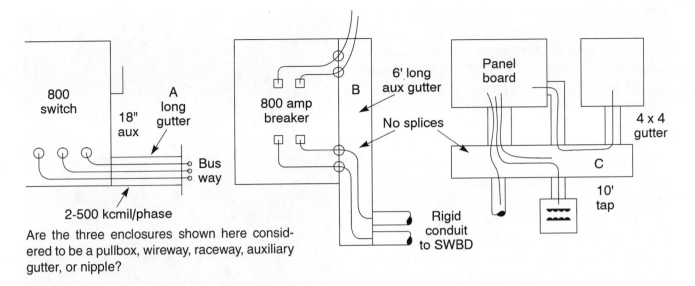

Are the three enclosures shown here considered to be a pullbox, wireway, raceway, auxiliary gutter, or nipple?

Figure 6-15 This question was asked of a panel of experts, and the consensus was that only the center sketch is an "Auxiliary Gutter"; the other two are "Wireways."

raintight construction. They are not permitted to be installed where subject to severe physical damage or corrosive vapors, or in hazardous locations except as permitted in *NEC® 501-4(b), 502-4(b), 504-20* and *505-15*. Installation of wireways shall also comply with *Article 300*. Wireways are not permitted to contain more than 30 current-carrying conductors at any cross section.

Conductors for signaling circuits and control circuits between motor and starter that are used only for starting duty shall not be considered current-carrying conductors for this purpose. The sum of a cross-sectional area of all contained conductors in any cross section of the wire shall not exceed 20% of the interior cross-section area of the wireway. The derating factors specified in *NEC® Section 310-15, Table 310-15(b)(2)(a)*, are not applicable to the 30 current-carrying conductors at the 20% level specified previously, except where the derating factors in *Article 310*, are applied. The number of current-carrying conductors shall not be limited, but the sum of all the contained conductors in the cross-sectional area still may not exceed 20% of the cross-sectional area of the wireway.

Where insulated conductors are deflected within a wireway, either at the ends or where the conduit fittings or raceways enter or leave the wireway, *NEC® Section 373-6* applies if the deflection is greater than 30°. Splices and taps are permitted within wireways, provided they are accessible and do not exceed 75% of the area at the point in which the taps or splices occur. Wireways are required to be

supported at intervals not to exceed 5 feet, or for individual lengths longer than 5 feet at each end or joint. In no case shall the distance between supports exceed 10 feet. Unbroken lengths are permitted to pass transversely through the walls if in unbroken lengths where passing through.

All dead ends of wireways must be closed, and extensions for wireways are permitted with any of the wiring methods in *Chapter 3* must include a means for equipment grounding. Where a separate equipment grounding conductor is employed, the connection of the equipment-grounding conductor and the wiring method to the wireway shall comply with *NEC® Section 250-8* and *Section 250-12*. Where rigid nonmetallic conduit, electrical nonmetallic tubing, or liquidtight, flexible nonmetallic conduit are used, connection to the equipment-grounding conductor and the nonmetallic raceway to the metal wireway shall comply with these same articles. Wireways must be marked so that the manufacturer's name and trademark are visible after the installation. Grounding of metal wireways must be done in accordance with *Article 250*.

Auxiliary Gutter

Auxiliary gutters, which are not raceways, are intended to be installed in accordance with *NEC® Article 374* and UL Std 870. They are permitted to supplement the wiring spaces at meter centers, distribution centers, switchboards, and similar points of wiring systems and may include conductors or bus bars, but they shall not be used to enclose switches, overcurrent devices, appliances, or other similar equipment. Auxiliary gutters shall not extend more than 30 feet beyond the equipment that it supplements. Gutters shall be supported throughout their entire length at intervals not to exceed 5 feet. There are no exceptions. Covers must be securely fastened to the gutter. Auxiliary gutters shall contain no more than 30 current-carrying conductors, and the same provisions as for wireways and exceptions apply.

Ampacity limitations for auxiliary gutters state that the current, carried continuously in bare copper bars in auxiliary gutters, shall not exceed 1000 amperes per square inch of cross section of the conductor. For aluminum bars, the current carried continuously shall not exceed 700 amperes per square inch of cross section of the conductor. Other current-carrying capacities within auxiliary gutters shall be in accordance with *NEC® Article 310*.

Bare conductors within auxiliary gutters must be securely and rigidly supported so that the minimum clearance between the bare current-carrying metal parts of opposite polarity mounted on the same surface will not be less than 2 inches, or less than 1 inch for parts that are held free in the air. Clearance of not less than 1 inch must be secured between bare current-carrying metal parts and any metal surface. Adequate provisions must be made for the expansion and contraction of bus bars. Splices and taps are permitted where they are accessible by means of removable covers or doors but shall not be permitted to fill more than 75% of the cross-sectional area where the taps or splices are made. Taps to the bare conductors are permitted and shall leave the gutter opposite the terminal connections. The conductors shall not be brought in contact with uninsulated current-carrying parts of opposite polarity. All taps shall be suitably identified in the gutter as to the circuit or equipment it is supplying. Overcurrent protection shall be as required in *NEC® Section 240-21*.

TYPES OF NONMETALLIC RACEWAYS

Rigid Nonmetallic Conduit Schedule 40 & Schedule 80

Rigid nonmetallic conduit (PVC) (Schedule 40 and Schedule 80), is required to be "listed" and is

Figure 6-16 Underground nonmetallic duct bank installation. (Courtesy of Carlon, A Lamson & Sessions Company)

intended for installation in accordance with *NEC® Article 347* and UL 651, was originally introduced as an underground duct for use by electrical utilities in the late 1950s. At that time the duct was produced from high-impact polystyrene, which had excellent physical properties and corrosion resistance. However, the styrene material did not have good fire-resistance properties, and by the early 1960s polyvinyl chloride (PVC) became the preferred material. UL first listed PVC Schedule 40 rigid nonmetallic conduit in 1962. The *NEC®* first recognized its use only for underground installations in 1968. In 1971 the *NEC®* expanded approved uses to include above-ground and building applications. *Article 347* as it appears in the 1996 *NEC®* remains virtually unchanged from the original 1971 edition.

Rigid nonmetallic conduit, fittings, and accessories also referred to as RNMC are manufactured according to NEMA TC-2, federal specifications WC1094A, and UL 651 specifications, and carry respective UL listings and UL labels. RNMC is a nonconductive, sunlight-resistant, UL-listed product for exposed or outdoor usage (the use of expansion fittings allows the system to expand and contract with temperature variations, and it will not rust or corrode). The most commonly used RNMC is

For underground applications encased in concrete or direct burial. Also for use in exposed or concealed applications aboveground.
- UL Listed
- Sunlight resistant
- Rated for use with 90°C conductors
- Superior weathering characteristics

For use in aboveground and belowground applications that are subject to physical damage.
- UL Listed
- Sunlight resistant
- Rated for use with 90°C conductors
- Superior weathering characteristics

Figure 6-17 Installation as per *NEC® Article 347* rigid nonmetallic conduit (Courtesy of Carlon, A Lamson & Sessions Company)

Schedule 40 rigid heavy wall PVC for underground applications encased in concrete or direct burial, exposed and concealed, which is rated for use with 90° C-conductors and is manufactured in nominal sizes 1/2 inch through 6 inches. It is produced in 10-foot lengths but may be produced in lengths shorter or longer than 10 feet, with or without belled ends. Where subject to physical damage for aboveground or below-ground applications, RNMC Schedule 80 extra heavy wall PVC-80 must be used. PVC Schedule 80 is also produced in 1/2-inch through 6-inch sizes, with or without belled ends. *NEC® Article 347* states the conditions under which rigid nonmetallic conduit and fittings are to be used (an informational fine print note advises that extreme cold may cause some nonmetallic conduits to become brittle and therefore more susceptible to damage from physical contact).

The following are permitted uses (varies by UL listing—some nonmetallic conduits are listed and marked for underground use only): concealed in walls, floors, and ceilings in locations subject to severe corrosive influences, as covered by *NEC® Section 300-6*, or where subject to chemicals for which the materials are specifically approved. For chemicals for which PVC is not acceptable, the designer or installer should select an appropriate writing method from *Chapter 3* of the *National Electrical Code®*

NOTE: See Appendix, Figure A-12.

The installations must comply with the specific *NEC®* article and in addition must be installed in accordance with the wiring methods specific listing instructions.

In dairies, laundries, canneries, and other wet locations or where walls are frequently washed, the entire conduit system shall be so installed and equipped as to prevent water from entering the system. All supports, bolts, straps, and screws shall be made of a corrosion-resistant material or protected against corrosion by corrosion-resistant materials. RNMC can be used in dry and damp locations and can be exposed where not subject to physical damage (Schedule 80, however, may be installed in locations subject to physical damage). Rigid nonmetallic conduit may be installed underground as per *NEC® Sections 300-5* and *300-50*.

NOTE: Schedule 80 has less interior space and a reduction in wire fill is necessary. See *NEC® Chapter 9*, and *Tables*. Underground installations are to be made as per *Sections 300-5* and *300-50*. (Additional information may be found in appendix, you may also consult the manufacturer for wire-fill instructions.)

Rigid nonmetallic conduit is not permitted in hazardous locations except as per *NEC® Sections 503-3(a), 504-20, 505-15, 514-8,* and *515-5,* and Class I, Division 2 locations as permitted in the *Exception* to *Section 501-4(b)*. It is not permitted for the

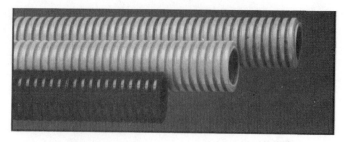

COLOR CODED
For Easy Circuit Identification of
Electrical Systems

Figure 6-18   Installation as per *NEC® Article 331* Electrical Nonmetallic Tubing (Courtesy of Carlon, A Lamson & Sessions Company)

support of fixtures or other equipment unless identi-fied specifically for such use. It is not permitted where subject to ambient temperatures exceeding 50°C (122°F). It is not permitted in theaters or sim-ilar locations, except under limited conditions as pro-vided in *Articles 518* and *520*. Rigid nonmetallic conduit also cannot be used in environmental air spaces as detailed in *Article 300-22*.

Rigid nonmetallic conduit products are joined by solvent cement. Sizes 1/2 inch through 1 1/2 inches should be cut square with a fine-toothed saw and deburred. For sizes 2 inches through 6 inches, a miter box or similar saw guide should be utilized to steady the material. After cutting and deburring, wipe the ends clean of dust, dirt, and shavings.

The joining process is as follows. Be sure the conduit end is clean and dry. Place a coating of cement on the conduit end and then on its mating part with a dauber. Thoroughly coat the surfaces to be mated. Allow the primer/cleaner a few seconds to soften the PVC surface (the time may need to be

adjusted, depending on the temperature). Push con-duit firmly into fitting while rotating conduit slightly about 1/4 turn to spread cement evenly. Allow the joint to set approximately ten minutes. Most manu-facturers recommend specific solvent cement. The cement is prepared for their products, compounds, and tolerances, and substitutions should not be made, because adverse effects can occur. In situations re-quiring extremely fast setting, or in low-temperature or difficult installation conditions, all-weather, quick-set cement should be used.

Rigid nonmetallic conduit may be bent in the field. *NEC® Sections 347-13* and *347-14* specify how to make the bends and the number permitted in a run. They must be made so that the conduit is not damaged and its internal diameter is not effectively reduced. It shall be made only with bending equip-ment identified for the purpose. For more informa-tion on bending, see *Chapter 3*. The number of conductors permitted shall not exceed that permitted by the percentage fill in *Table 1, Chapter 9*.

Electrical Nonmetallic Tubing

Electrical nonmetallic tubing (ENT) is required to be "listed" and is intended for installation in accordance with *NEC® Article 331.*

NOTE: ENT is listed for its intended purpose by UL standard CSA/UL 1653, which has recently been published for this product.

ENT is a pliable corrugated raceway with a circular cross section. Even though ENT was not commercially introduced in the United States until the late 1970s, the product has been used successfully in a lightweight form in European countries since the 1950s, where it is generally embedded in plaster. ENT was designed as a branch-circuit wiring method for use only within buildings. It is not sunlight resistant and is not permitted for outdoors installation. ENT electrical nonmetallic tubing, fittings, and accessories, as covered in *Article 331*, are available in various colors for easy circuit identification of electrical systems (ENT is commonly produced in blue, red, and yellow). It is a pliable material for which no bending equipment is necessary and is of a corrugated design. It is available in 1/2-inch through 2-inch sizes.

ENT is recognized by the three model building codes—BOCA, ICBO, and SBCCI—in their National Evaluation Report, NER 290. A National Evaluation Report is necessary where a product is normally not permitted in some types of construction. Because it is combustible, ENT is normally not allowed in noncombustible construction. NER-290 permits its use only in non-load-bearing partitions of specific construction with a two-hour fire-resistance rating (it is not approved for load-bearing walls). This application is more restrictive than the *NEC®* and is limited to a maximum of three runs in any 6-foot length of wall, with a maximum of two tubes or conduits in any one stud cavity.

It is also permitted in a fire-resistant, floor-ceiling assembly with a rating of three hours or fewer under limited conditions. The total volume of the nonmetallic installation in the rated floor-ceiling assembly cannot exceed 380 cubic inches per 100 square feet of ceiling area; this limit is any combination of ENT, rigid nonmetallic conduit, and liquidtight conduit. Computation of the 380 cubic inches is required on a cubic-inch-per-linear-foot basis. The conduits are not permitted to penetrate the dropped ceiling membrane. The ceiling membrane must carry the mark of a recognized laboratory to indicate that it is fire rated for the purpose. Also, there must be a distance of not fewer than 16 3/8 inches between the ceiling and floor above for ENT to be used in fire resistant, rated floor-ceiling assemblies. One final caution—not all PVC conduit is approved for this application. Only those manufacturers who have received an evaluation report are approved. You should check with your building department before installing.

ENT must always be concealed in walls, floors, or ceilings when it is used in any building that is higher than three floors above grade.

NOTE: Three floors above grade is defined in *NEC® Article 336-5(a)(1)* as follows: ". . . the first floor of a building shall be that floor that has fifty percent or more of the exterior wall surface area level with or above finished grade. One additional level that is the first level and not designed for human habitation and used only for vehicle parking, storage, or similar use shall be permitted."*

Additionally, the walls, floors, and ceilings behind which ENT is concealed must provide a thermal barrier of material having at least a 15-minute finish rating as identified in listings of fire-rated assemblies. The UL *Fire-Resistance Directory* advises that a finish rating is established for fire-resistant rated assemblies containing combustible supports. It is described as follows: "The finish rating is the time at which the wood stud or joist reaches an average temperature rise of 250°F or an individual temperature rise of 325°F as measured on the plane of the wood nearest the fire." A finish rating is different from a fire-resistant rating; UL must establish finish ratings for metal and wood stud assemblies.

The space above suspended ceilings is considered an accessible location by *NEC®* definition. *Article 331-3(5)* permits ENT when the suspended

*(Reprinted with permission from NFPA 70-1999.)

ceiling is part of a fire-rated assembly and meets the 15-minute thermal barrier requirement. The actual assembly should be verified as to its rating, as grid-type ceilings frequently do not meet these ratings. *Article 331-3(1)* permits ENT for other exposed work in buildings that do not exceed three floors above grade, if it is not subject to physical damage.

ENT is also permitted in locations subject to severe corrosive influences as covered in *NEC® Section 300-6,* where subject to chemicals for which the materials are specifically approved. It is permitted to be embedded in poured concrete, provided that fittings identified for this purpose are used for connections. This fitting identification is located by a marking on the carton.

The UL Green Book further tells us that some fittings can be used only with ENT produced by a specific manufacturer and are not interchangeable with other brands. In such case the carton is marked, "Suitable for use only with [brand X] ENT."

As with other nonmetallic conduit in the *NEC®, Article 331* also carries a *Fine Print Note* stating "extreme cold may cause some types of nonmetallic conduits to become brittle, and therefore, more susceptible to damage from physical contact."*

Electrical nonmetallic tubing is not permitted in the following conditions: in hazardous (classified) locations, for support of fixtures or other equipment, where subject to ambient temperatures exceeding 50°C; (122°F), for conductors whose insulation limitations exceed those for which the tubing is listed, for direct burial, where the voltage is over 600 volts, or in theaters and similar locations, except as provided in *NEC® Articles 518* and *520.*

ENT is usually joined with listed snap fittings. This may be a one-piece snap fitting, or a fitting integral to the box. Glue-on, rigid nonmetallic couplings are permitted when both coupling and tubing are PVC.

ENT shall be installed as a complete system, as provided in *NEC® Article 300*, and securely fastened in place within 3 feet of each outlet box, junction box, cabinet, or fitting. The tubing must also be secured at least every 3 feet. Where the tubing enters a box or fitting or other enclosure, a bushing or adapter

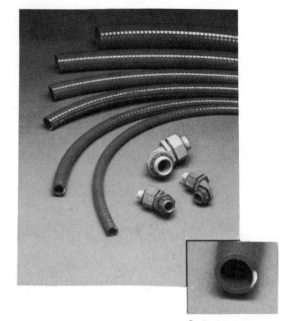

Carflex® has a hard PVC spiral completely surrounded by flexible PVC.

Figure 6-19 Installation as per *NEC® Article 351* Flexible Nonmetallic Conduit (Courtesy of Carlon, a Lamson & Sessions Company)

must be provided to protect the wire from abrasion unless the box, fitting, or enclosure design provides equivalent protection. The number of conductors permitted in ENT shall not exceed that permitted by the percentage fill in *Table 1, Chapter 9.*

> **NOTE:** When using solvent weld fittings for concrete-tight performance, do not use primer or cleaner. Apply only a light uniform coat of the cement labeled for use with ENT. Do not use a dauber, and brush excess cement out of the ENT grooves. The concern is that excessive solvent may greatly reduce the physical properties of ENT at the joint.

Sizes ½ through 1 inch trade size are permitted as a listed, manufactured, prewired assembly. Although this new permitted use will allow any number and combination of conductors that comply with *Table 1, Chapter 9*, of the *NEC®* it will require that combinations be ordered from the factory as listed

*Reprinted with permission from NFPA 70-1999.

assemblies, and therefore it is thought that the primary use will be for two- and three-wire branch circuit wiring. These common wire configurations will be available in coils and reels. This will allow ENT to compete with type MC and type AC cables. The same cautions must be observed when cutting prewired ENT: the cutters designed for the purpose should be used when cutting so that conductor insulation is not damaged.

Flexible Nonmetallic Liquidtight Conduit

Liquidtight Flexible Nonmetallic Conduit (LFNC) is covered in NEC Article 351, Part B, and evaluated by UL1660. This product was originally developed in the late 1970s as a raceway for industrial equipment where flexibility was required and where protection of conductors from liquids was also necessary. LFNC was first included in the 1981 *NEC*® as *Article 351, Part B*. The product was limited to industrial installations, and unless special permission was granted, installations were limited to six feet in length. The 1987 *NEC*® deleted the industrial limitation and recognized outdoor use where listed. The 1990 *NEC*® added direct burial when so listed as a permitted use. Several varieties of LFNC have been introduced over the years. The product first introduced, commonly referred to as "hose," consisted of an inner and outer layer of neoprene with a nylon reinforcing web between the layers. The 1984 *NEC*® recognized a second-generation product which consisted of a smooth wall of flexible PVC with a rigid PVC integral reinforcement rod. The 1990 *NEC*® recognized a nylon corrugated shape without any integral reinforcements. These three types are defined in *Section 351-22(a), 351-22(b)* and *351-22(c)*. The type LFNC-B is most common and is allowed to be used in lengths longer than 6 feet and must be flame resistant with fittings approved for the installation of electrical conductors. These three LTFNMC raceways are limited to 600-volt installations or less. They are sunlight resistant and suitable for use at conduit temperatures of 80°C dry and 60°C wet, and 60° oil resistance as required by Section 15-6 of ANSI/NFPA 79-1985 and UL 1660. LTNFNMC may be manufactured in sizes 3/8 inch through 4 inch trade size. The number of conductors

Figure 6-20 Installation as per *NEC*® *Article 362* nonmetallic wireway (Courtesy of Carlon, a Lamson & Sessions Company)

permitted in a single conduit shall be in accordance with the percentage fill specified in NEC Table 1, Chapter 9. The number of bends permitted in liquid tight flexible nonmetallic conduit shall not exceed 360° in one run, the equivalent of four quarter bends between foe points, e.g., conduit bodies and boxes. Angle connectors shall not be used for concealed raceways installations.

Liquidtight flexible nonmetallic conduit (LTFNMC) is defined as a raceway of circular cross section. There are presently three types, (1) A smooth seamless inner core and cover bonded together and having one or more reinforcement layers between the core and cover. (2) A smooth inner surface with integral reinforcement within the conduit wall. (3) A corrugated internal and external surface without integral reinforcement within the conduit wall. The code requires it to be flame-resistant and, with fittings listed for the purpose. UL (Underwriters Laboratories) presently lists nonmetallic connectors for these products and some liquid tight metallic flexible conduit connectors are dual listed for both metallic and nonmetallic liquid tight flexible conduit.

This raceway is permitted to be used in exposed or concealed locations however, cold may cause some types of nonmetallic conduits to become brittle and therefore more susceptible to damage from physical contact. It is designed and permitted where flexibility is required for installation, operation, or maintenance, and for the protection of the contained conductors is required from vapors, liquids, or solids. Where listed and marked as suitable for the purpose it can be used for outdoor locations and for direct burial.

Liquidtight flexible nonmetallic conduit shall not be used where subject to physical damage or where any combination of ambient and conductor temperatures are in excess of that for which it is approved.

Only Type LFNC-B, may be used generally in lengths longer than six feet but when it is used in lengths longer than six feet it must be supported and secured every 4 ½ feet and within 12 inches of all terminations. Sizes ½ through 1 inches trade size are permitted as a listed manufactured prewired assembly. Although this new permitted use will allow any number and combination of conductors that comply with *Table 1, Chapter 9* of the *NEC®*. The permitted use will require that combinations be ordered from the factory as listed assemblies. Common wire configurations will be available in coils and reels. This will allow the installer to cut whips to length and install AC hookups with a minimum of effort. The same cautions must be observed when cutting prewired LFNC. The cutters designed for the purpose should be used when cutting so that conductor insulation is not damaged. Although this new permitted use will allow any number and combination of conductors that comply with *Table 1, Chapter 9* of the *NEC®*. The permitted use will require that combinations be ordered from the factory as listed assemblies.

Generally Types LFNC-A and LFNC-C are not permitted in lengths longer than 6 feet except where longer length is essential for a required degree of flexibility, such as connections to light mixers that are regularly moved from vessel to vessel for mixing and other such applications.

> **NOTE:** for Chemical Resistance Chart see Figure A-12

Nonmetallic Wireways

This wiring method was accepted into the 1993 *NEC®* as "*Part B*" for *Article 362*. It is approved for exposed installations, with one exception, and may be used in wet locations and where subject to

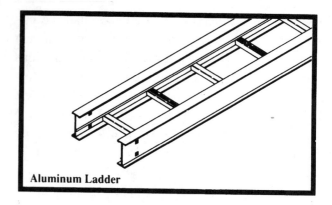

Aluminum Ladder

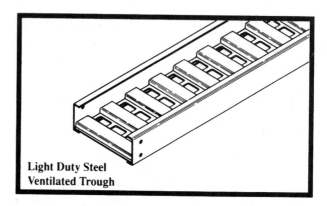

Light Duty Steel Ventilated Trough

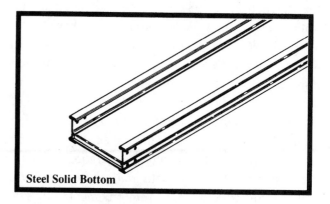

Steel Solid Bottom

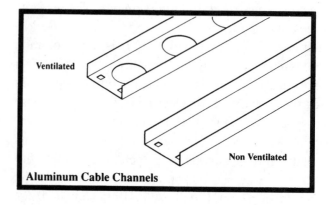

Ventilated

Non Ventilated

Aluminum Cable Channels

Figure 6-21 Basic cable tray support systems designs (Courtesy of B-Line Systems Inc.)

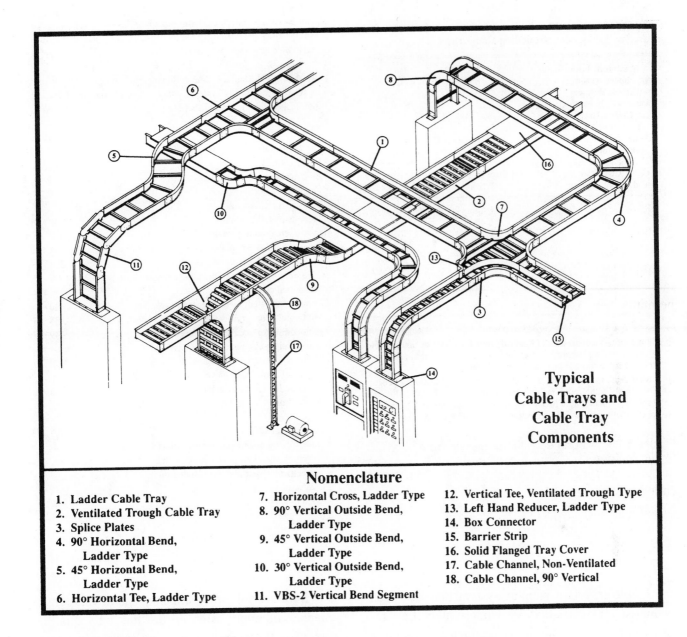

Figure 6-22 Typical cable trays and components (Courtesy of B-Line Systems Inc.)

Nomenclature

1. Ladder Cable Tray	7. Horizontal Cross, Ladder Type	12. Vertical Tee, Ventilated Trough Type
2. Ventilated Trough Cable Tray	8. 90° Vertical Outside Bend, Ladder Type	13. Left Hand Reducer, Ladder Type
3. Splice Plates	9. 45° Vertical Outside Bend, Ladder Type	14. Box Connector
4. 90° Horizontal Bend, Ladder Type	10. 30° Vertical Outside Bend, Ladder Type	15. Barrier Strip
5. 45° Horizontal Bend, Ladder Type	11. VBS-2 Vertical Bend Segment	16. Solid Flanged Tray Cover
6. Horizontal Tee, Ladder Type		17. Cable Channel, Non-Ventilated
		18. Cable Channel, 90° Vertical

corrosive vapors. Nonmetallic wireway may not be installed where subject to physical damage, exposed to sunlight unless listed as sunlight resistant, or in hazardous locations, and is limited to ambient temperatures for which the product is "Listed." One very important difference between metal wireway and nonmetallic wireway is that although the derating factors in *Article 310 do apply* for the nonmetallic wireway, they *do not apply* to the metal wireway. The support and expansion requirements are also different and must be strictly adhered to when using this wiring method. See *NEC® Article 362, Part B* for the complete installation requirements.

CABLE TRAY SUPPORT SYSTEMS

Cable trays are intended to be field assembled and installed in accordance with *NEC® Article 318.*

> **Note:** Cable tray is not a raceway; to refer to it as a raceway is an error.

Table 318-7(b)(2). Metal Area Requirements for Cable Trays Used as Equipment Grounding Conductors

Maximum Fuse Ampere Rating, Circuit Breaker Ampere Trip Setting, or Circuit Breaker Protective Relay Ampere Trip Setting for Ground-Fault Protection of Any Cable Circuit in the Cable Tray System	Minimum Cross-Sectional Area of Metal[a] (in.^{2})	
	Steel Cable Trays	Aluminum Cable Trays
60	0.20	0.20
100	0.40	0.20
200	0.70	0.20
400	1.00	0.40
600	1.50[b]	0.40
1000	—	0.60
1200	—	1.00
1600	—	1.50
2000	—	2.00[b]

Note: For SI units, 1 in.^{2} = 645 sq mm^2.

[a]Total cross-sectional area of both side rails for ladder or trough cable trays; or the minimum cross-sectional area of metal in channel cable trays or cable trays of one-piece construction.

[b]Steel cable trays shall not be used as equipment grounding conductors for circuits with ground-fault protection above 600 amperes. Aluminum cable trays shall not be used as equipment grounding conductors for circuits with ground-fault protection above 2000 amperes.

Figure 6-23 Reprinted with permission from NFPA 70-1999.

318-3. Uses Permitted. Cable tray installations shall not be limited to industrial establishments.

(a) Wiring Methods. The following shall be permitted to be installed in cable tray systems under the conditions described in their respective articles and sections:

	Section	Article
Armored cable		333
Electrical metallic tubing		348
Electrical nonmetallic tubing		331
Fire alarm cables		760
Flexible metal conduit		350
Flexible metallic tubing		349
Instrumentation tray cable		727
Intermediate metal conduit		345
Liquidtight flexible metal conduit and liquidtight flexible nonmetallic conduit		351
Metal-clad cable		334
Mineral-insulated, metal-sheathed cable		330
Multiconductor service-entrance cable		338
Multiconductor underground feeder and branch-circuit cable		339
Multipurpose and communications cables		800
Nonmetallic-sheathed cable		336
Power and control tray cable		340
Power-limited tray cable	725-61(c) and 725-71(e)	
Optical fiber cables		770
Other factory-assembled, multiconductor control, signal, or power cables that are specifically approved for installation in cable trays		
Rigid metal conduit		346
Rigid nonmetallic conduit		347

Reprinted with permission from NFPA 70-1999.

Cable tray is a support system that has been evaluated and classified by Underwriters Laboratories as to their suitability as equipment-grounding conductors only. In the 1971 *National Electrical Code*® and previous editions, *Article 318* governing cable tray was referred to as "Continuous Rigid Cable Supports." *Section 318-3(c)* limits cable tray to be used as an equipment-grounding conductor in commercial and industrial establishments only where conditions of maintenance and supervision provide qualified persons for servicing the installed cable tray system.

NEC® *Article 318* covers cable tray systems including ladders, troughs, channels, solid-bottom trays, and other similar structures. A cable tray system is a unit or assembly of units, sections, and associated fittings forming a rigid structural system to support cables.

In industrial establishments, again where maintenance and supervision ensure that only qualified persons will service the installed cable tray system, any of the cables listed in the following shall be permitted to be installed in ladder; ventilated trough; 4-inch (102-mm), ventilated, channel-type cable trays; or 6-inch (152-mm), ventilated, channel-type cable trays: (a) Single conductor cable shall be 1/0 or larger and shall be of a type listed and marked on the surface for use in cable trays. Single conductor cable 1/0 through 4/0 must be installed in a ladder-type cable tray with a maximum rung spacing of 9 inches (229 mm) or in a ventilated trough cable tray. Where exposed to direct rays of the sun, cables shall be identified as sunlight resistant welding cables as permitted in *Article 630. (b) Multiconductor cables of Type MV (Article 326)* where exposed to direct rays of the sun shall be identified as sunlight resistant.

Table 318-9. Allowable Cable Fill Area for Multiconductor Cables in Ladder, Ventilated Trough, or Solid Bottom Cable Trays for Cables Rated 2000 Volts or Less

| | Maximum Allowable Fill Area for Multiconductor Cables | | | |
| | Ladder or Ventilated Trough Cable Trays, Section 318-9(a) | | Solid Bottom Cable Trays, Section 318-9(c) | |
Inside Width of Cable Tray (in.)	Column 1 Applicable for Section 318-9(a)(2) Only (in.²)	Column 2[a] Applicable for Section 318-9(a)(3) Only (in.²)	Column 3 Applicable for Section 318-9(c)(2) Only (in.²)	Column 4[a] Applicable for Section 318-9(c)(3) Only (in.²)
6.0	7.0	7–(1.2 Sd)[b]	5.5	5.5–Sd[b]
9.0	10.5	10.5–(1.2 Sd)	8.0	8.0–Sd
12.0	14.0	14–(1.2 Sd)	11.0	11.0–Sd
18.0	21.0	21–(1.2 Sd)	16.5	16.5–Sd
24.0	28.0	28–(1.2 Sd)	22.0	22.0–Sd
30.0	35.0	35–(1.2 Sd)	27.5	27.5–Sd
36.0	42.0	42–(1.2 Sd)	33.0	33.0–Sd

Note: For SI units, 1 in.² = 645 mm².

[a]The maximum allowable fill areas in Columns 2 and 4 shall be computed. For example, the maximum allowable fill, in square inches, for a 6-in. (152-mm) wide cable tray in Column 2 shall be 7 minus (1.2 multiplied by Sd).

[b]The term Sd in Columns 2 and 4 is equal to the sum of the diameters, in inches, of all Nos. 4/0 and larger multiconductor cables in the same cable tray with smaller cables.

Table 318-9(e). Allowable Cable Fill Area for Multiconductor Cables in Ventilated Channel Cable Trays for Cables Rated 2000 Volts or Less

| | Maximum Allowable Fill Area for Multiconductor Cables (in.²) | |
Inside Width of Cable Tray (in.)	Column 1 One Cable	Column 2 More than One Cable
3	2.3	1.3
4	4.5	2.5
6	7.0	3.8

Note: For SI units, 1 in.² = 645 mm².

Table 318-10. Allowable Cable Fill Area for Single Conductor Cables in Ladder or Ventilated Trough Cable Trays for Cables Rated 2000 Volts or Less

| | Maximum Allowable Fill Area for Single Conductor Cables in Ladder or Ventilated Trough Cable Trays | |
Inside Width of Cable Tray (in.)	Column 1 Applicable for Section 318-10(a)(2) Only (in.²)	Column 2[a] Applicable for Section 318-10(a)(3) Only (in.²)
6	6.5	6.5–(1.1 Sd)[b]
9	9.5	9.5–(1.1 Sd)
12	13.0	13.0–(1.1 Sd)
18	19.5	19.5–(1.1 Sd)
24	26.0	26.0–(1.1 Sd)
30	32.5	32.5–(1.1 Sd)
36	39.0	39.0–(1.1 Sd)

Note: For SI units, 1 in.² = 645 mm².

[a]The maximum allowable fill areas in Column 2 shall be computed. For example, the maximum allowable fill, in square inches, for a 6-in. (152-mm) wide cable tray shall be 6.5 minus (1.1 multiplied by Sd).

[b]The term Sd in Column 2 is equal to the sum of the diameters, in inches, of all 1000 kcmil and larger single conductor cables in the same ladder or ventilated trough cable tray with small cables.

Figure 6-24 Reprinted with permission from NFPA 70-1999.

Cable trays shall be installed as a complete system. All field bends and modifications shall be made so that electrical continuity and support of cables is maintained. Each run of cable tray must be completely installed before the installation of cables. Supports must be provided to prevent stress on cables where they enter the raceways or other enclosures from the cable tray system. Where additional protec-tion is required, covers and enclosures providing that protection shall be of a material compatible with the cable tray. Multiconductor cables rated 600 volts or fewer shall be permitted in the same cable tray. Cables rated over 600 volts shall not be installed with cables rated 600 volts or fewer, unless separated by a solid fixed barrier of material compatible with the cable tray, or where cables over 600 volts are Type MC.

a. Underfloor raceways

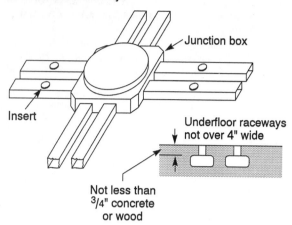

Figure 6-25a Underfloor raceways must be installed in accordance with *NEC*® *Article 354* (Courtesy of American Iron & Steel Institute)

Cable trays are permitted to extend transversely through partitions and walls or vertically through platforms and floors in wet or dry locations where the installations, complete with installed cables, are made in accordance with the requirements of *NEC*® *Section 300-21*. Cable trays must be exposed and accessible except as permitted by *Section 318-6(g)*.

Cable installation shall be as permitted in *NEC*® *Sections 318-8* through *318-13*.

Nonmetallic cable tray is permitted in corrosive areas and in areas requiring voltage isolation.

As per *NEC*® *Section 318-5*, cable trays shall have suitable strength and rigidity to provide adequate support for all contained wiring. They shall not have sharp edges, burrs, or projections that may damage the insulation or jackets of the wiring. They shall be made of a corrosion-resistant material or, if made of metal, shall be adequately protected against corrosion. They shall have side rails or equivalent structural members. Cable trays shall include fittings for changes in direction and elevation of runs. Nonmetallic cable trays shall be permitted in corrosive areas and in areas requiring voltage isolation and shall be made of a flame-retardant material.

OTHER TYPES OF RACEWAYS

Underfloor Raceways

Underfloor raceways are described in *NEC*® *Article 354* and UL 884. Underfloor raceways shall

comply with *Article 300* and *Article 354*. Underfloor raceways are permitted beneath the surface of concrete or other flooring material or in office occupancies where laid flush with a concrete floor and covered with linoleum or an equivalent floor covering. Underfloor raceways shall not be installed where subject to corrosive vapors or in any hazardous location except as permitted in *Article 504*, unless made of a material judged suitable for the condition, or unless corrosion protection approved for the condition is provided.

Underfloor raceways are half-round and flat-top material not over 4 inches in width, and not less than 3/4 inch of concrete or wood is to be placed above the raceway. If raceways are over 4 inches, but not over 8 inches, in width, a minimum 1-inch spacing between raceways shall be covered with concrete to a depth of not less than 1 inch. Raceways spaced less than 1 inch apart must be covered with concrete to a depth of 1 1/2 inches. Trench-type raceways with removable covers are permitted to be laid flush with the floor surface, and the cover plates must provide adequate mechanical protection and rigidity equivalent to junction box covers. For a specific application, refer to *Article 354* of the *National Electrical Code*®.

Cellular Metal Floor Raceways

Cellular metal floor raceways are intended to be installed in accordance with *NEC*® *Article 356* and UL 209. This wiring method consists of hollow spaces of cellular metal floors, together with suitable fittings for electrical conductors. It is not permitted to be installed where subject to corrosive vapors or any hazardous classified location, except as described in *NEC*® *Article 504* and *501-4 (b)*, or in commercial garages other than for supplying ceiling outlets or extensions in an area below the floor but not above.

Cellular metal floor raceways must be installed in compliance with the applicable provisions in *NEC*® *Article 300*. It is limited to No. 1/0 conductors, except by special permission. It must be constructed so that adequate electrical mechanical continuity of a complete system is secured, and it shall provide a complete enclosure for conductors. The interior surfaces must be free from burrs and sharp edges, and the surfaces over which the conductors are drawn must be smooth. Suitable bushings or fittings having smooth, round edges shall be provided where the conductors must pass. For a specific installation, the complete requirements are found in *Article 356*.

Cellular Concrete Floor Raceways

The installation requirements are covered by *NEC® Article 358*. No UL standard covers this product. A cellular concrete floor raceway consists of the hollow spaces in floors constructed of precast cellular concrete slabs, together with suitable metal fittings to provide access to the floor cells. A cell is defined as a single enclosed tubular space in the floor made of precast cellular concrete slabs, the direction of the cell being parallel to the direction of the floor member. A header shall be designed as a transverse metal raceway for electrical conductors providing access to predetermined cells of precast cellular floor, thereby permitting the installation of electrical conductors from the distribution point to the floor cells.

This wiring method must also comply with the applicable provisions of *NEC® Article 300*. It is not permitted where subject to corrosive vapors or in hazardous locations with exceptions (e.g., in commercial garages). It is a limited wiring method for specific use. Connections from the headers to cabinets and other enclosures shall be made by means of metal raceways and approved fittings. All junction boxes must be leveled at floor grade and sealed against the concrete. All junction boxes must be metal and be mechanically and electrically continuous to the headers. No conductors larger than 1/0 shall be installed except by special permission.

NOTE: special permission is defined in *NEC® Article 100* and requires written permission from the authority having jurisdiction.

Raceways shall be limited so that the number of conductors does not exceed 40% of the cross-sectional area of the cell or the header. Splices and taps shall be made only in access units or junction boxes.

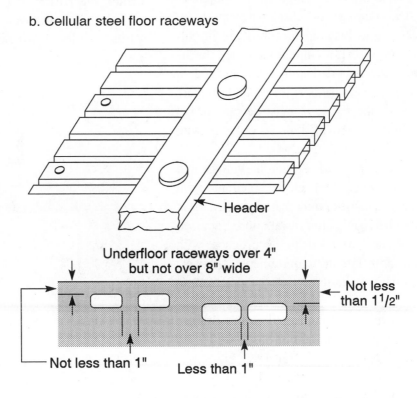

b. Cellular steel floor raceways

Figure 6-25b Cellular steel floor raceways must be installed in accordance with *NEC® Article 356*. (Courtesy of American Iron & Steel Institute)

Where an outlet is abandoned, discontinued, or removed sections of circuit conductors supplying the outlet shall be removed. No splices or reinsulated conductors such as would be the case with abandoned outlets shall be allowed in the raceways.

Busways

Busways are intended to be installed in accordance with *NEC® Article 364* and UL STD 857. A busway is considered to be a grounded metal enclosure containing factory-mounted, bare, or insulated conductors that are usually copper or aluminum bars, rods, or tubes. Busways shall comply with *Article 364* and the applicable provisions of *Article 300*. Busways must be installed only where located in the open and must be visible, except behind panels if access is provided, if no overcurrent devices are installed other than for an individual fixture, if the space behind the access panel is not used for air-handling purposes, if they are totally enclosed non-ventilating types, or if they are installed so that the joints between sections and fittings are accessible, or in accordance with *Section 300-22 (c)*.

Busway shall not be installed where subject to severe physical damage or corrosive vapors, in hoistways, hazardous locations generally, or in outdoor, wet, or damp locations unless identified for the purpose. It must be supported at intervals not exceeding 5 feet unless otherwise designated and marked. All dead ends of the busways must be closed and overcurrent must be provided in accordance with *NEC® Articles 364-10* through *364-14*.

Busways must be marked with the voltage and current rating for which they are designed and with the manufacturer's name or trademark in such a manner that it shall be visible after installation. Enclosed busways must be grounded in accordance with *NEC® Article 250*. For the requirements for over-600-volt nominal installations, see *Part B of Article 364*.

Busways are a preengineered wired system consisting of copper or aluminum buses supported in sheet steel enclosures. Branches from bus phase must be made with busways, rigid metal conduit, intermediate metal conduit, rigid nonmetallic conduit, electrical metallic tubing, flexible metal conduit, or with suitable cord assemblies

---

364–11. Reduction in Ampacity Size of Busway. Overcurrent protection shall be required where busways are reduced in ampacity.

Exception: For industrial establishments only, omission of overcurrent protection shall be permitted at points where busways are reduced in ampacity, provided that the length of the busway having the smaller ampacity does not exceed 50 feet (15.2 m) and has an ampacity at least equal to 1/3 the rating or setting of the overcurrent device next back on the line, and provided further that such busway is free from contact with combustible material.

(Reprinted with permission from NFPA 70-1999.)

---

approved for extra-hard usage for the connection of portable equipment or for the connection of stationary equipment to facilitate their interchange. (For flexible cord installation requirements, see *NEC® 364-8*.)

Busways are available in several general categories, some of which are suitable for locations

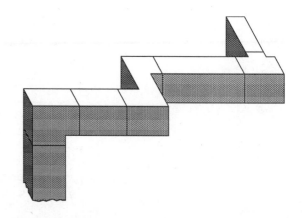

Of sheet metal with bus bars permanently positioned within for mechanical and electrical continuity.
May be used only for exposed work.
Securely supported at intervals not exceeding five feet unless approved for supports at greater intervals.
Branches from busways shall be as permitted by *NEC® Section 364-8*.

Figure 6-26 Busways (Courtesy of American Iron & Steel Institute)

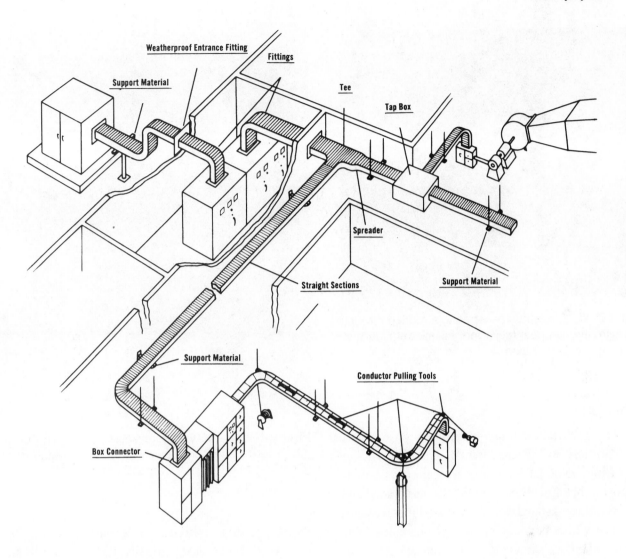

Figure 6-27 Cable bus systems are used for incoming primary feeders and for secondary distribution. (Courtesy of MP Husky Corporation)

where weatherproof, totally enclosed, or ventilated design features are required. Low-impedance feeder buses are generally ventilated and used for long heavy-duty feeders, and in ampacities of 4000 or more, single-phase, three-phase, and three-phase, four-wire systems are available. Busways of other than the low-impedance types usually are not suitable for heavy currents over long distances because their reactance is too high. On runs of 25 feet or fewer, however, they may often be used to advantage. Plug-in distribution busways are used for circuit distribution. They may have covered openings on about 10-inch to 12-inch centers for plug-in type combinations, switch overcurrent units, motors, and other electrical equipment. They are available in ampacities up to 1350 or more.

Trolley-type busways provide continuous contact with enclosed bus bars by means of mobile current collector taps for operation of portable tools and devices. They are available for lighting circuits up to 50 amperes, 250 volts, and for power circuits up to 500 amperes, 575 volts. The length in feet should not exceed three times the ampere rating of the circuit, unless plug and equipment are individually protected by overcurrent devices. The construction of plug-in and trolley-type busways makes them extremely flexible and suitable where changes in layout may be frequent. Sections usually can be reassembled when the busway is relocated. Feeder-type busways should never be installed vertically unless specifically approved for this purpose. Vertical busways require special internal as well as external supports.

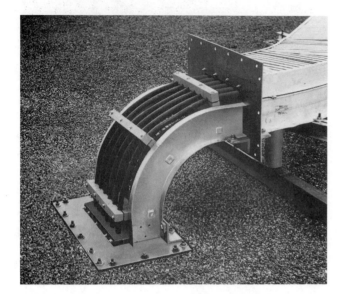

Figure 6-28 Cable bus penetrating a rooftop through an environmental seal (cover not installed). (Courtesy of MP Husky Corporation)

Figure 6-29 4,000 ampere cable bus turning down into a termination enclosure on top of a switch gear. (Courtesy of MP Husky Corporation)

Cable Bus

The installation requirements are covered by *NEC*® Article 365. There is no UL standard to cover this wiring method. Cable bus is a means of distributing power that utilizes conductors that are braced and properly phase-arranged in a ventilated metal enclosure into which conductors are field-installed. The housing includes conductor supports, usually consisting of wood or fiberglass blocks that also provide controlled spacing for the conductors. These support blocks brace these cables in the event of a short circuit and are placed every 3 feet horizontally and every 18 inches vertically. Cable bus is typically used for 800- to 8000-ampere installations and all voltages up to 69kV. It is approved for exposed locations only, and may be used in corrosive and wet locations only if approved for the purpose. The conductors are required to have an insulation rating of 75° C or higher and be #1/0 or larger in size. It is ordinarily assembled at the point of installation from components furnished and/or specified and engineered by the manufacturer in accordance with the instructions (job-site plans and specifications) for a specific job. The allowable ampacity of the conductors shall be in accordance with the standard rating of the overcurrent device or the next ampere rated overcurrent device where the conductor does not correspond with a standard overcurrent device. Except where fire stops are required, it shall be permissible to extend cable bus vertically through floors and platforms in wet locations. Curbs or other suitable means to prevent water flow through the floor or platform openings must be provided and the cable bus is totally enclosed at the point where it passes through the floor or platform for a distance of six feet above the floor or platform. Fire-stops and environmental needs are provided by the manufacturer as a part of the cable bus system when these penetrations are necessary.

Grounding requirements for cable bus are that it shall be electrically bonded by inherent design of mechanical joints or by applying a bonding means, in accordance with *NEC*® *250-96*. The cable bus installation must be grounded in accordance with *NEC*® *Sections 250-80* and *250-86*. Each section of cable bus must be marked with the manufacturer's name or trade design, and the maximum diameter number, voltage rating and ampacity of the conductors to be installed. All markings must be visible after the installation is made.

REVIEW QUESTIONS

1. Name two restrictions that apply when installing exposed electrical nonmetallic tubing (ENT).

2. Is metal cable tray "Listed"? Is nonmetallic tray "Listed"?

3. Name four types of raceways approved for installation in areas subject to physical damage.

4. Under what conditions can liquidtight flexible nonmetallic conduit be installed in lengths longer than 6 feet?

5. Name two types of raceways that may be installed in locations subject to physical damage and severe corrosion.

6. Name three wiring methods approved for installation in hazardous Class I, Division 1 locations.

7. What is the maximum permitted size of flexible metal tubing?

8. What is the maximum permitted length for an auxiliary gutter?

9. Some types of raceways and cables are acceptable grounding conductors, and a separate grounding conductor is not required to be run (included) with the circuit conductors. What *NEC*® article and section lists these types?

10. Is cable tray a raceway?

Chapter 7

Other Wiring Methods

OPEN WIRING ON INSULATORS

Open wiring on insulators shall be installed in accordance with *NEC® Article 320*. No UL standard covers this wiring method. Open wiring on insulators is the oldest wiring method, using cleats, knobs, tubes, and flexible tubing for the protection and support of single insulated conductors run in or on buildings and not concealed by the building structure. It is permitted on systems of 600 nominal volts or fewer in industrial or agricultural establishments, indoor or outdoor, wet or dry locations, and is one of the conductor types specified in *Article 310*. The ampacity of these conductors shall comply with *Section 310-15*. The conductors shall be supported on noncombustible, nonabsorbent insulating materials and shall not be in contact with any other objects.

The conductor supports shall be installed as follows: (1) within 6 inches of a tap or splice, (2) within 12 inches of a dead-end connection to a lampholder or receptacle, and (3) at intervals not exceeding 4 1/2 feet and at closer intervals sufficient to provide adequate support where likely to be disturbed, with the exception that supports for No. 8 and larger across open spaces shall be permitted up to 15 feet apart provided that spacers of a noncombustible, nonabsorbent material are installed between the cables at least every 4 1/2 feet to maintain a 2 1/2-inch space between the conductors.

Where not likely to be disturbed in mill construction, No. 8 and larger are permitted to be run across open spaces if supported from each wood cross member on approved insulators, maintaining 6 inches between conductors, and in industrial establishments where the conditions of maintenance and supervision ensure that only qualified persons service the system, conductors of size 250 kcmil and larger shall be permitted across open spaces supported at intervals up to 30 feet in industrial establishments where serviced by qualified persons. Open conductors shall be separated by at least 2 inches from metal raceways, piping, and other conducting material, and from exposed lighting, power, or signaling conductors, and shall be separated therefrom firmly and continuously. Whenever practical, conductors shall pass over rather than under piping subject to leakage, accumulations, or moisture. Conductors entering or leaving locations because of dampness, wetness, or corrosive vapors shall have drip loops formed on them and shall pass upward on the outside of the building. Conductors within 7 feet of the floor shall be considered exposed to physical damage, and where conductors cross ceiling joists, wall studs, or are exposed to physical damage, they shall be protected in accordance with *NEC® Section 320-14*.

MESSENGER-SUPPORTED WIRING

Messenger-supported wiring shall be installed in accordance with *NEC® Article 321*. No UL standard covers this wiring method. Messenger-supported wiring is an exposed wiring support system using a messenger wire to support insulated conductors with rings and saddles for conductor support,

143

field-installed lashing material, factory-assembled aerial cable, and multiplex cables utilizing bare cable that are factory-assembled and twisted with one or more insulated conductors, such as duplex-, triplex-, or quadruplex-type construction. The following types of cables are permitted in messenger-supported wiring under the conditions described in the article for each.

> **NOTE:** When using a cable type or conductor type listed here, caution must be observed because both this article and the article covering that cable type must be considered and complied with.

The acceptable types are mineral-insulated sheath cable, as per *NEC® Article 330*; metal-clad cable, as per *Article 334*; multiconductor service-entrance cable, as per *Article 338*; multiconductor underground feeder and branch-circuit cable, as per *Article 339*; power and control tray cable, as per *Article 340*; power-limited tray cables, in accordance with *Sections 725-61(c)* and *725-71(e)*; and other factory-assembled, multiconductor control, signal, and power cables identified for this use.

In industrial establishments where conditions of maintenance and supervision ensure that only competent individuals service the messenger-supported wiring, any conductor types given in *NEC® Tables 310-13* or *310-16* or Type MV cable may be used. Where exposed to weather, conductors shall be listed for use in wet locations. Where exposed to the rays of the sun, conductors or cables shall be sunlight resistant. Messenger-supported wiring is permitted in hazardous (classified) locations where the contained cables are permitted for such use in *Sections 501-4, 502-4, 503-4, 504-20* and *505-15*. Messenger-supported cable is not permitted in hoistways or where subject to physical damage. The messenger shall be supported at dead ends and at intermediate locations to eliminate the tension on conductors. Conductors shall not be permitted to come into contact with messenger supports or any of the structural members, walls, or pipes. Messenger wire must be grounded in accordance with

Article 250. All conductor splices and taps must be made and insulated by approved methods.

CONCEALED KNOB-AND-TUBE WIRING

Concealed knob-and-tube wiring shall be installed in accordance with *NEC® Article 324*. No UL standard covers this wiring method. Concealed knob-and-tube wiring is a wiring method using knobs, tubes, and flexible nonmetallic lume or tubing for protecting single-support insulated conductors concealed in hollow spaces of walls and ceilings. This is one of the oldest types of wiring method. This method served a very useful purpose in the early days of the electrical industry, but new wiring methods are far superior to this method and provide superior protection from fire and physical damage. Therefore, concealed knob-and-tube wiring is now allowed only for extensions of existing installations or by special permission (see *Definition, Article 100*).

INTEGRATED GAS SPACER CABLE

Type IGS cable shall be installed in accordance with *NEC® Article 325*. No UL standard covers this wiring method. Type IGS cable is a factory assembly of one or more conductors, each individually insulated and enclosed in a loose-fit, nonmetallic flexible conduit as an integrated gas spacer cable rated 0-600 volts, which is permitted for use underground, including direct burial in the earth, as service-entrance conductors, or as feeder or branch-circuit conductors. It is not permitted as an interior wiring or exposed in contact with buildings. The conductors are solid aluminum rods, laid parallel, consisting of one to nineteen 1/2-inch-diameter rods. The minimum conductor size shall be 250 kcmil and the maximum size, 4750 kcmil. Type IGS cable shall be identified as a type suitable for maintaining the gas pressure within the conduit; a valve or cap shall be provided at each length of cable or conduit to check the gas pressure and inject gas into the content. The ampacity of Type IGS cable and conduit shall not exceed the values in *Table 325-14*.

325-11. Bending Radius. Where the coilable nonmetallic conduit and cable is bent for installation purposes or is flexed or bent during shipment or installation, the radii of bends measured to the inside of the bend shall not be less than specified in Table 325-11.

Table 325-11. Minimum Radii of Bends

Conduit Trade Size (in.)	Minimum Radii	
	in.	mm
2	24	610
3	35	889
4	45	1143

Reprinted with permission from NFPA 70-1999

325-14. Ampacity. The ampacity of Type IGS cable and conduit shall not exceed values shown in Table 325-14 for single conductor or multiconductor cable.

Table 325-14. Ampacity of Type IGS Cable

Size (kcmil)	Amperes	Size (kcmil)	Amperes
250	119	2500	376
500	168	3000	412
750	206	3250	429
1000	238	3500	445
1250	266	3750	461
1500	292	4000	476
1750	344	4250	491
2000	336	4500	505
2250	357	4750	519

Reprinted with permission from NFPA 70-1999

MEDIUM-VOLTAGE CABLE

Medium-voltage cable shall be installed in accordance with *NEC® Article 326* and *UL 1072*. Type MV cable is a single- or multiconductor, solid, dielectric, insulated cable rated 2001 volts or higher. Type MV cable is permitted in power systems up to 35,000 volts, nominal, in wet or dry locations, in raceways, in cable trays as specified in *Section 318-3,* directly buried in accordance with *Section 300-50,* and in messenger-supported wiring. It is not permitted to be exposed to direct sun or in cable trays, unless identified for the use. The ampacity of Type MV cable shall be in accordance with *Section 310-15.* Marking requirements shall be as required in *Section 310-11.* Ampacity shall be in accordance with *Section 310-15* except when installed in cable tray the ampacity shall be in accordance with *Section 318-13.*

325-21. Insulation. The insulation shall be dry kraft paper tapes and a pressurized sulfur hexafluoride gas (SF_6), both approved for electrical use. The nominal gas pressure shall be 20 pounds per square inch gauge (psig) (138 kPa gauge).

The thickness of the paper spacer shall be as specified in Table 325-21.

Table 325-21. Paper Spacer Thickness

Size (kcmil)	Thickness	
	in.	mm
250–1000	0.040	1.02
1250–4750	0.060	1.52

Reprinted with permission from NFPA 70-1999

325-22. Conduit. The conduit shall be a medium density polyethylene identified as suitable for use with natural gas rated pipe in 2-in., 3-in., or 4-in. trade size. The percent fill dimensions for the conduit are shown in Table 325-22.

The size of the conduit permitted for each conductor size shall be calculated for a percent fill not to exceed those found in Table 1, Chapter 9.

Table 325-22. Conduit Dimensions

Conduit Trade Size (in.)	Outside Diameter		Inside Diameter	
	in.	mm	in.	mm
2	2.375	60	1.947	49.46
3	3.500	89	2.886	73.30
4	4.500	114	3.710	94.23

Reprinted with permission from NFPA 70-1999

FLAT CONDUCTOR CABLE

Type FCC cable shall be installed in accordance with *NEC® Article 328.* No UL standard covers this wiring method. Type FCC cable was first accepted by tentative interim amendment (as defined in *Chapter 1*) to the 1978 *NEC®* as a new article. Type FCC cable consists of three or more flat copper conductors placed edge to edge, separated, and enclosed within an insulated assembly. It is a complete wiring system for branch circuits designed for installation under carpet squares. These carpet squares must be installed on hard, sound, smooth, continuous floor surfaces of concrete, ceramic or composite flooring, wood, or similar materials, or on wall surfaces in surface metal raceways. They are not permitted to be used outdoors or in wet locations or where subject to corrosive vapors. They are not permitted in residential, school, or hospital buildings. The carpet squares concealing Type FCC cables must be no larger than 36 inches square and must be adhered to the floor with a release-type adhesive. For the specific installation and construction specifications, see *Article 328* of the *National Electrical Code®*.

MINERAL-INSULATED, METAL-SHEATHED CABLE

MI cable was developed by a French firm in 1934. They produced the first fire resistant mineral insulated copper sheathed for commercial production. It caught on and there was a ready market in ships as shipboard cable during the war (World War II) and the uses spread to industrial and large commercial facilities. By 1940 this product was being produced in North America.

Type MI cable shall be installed in accordance with *NEC® Article 330*. In a 1951 fact-finding investigation, Underwriters Laboratories concluded that MI is a viable wiring method. UL lists any MI-type cable that meets criteria in the investigation. Type MI cable is a factory assembly of one or more conductors insulated with a highly compressed, refractory mineral insulation enclosed in a liquidtight and gas-tight continuous copper or alloy steel sheath. It is presently labeled in one-conductor sizes 16 AWG through 500 kcmil, two- and three-conductor sizes 16 AWG through 4 AWG, four-conductor sizes 16

AWG through 6 AWG, and seven-conductor sizes 16 AWG through 10 AWG. It is widely permitted for services, feeders, and branch circuits in many locations and under a wide variety of conditions, as permitted in *Section 300-3*. It is not permitted where exposed to destructive corrosive conditions, except where protected by materials suitable for the conditions, such as solid copper or nickel-clad copper with a resistance corresponding to standard AWG sizes.

MI cable shall be supported securely at intervals not exceeding 6 ft by straps, staples, hangers, or similar fittings so designed and installed that the cable is not damaged, or installed in cable tray. When bending MI cable, the bend shall be so made as not to damage the cable. The radius of the inner edge of any bend shall not be less than five times the external diameter of the ¾-inch or smaller cable and not less than ten times the diameter of cable greater than 3 / 4 inch but not more than 1 inch.

Fittings used for connecting MI cable to boxes or other equipment shall be identified for such use. Where single-conductor cables enter ferrous metal boxes or cabinets, the installation shall comply with *Section 300-20* to prevent inductive heating. Where single-conductor cables are used, all phase conductors and, where used, the neutral conductor shall be grouped together to minimize induced voltage on the sheath. Where Type MI cable terminates, a seal shall be provided immediately after stripping to prevent the entrance of moisture into the insulation. The con-

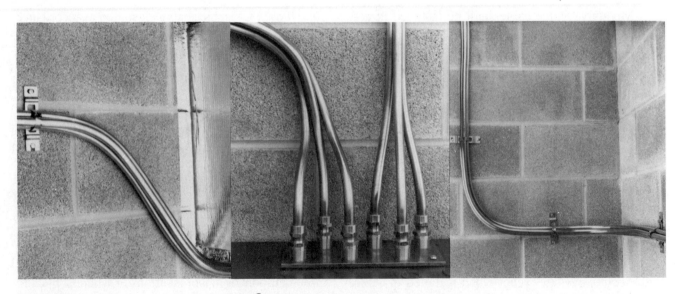

Figure 7-1   MI cable installed as per *NEC® Article 330* (Courtesy of Pyrotenax USA Inc., photos by Robert Stewart Associates)

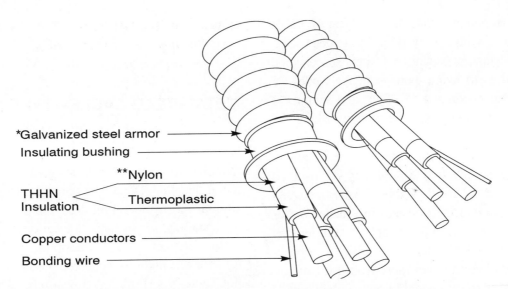

*Galvanized steel armor

Insulating bushing

**Nylon

THHN Insulation — Thermoplastic

Copper conductors

Bonding wire

*Armor must be acceptable as an equipment-grounding conductor in accordance with *NEC® 250-118*
**Conductors individually wrapped in kraft paper or other acceptable materials

Figure 7-2 Type AC cable; the armor is acceptable as an equipment-grounding conductor in accordance with *NEC® 250-91(b)* and is identifiable from Type MC cable, the enclosed conductors are individually wrapped with kraft paper or other suitable materials. (Courtesy of AFC/A Monogram Company)

ductors extending beyond the sheath shall be individually provided with an insulating material.

> **Note:** It is very important that the ends be sealed at all times, including during storage, to prevent moisture from entering

It is a highly compressed refractory mineral that provides proper spacing for conductors. The outer sheath shall be of a continuous construction to provide mechanical protection and moisture seal. Where made of copper, it shall provide an adequate path for equipment-grounding purposes. Where made of steel, an equipment-grounding conductor in accordance with *NEC® Article 250* shall be provided.

ARMORED CABLE—TYPE AC

Armored cable is intended for use in accordance with *NEC® Article 333* and UL 4. Armored cable is the oldest cable assembly developed in this country. It was listed by Sprague Electric before the turn of the century (1899) and is still commonly referred to today as BX cable, the *B* standing for one of the two types made by Sprague Electric, the *X* standing for experimental. This designation "BX," a

registered trade name of General Electric Co., has stayed with this cable through all these years. Type AC cable first appeared in the *NEC®* in the 1903 edition. It consists of an armor of flexible metal tape with insulated conductors enclosed. It must have an internal bonding strip of copper or aluminum in intimate contact with the armor for its entire length. Type AC cable is made in both steel-armored sheathing and aluminum-armored sheathing. The aluminum-armored sheathing is manufactured with conductor sizes No. 14 through No. 1 AWG up to 600 volts, AC phase-to-phase (277 volts-to-ground or fewer). It is suitable for alternating-current (AC) circuits only. AC cable is listed in sizes No. 14 through No. 1 AWG copper, or No. 12 through No. 1 AWG aluminum or copper-clad aluminum, and is rated for not over 600 volts.

> **Note:** Type AC armored cable shall provide an adequate path for equipment grounding as required by *Sections 250-2(d)* and *250-118(g)*.

Type AC cable is distinguishable from Type MC cable by the internal, continuous bonding strip enclosed, and the enclosed conductors are individually wrapped with a kraft paper, whereas Type MC cable's conductors are aggregately enclosed in a mylar-type wrapping. Installations of Type AC cable shall be secured with approved fastening means: staples, straps, hangers, or similar fittings so designed and installed as to avoid damage to the cables, at intervals not exceeding 4 1/2 feet, and within 12 inches of every outlet box, junction box, cabinet, or fitting.

AC cable may be supported by cable tray in accordance with *Article 318*, and may be installed without support: where the cable is fished between access points in concealed spaces in finished buildings and supporting is impracticable; where lengths of not more than 2 ft at terminals where flexibility is necessary; where it is not more than 6 ft from an outlet lighting fixture; or within an accessible ceiling.

Bends shall be made to avoid damaging the cable, and the radius of the inner edge of any bend shall not be fewer than five times the diameter of the cable. At all termination points Type AC cable shall terminate in fittings approved to protect the wires from abrasion. An approved insulated bushing or its equivalent approved for this protection shall be provided between the conductors and the armor. The fittings shall be listed as required in *Section 300-15(c)*.

Where installed through studs, joists, or rafters, the installation shall be made in compliance with *NEC® Section 300-4*. Where exposed the cable shall follow the surface of the building finish or be on running boards, generally. Installations of Type AC cable are permitted for branch circuits and feeders in both exposed and concealed work and in cable trays. AC cable marked with the suffix LS is flame retardant and has limited smoke characteristics. It is permitted in dry locations and for embedding in plaster finish on brick or other masonry, except in damp or wet locations. AC cable is permitted to be fished in air voids of masonry, block, or tile walls. Type AC cable shall not be used where exposed to weather or continuous moisture for underground runs in raceways embedded in masonry, concrete, or fill, or where exposed to oil or other conditions having a deteriorating effect on the insulation. AC cable is not permitted in theaters and similar locations except as provided in *Article 518* for places of assembly. It is not permitted in motion picture studios, hazardous locations (except as permitted in *NEC® Chapter 5*), where exposed to corrosive fumes or vapors, or on

cranes or hoists. Generally a Type AC cable is not permitted in storage battery rooms.

METAL-CLAD CABLE—TYPE MC

Type MC cable shall be installed in accordance with *NEC® Article 334* and UL 1569. MC cable is manufactured in several designs: (a) interlocked metal sheathing tape, (b) corrugated tube, and (c) smooth metal tube. Type MC cable that is suitable for installation in hazardous locations is not covered in this chapter. A full description and recognized uses for this highly specialized product can be found in Chapter 9, "Hazardous Locations." The type with an interlocking metal tape must be provided with an internal equipment-grounding conductor. The sheathing is not permitted as the grounding conductor, but the smooth or corrugated tube type may be acceptable as an equipment-grounding conductor. An internal equipment-grounding conductor when provided shall be used in combination with the tube for suitable grounding. Cables suitable for use in cable trays, direct sunlight, or direct burial applications shall be so marked.

This wiring method is acceptable and permitted to be used for services, feeders and branch circuits, power lighting, control and signal circuits, indoors or outdoors, exposed or concealed, in any approved raceway, as open runs, as an aerial cable on a messenger, in hazardous locations where permitted in *Chapter 5*, any wet locations (provided the cable is approved for that location), with a metallic covering that is impervious to moisture, or with insulated conductors under the metallic covering approved for wet locations. This wiring method shall not be used where exposed to destructive, corrosive conditions, except where protected by a material suitable for those atmospheric conditions. It shall be secured and supported at least every 6 feet. Cables containing four or fewer conductors sized no larger than No. 10 shall be secured within 12 inches of every outlet, junction box, cabinet, or fitting. When used as service-entrance cable, it must comply with *NEC® Article 230*. MC cable shall be installed, and all bends made, in a manner that does not damage the cable. The radius curve of any bend shall not be less than as shown in *Section 334-11*. Fittings and connections for MC cable shall be identified for use and installed in compliance with *Article 300*.

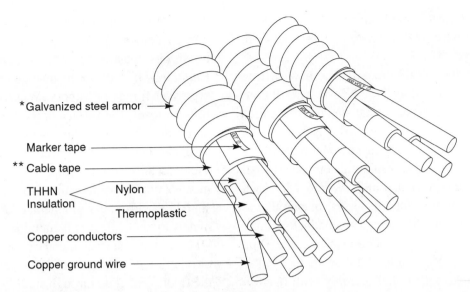

*Galvanized steel armor

Marker tape

** Cable tape

THHN Insulation — Nylon

Thermoplastic

Copper conductors

Copper ground wire

*Armor not suitable as a grounding conductor. An equipment-grounding conductor must be run enclosed with conductors.
**All conductors wrapped together with nylon or other suitable materials.

Figure 7-3 Type MC Cable; Armor is not suitable as an equipment-grounding conductor. An equipment grounding conductor must be enclosed, and is identifiable from Type AC cable, the enclosed conductors are wrapped together with mylar or other suitable materials. (Courtesy of AFC/A Monogram Company)

Caution: When used for service conductors, Type MC cable should comply with *NEC® Article 250* for sizing both the neutral (grounded conductor) and equipment-grounding conductor within the cable. Where conductors are run in parallel in multiple-cable assemblies as permitted in *Section 310-4*, the equipment-grounding conductor where used shall be run in parallel and each equipment-grounding conductor shall be sized on the basis of the ampere rating of the overcurrent device protecting the circuit conductors in accordance with *Table 250-118(11)*. This may pose a problem with the manufacture of Type MC cable. However, it is imperative, following good engineering practices, that each equipment-grounding conductor be large enough to handle any encountered fault. The grounded (neutral) conductor may not pose a problem but must comply with *NEC® Article 250*.

NONMETALLIC-SHEATHED CABLE—TYPES NM AND NMC

Nonmetallic-sheathed cable shall be installed and used in accordance with *NEC® Article 336* and UL Standard 1719. Type NM cable, developed by the General Cable Co. first appeared in the *NEC®* in the 1926 supplement and was listed the same year. Type NM cable is commonly referred to as "Romex" by most field installers; "Romex" is a registered trade name of Rome Cable Co. When Type NM cable was first developed, it was primarily a 2-wire, branch-circuit material; today it is manufactured in sizes 14 through #2 AWG copper and may be made in aluminum or copper-clad aluminum sizes 12 through #2 AWG. Until the 1984 edition of the *NEC®*, NM cable was manufactured as a 60°C wiring method. In 1984 the *NEC®* changed the requirement, owing to the energy conservation practices for new homes, which require much greater insulation R-values in residential installations. This insulation and tighter-

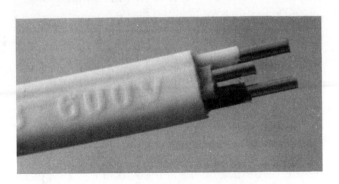

Figure 7-4 Type NM Cable, commonly known as "Romex" (Courtesy of Southwire Company)

built structures created the need for higher-temperature wire because of the higher-temperature attics in these structures.

In 1984 the *NEC*® required a new wiring method, which is listed as Type NM-B wire. It contains 90°C conductors; however, the *NEC*® requires that even though they have 90° insulation, the ampacity must be calculated as a 60° wiring method. *Section 336-26* states that insulating conductors shall be of one of the types listed in *Table 310-13*, which are suitable for branch-circuit wiring, or one that is identified for use in NM cable. The conductors must be rated at 90°C (194°F). The ampacity of the Type NM and Type NMC cable shall be that of 60°C (140°F) conductors and shall comply with *Section 310-15*, which covers the ampacity of conductors.

Many manufacturers produce other cable types permitted to be installed indoors as per *NEC*® *Article 336* and also manufacture cable to meet these requirements, such as Type UF-B cable, which is readily available today. Type NM cable is most commonly used for residential installations. However, *Article 336* has no limitation as to residential work. It is permitted for use in one- and two-family dwellings, multifamily dwellings, and other structures except as prohibited elsewhere in the Code. It is permitted for use both exposed and concealed in normally dry locations. It is permitted to be installed or fished in air voids of masonry, block, or tile where such walls are not exposed or subject to excessive moisture. It may not be used, however, in a structure exceeding three floors above grade, other than one- and two-family dwellings.

NOTE: For the purpose of this article, the first floor of a building shall be that floor which has 50% or more of the exterior wall-surface area level with or above finished grade.

One additional level, that is, the first level, which cannot be used for human habitation but only for vehicle parking, storage, or similar use, shall be permitted. It cannot be used as service-entrance cable or in commercial garages having hazardous (classified) locations as provided in *NEC*® *Section 511-3*, or in theaters or similar locations except as provided in *Article 518* for places of assembly. *NEC*® cable is not permitted in motion picture stu-

dios, storage battery rooms, hoistways, embedded in poured cement, concrete, or aggregate, or in any hazardous location except as permitted in *NEC*® *Chapter 5*. It shall not be installed where exposed to corrosive fumes or vapors. When exposed, it must follow closely the surface of the building finish or be on running boards. It must always be protected from physical damage where necessary by conduit, electrical metallic tubing, pipe, guard strips, or other means. Where passing through the floor, the cable shall be enclosed in rigid metal conduit, intermediate metal conduit, electrical metallic tubing, Schedule 80 PVC, listed surface metal or nonmetallic raceway, or other metal pipe extending at least 6 inches above the floor. When run through studs, joists, or rafters, it must be protected in accordance with *Section 300-4*. In accessible attics, it must be protected in accordance with *Section 333-12*. It shall be permitted to run at angles in unfinished basements where not smaller than #6 or #8 conductors are used. Where run parallel to the joists, cable of any size shall be secured to the sides or faces of the joists.

Type NM cable shall be secured by staples, straps, or similar fittings so designed and installed as not to damage the cable and shall be secured at intervals not exceeding 4 1/2 feet and within 12 inches of every cabinet, box, or fitting. Some exceptions to this rule are found in *NEC*® *Section 336-18*. NM and NMC non-metallic-sheathed cables are rated 600 volts and are listed in sizes 14 through 2 AWG. These cables are surface marked with a suffix letter *B*, immediately following the type letters, to designate the use of conductors with 90°C insulation. The ampacities of these cables are the same as 60° conductors as specified in the latest edition of the *National Electrical Code*®, *Section 336-26*.

SERVICE-ENTRANCE CABLE— TYPES SE AND USE

Service-entrance cable Types SE and USE are intended to be installed in accordance with *NEC*® *Article 338* and UL Standard 854. Service-entrance cable is a single- or multiconductor assembly with or without an overall covering. It has long been used in the construction industry for service drops on residential and small commercial installations where permitted by local codes, for service entrance from

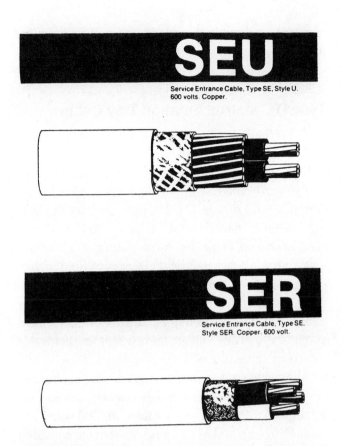

Figure 7-5 Type SER and USE cable available with copper or aluminum conductors (Courtesy of Southwire Company)

the service drop to the meter fitting, for the service entrance from the meter to the main disconnecting means, from the main disconnecting means for feeders to additional subpanels, and for branch circuits where approved. Type SE has a flame-retardant, moisture-resistant covering. Type USE is identified for underground use.

Caution: It is not required to have a flame-retardant covering and therefore may not be permitted within the building. However, manufacturers today may manufacture USE with the flame-retardant covering.

SE and USE are permitted to have a bare grounded conductor. However, some SE cables, Type SER, are manufactured with three insulated conductors and a bare, equipment-grounding conductor. It is permitted to be installed as per *NEC® Article 230* and shall be permitted in interior wiring systems where all the circuit conductors of the cable are rubber covered

and thermoplastic. SE cable without individual insulation on the grounded conductor may not be used as a branch-circuit or feeder conductor except as a final nonmetallic outer covering supplied by a circuit of not over 150 volts-to-ground. *Article 250-140* permits this conductor to supply existing range, wall-mounted oven, counter-mounted cooking units, or clothes dryers where the branch circuit originates at the service panel. Not a recommended engineering practice, it is no longer permitted for new installations. When Type SE service-entrance cable is used for interior wiring, it must be installed in accordance with the provisions of *Article 336* and the applicable provisions of *Article 300*.

UNDERGROUND FEEDER AND BRANCH-CIRCUIT CABLE —TYPE UF

Underground feeder and branch-circuit cable shall be installed in accordance with *NEC® Article 339* and UL Standard 493 and meets or exceeds UL Standard 83 for nonmetallic sheath cable (NMB). This wiring method, designed for direct burial, is manufactured in sizes 14 through 4/0 conductors and may be manufactured as a single conductor or as a cable assembly of multiple conductors. Where single conductor cables are installed, all cables of the feeder, subfeeder circuit, or branch circuit, including the neutral and equipment-grounding conductor, shall be run together in the same trench or raceway and shall meet the underground requirements in *Section 300-5* for burial depths. Installed as an interior wire method in wet, dry, or corrosive locations under the recognized wiring methods of the *NEC®*, or as nonmetallic sheath cable, the installation and conductor requirements shall comply with the provisions of *Article 336* and be only of the multiconductor types. One exception can be found in *NEC® Article 339-3*.

Type UF cable is not permitted for use as service-entrance cables, in commercial garages, theaters, motion picture studios, storage battery rooms, hoistways, hazardous locations, installed in poured concrete or aggregate, or where exposed to the direct rays of the sun unless specifically identified as sunlight resistant. This wiring method is most com-

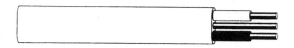

Figure 7-6 Type UF cable jacketed with sunlight-, moisture-, and fungus-resistant Polyvinyl chloride (PVC). (Courtesy of Southwire Company)

monly used as branch-circuit or feeder conductors outside; it is used to supply loads such as post lamps, pumps, and other loads or apparatuses fed from a distribution point. Residential buildings and farmsteads are the most common use for this product.

POWER AND CONTROL TRAY CABLE—TYPE TC

This wiring method is manufactured for installation in accordance with *NEC® Articles 340* and *318*. It is manufactured to meet or exceed UL Standards 44, 83, and 1277, and when used in hazardous locations, *Chapter 5* of the *National Electrical Code®*. It is available in sizes 8 through 750 kcmil, two conductors or greater; as power and control cable #16 through #10, two conductors or more; and as combination power and control cable power sizes No. 12 through No. 2 AWG and control sizes No. 14 and No. 12 AWG.

Type TC tray cable may be used for power, lighting, control and signaling circuits, cable trays or raceways, or where supported by messenger wire. Type TC cable is also permitted in classified locations as per *NEC® Article 318* and *Chapter 5* and in industrial establishments where the conditions of maintenance and supervision ensure that only qualified persons service the installation. It is also permitted for Class I circuits, as per *Article 725*.

This wiring method cannot be used where exposed to physical damage or where installed as an open cable on brackets. It is not permitted to be exposed to direct sun unless identified as sunlight

resistant, or directly buried unless identified for that use.

Type ITC Instrumentation Tray Cable

ITC cable was added to the 1996 *NEC®* as a new *Article 727*. Although this wiring method is not new to industrial applications, until now it has been installed as a power-limited wiring method and installed in accordance with *Article 725*. The proponents of this wiring method have stated a need for this new state-of-the-art cable for instrumentation technologies for low energy electrical/electronic circuitry, like process control, process monitoring and measurement applications (involving intrinsically safe apparatus, non-incendive circuits, building management systems, etc.). The voltage and current levels for this cable is a maximum of 150 volts, and 5 amps or less; the allowable sizes are no. 22 through no. 12. The conductor material is copper or a thermocouple alloy. Insulation on the conductors are rated for 300 volts. The cable shall be listed as being resistant to the spread of fire, the outer jacket shall be sunlight- and moisture-resistant. Where a smooth metallic sheath, continuous corrugated metallic sheath, or interlocking tape armor is applied over the nonmetallic sheath, an overall nonmetallic jacket shall be permitted but not required. Shielding is permitted and bends shall be made so as not to damage the cable. Type PLTC (power-limited tray cable) where installed in accordance with the requirements of Article 727 shall be permitted to be substituted; however, it must be permanently marked at the terminations.

Type ITC cable is permitted:

1. In industrial establishments where the conditions of maintenance and supervision ensure that only qualified persons will service the installation In cable trays, In raceways.
2. In hazardous locations as permitted in *Articles 501, 502, 503, 504,* and *505*. It may also be used as open wiring where equipped with a smooth metallic sheath, continuous corrugated metallic sheath, or interlocking tape armor applied over the nonmetallic sheath. It must be supported and secured at intervals not exceeding 6 feet.
3. Without a metallic sheath or armor—it shall

TYPE TC TRAY CABLES
POWER & CONTROL 600-VOLT
DIRECT BURIAL— SUNLIGHT RESISTANT

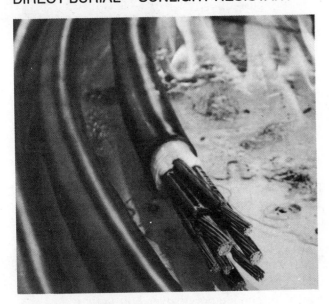

AVAILABLE CONSTRUCTIONS (Jackets)
Polyvinyl Chloride (PVC)
Chlorosulfonated Polyethylene (Hypalon)
Polychloroprene (Neoprene)
Thermoplastic Elastomer (TPE)
Chlorinated Polyethylene (CPE)

AVAILABLE CONSTRUCTIONS (Design Options)
Stranding class
Conductor coating
Color coding
Grounding conductors
Shielding systems
VW - 1 rated individual conductors
"Gas Blocked"
210,000 BTU Rated Cables

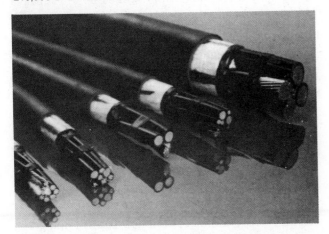

Figure 7-7   Type TC Cable: A factory assembly of two or more insulated conductors, with or without associated bare or covered grounding conductors under nonmetallic sheath, approved installation in cable trays, in raceways, or where supported in outdoor locations by messenger wire. (Courtesy of Triangle Wire and Cable Company)

be permitted to be installed as open wiring between cable tray and utilization equipment in lengths not to exceed 50 ft (15.24 m), where the cable is supported at 6 ft intervals and protected against physical damage using mechanical protection, such as dedicated struts, angles, or channels.

4. Where it complies with the crush and impact requirements of Type MC cable and is identified for such use, it shall be permitted as open wiring where supported at intervals not exceeding 6 feet between the cable tray and the utilization equipment in lengths not to exceed 50 feet.

5. As aerial cable on a messenger;

6. Direct buried where identified for the use;

7. Under raised floors in control rooms and rack rooms where arranged to prevent damage to the cable.

Type ITC cable is not permitted:

1. To be installed on circuits operating at more than 150 volts or more than 5 amperes

2. With other cables—it shall not be installed with power, lighting, Class 1, or nonpower-limited circuits.

3. Where the governing *NEC*® articles do not contain stated provisions for installation with Type ITC cable, the installation of Type ITC cable with other cables shall not be allowed.

4. Where terminated within equipment or junction boxes and separations are maintained by insulating barriers or other means.

5. Where a metallic sheath or armor is applied over the nonmetallic sheath of the Type ITC cable.

Type ITC, instrumentation tray cable, shall be permitted to be used in industrial establishments where the conditions of maintenance and supervision assure that only qualified persons will service the installation.

NONMETALLIC EXTENSIONS

Nonmetallic extensions are intended to be installed in accordance with *NEC*® Article 342. Nonmetallic extensions are an assembly of two insulated conductors within a nonmetallic jacket or extruded thermoplastic covering. The classification includes both surface extensions, direct mounting on the surface, and aerial cable containing a messenger-supporting cable as an integral part of the cable assembly. All the other applicable provisions of the *NEC®* apply to these installations. These extensions are permitted only from an existing outlet of a 15- or 20-ampere branch circuit that conforms to the requirements of *Article 310*, in exposed dry locations. Nonmetallic surface extensions are permitted for a building that is used for residential or office purposes and does not exceed the height limitations specified in *Article 336*. Aerial cable is permitted in industrial buildings and where the occupancy requires a highly flexible means for connecting equipment.

It is not permitted as aerial cable to substitute for one of the general wiring methods specified in the *NEC®*. It is not permitted in unfinished areas, such as basements, attics, and roof spaces. Nonmetallic extensions are not permitted where the voltage between conductors exceeds 150 volts (300 volts for aerial cable). It is also not permitted where corrosive vapors may be present or where run through floor or partitions or outside the room in which it originates.

UNDERGROUND CABLE IN NONMETALLIC CONDUIT

NEC® Article 343 was added in the 1993 *NEC®* to cover this wiring method, which is said to have been used by utility companies and for airport and highway lighting since the 1960s. It is required to be "listed" and its uses shall be limited to direct burial installations. It may not be installed in exposed locations nor shall it be permitted in buildings, except the conductor portion of the assembly may enter the building for termination purposes in accordance with *Section 300-3*. It is permitted in 1/2-inch through 4-inch sizes. Burial depths are in accordance with *NEC® Section 300-5* and *300-50* and *Tables 300-5* and *300-50*. Conductor terminations shall be in accordance with *Section 110-14* approved for the type of conductor or cable being used. All splices and taps shall be made in boxes or enclosures in accordance with *Article 370*.

Table 343-10. Radius of Conduit Bends

Trade Size (in.)	Minimum Bending Radius (in.)
½	10
¾	12
1	14
1¼	18
1½	20
2	26
2½	36
3	48
4	60

Note: For SI units, in. = 25.4 mm (radius).

Figure 7-8   Reprinted with permission from NFPA 70-1999.

The voltage is not limited; however, conductors of 600 volts or less shall not be run in the same raceway with conductors of over 600 volts.

Although the conduit material is not specified in the *NEC®*, the most common material used is Schedule 40 high-density polyethylene tube per ASTM 1248 with any type of conductor specified by the customer enclosed. They are pre-lubricated so they can be removed. The manufacturer recommends that the ends be sealed during storage and during the installation to keep the inside clean and free of debris. Before being unreeeled the cable should be cut free from the conduit, as it will pull back inside the conduit approximately 1½ feet per 100 feet. The conduit must be cut back to locate the conductors before completing the installation. Bends can be made by hand; it should be run as straight as possible without sharp bends. (See Figure 7-9.)

FLAT CABLE ASSEMBLIES—TYPE FC

Type FC assemblies are installed in accordance with *NEC® Article 363*. A flat cable assembly is an assembly of parallel conductors formed integrally with an insulating material web specifically designed for field installation in surface metal raceway. Flat cable is permitted only for branch circuits that supply suitable tap devices for lighting, small appliances, and small power loads. It must be installed for exposed work only, in locations where it

STEP #1: CORRECT

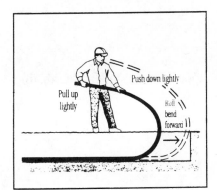

Pull up lightly
Push down lightly
Roll bend forward

STEP #2: CORRECT

Backfill to support under the bend

INCORRECT

Do not exceed minimum bend radius when installing underground cable in nonmetallic conduit. (Courtesy of Southwire Company)

will not be subject to severe physical damage. It is not permitted in locations subject to corrosive vapors, in hoistways, in hazardous locations, out-doors, or in wet or damp locations unless identified for the purpose. For installation requirements, see *Article 363.*

REVIEW QUESTIONS

1. Nonmetallic sheathed cable is limited to installation in buildings of three stories or fewer. What is the definition for a three-story building as related to this limitation?

2. Type UF cable is designed for installation underground but it may also be installed indoors in some locations and under specific conditions. Which *NEC*® article governs Type UF cable when installed indoors?

3. What types of wiring methods do not have to be protected from screws and nails when run parallel to framing members?

4. Type SE service-entrance cable is permitted for use in supplying existing ranges and dryers when the circuit originates in the service panel. What *NEC*® section governs this requirement, and how were the equipment enclosures (frames) permitted to be grounded? What *NEC*® section governs new installations of ranges and dryers?

5. May Type FCC flat conductor cable be installed under a full carpeted floor?

6. Is the sheath of all types of MI cable permitted as an acceptable grounding conductor?

7. Knob and tube wiring methods were used for many years as residential wiring methods, and many electricians still believe them to be good wiring methods. Are they permissible wiring methods for new installations?

8. Types AC and MC cable are very hard to distinguish by appearance. Name three distinct differences in these two wiring methods.

9. Is Type TC cable installation limited to cable trays?

10. When installing Type NM cable in accessible attics, what special precautions must be adhered to and what section governs these installations?

Chapter 8

Boxes, Enclosures, Fittings, and Accessories

After a wiring method has been selected, it is not complete without fittings designed for the purpose, such as connectors, couplings, bushings, elbows, junction boxes, pull boxes, metal enclosures for distribution panels, and similar equipment. Some of the items covered in this chapter may be considered installation details, and little, if anything, may be said about them on plans and in specifications. They are nevertheless an integral part of any good electrical system and should be adequately considered in the initial design stage and preparation of specifications. Failure to do this can result in misunderstanding, confusion, delay, and possible disagreement over extras. Fittings for each wiring method selected must be designed and listed for the purpose for which they are to be used, as per *NEC® Section 300-15 (c)*.

ENCLOSURE APPLICATION AND INSTALLATION GUIDELINES

The increasing use of electrical and electronic equipment for communication, control and security in commercial and industrial applications is making the proper installation of metallic and non-metallic enclosures ever more critical. It is important that enclosure products be applied, installed and used only by qualified engineers, technicians or electricians familiar with the standards, laws, regulations and ordinances associated with the given application.

The *NEC®* classifies areas according to the nature, probability and extent of ignitable flammable hazards that could exist where electrical equipment is installed. The intent of area classification is to

Figure 8-1 Typical junction box installation in compliance with *NEC® Article 370* (Courtesy of Allied Tube & Conduit)

157

prevent fires and explosions that could be caused by electrical equipment serving as an ignition source (arc, spark, high temperature, etc.). The *NEC®* divides applications in two broad categories: Normally Hazardous and Not Normally Hazardous. Since the Code further divides these categories into groups and considerable judgment must be applied, one should refer to the *NEC®* for detailed information on specific atmospheres.

The Canadian Standards Association (CSA), International Electrotechnical Commission (IEC), National Electrical Manufacturers Association (NEMA) and Underwriters Laboratories (UL) issue standards and use rating classifications to establish performance requirements for enclosures. The enclosure classifications and test requirements are specified in NEMA Standards Publication 250, *Enclosures for Electrical Equipment (1000 Volts Maximum)*; UL 50, *Enclosures for Electrical Equipment*; and CSA C22.2 Spec No. 25 and Spec No. 94. Enclosures are classified by standards according to the protection provided against various environmental conditions.

All enclosures are constructed to provide a degree of protection to personnel against incidental contact. General information for the degree of environmental protection provided by various ratings is provided in the table; for detailed information, the respective standards should be consulted. The ratings are applicable to both metallic and non-metallic enclosures.

Proper installation is just as critical as enclosure selection. Sensitive electronics and electrical equipment are routinely used to control or protect very expensive processes and equipment. In such applications enclosure grounding is extremely important for safe and reliable operation. In addition to personnel safety hazards from electrical shock, improper grounding can cause voltage transients and electric charges that result in equipment failure. Lack of or improper grounding in applications sensitive to electromagnetic or radio frequency interference (EMI/RH) could also lead to unwanted operation of equipment.

Anywhere wiring enters an enclosure via a conduit system, the conduit and equipment are required to be electrically bonded to the enclosure. Approved hubs or locknuts are utilized to securely bond the

conduit or cable to the enclosure. The hub or locknut connection must penetrate the coating system on metal enclosures to form a sufficient electrical bond. Note that hubs or locknuts can be provided to make an electrical connection to equipment that may be mounted on a separate ungrounded panel.

Nonmetallic enclosures are nonconductive and thus do not provide bonding between the enclosure and conduit connection. Fiberglass enclosures can be obtained with a stainless steel jam nut installed in the wall to provide a ground contact. Bonding bushings and bonding jumpers are approved connection methods to ground the fiberglass enclosure. The grounding screw provided on the hub can also be used to make an electrical connection between equipment mounted in the enclosure and the conduit. In a similar way the ground sheath of any incoming cables must be bonded with the equipment and enclosure grounds. Mounting panels should be bonded to the enclosure ground.

Note: After equipment selection, grounding is probably the most important element to the safe and successful operation of an electrical system, but it remains one of the least understood and most error prone parts of the electrical system.

Military specification (MIL-STD-285) is most commonly used to test the shielding effectiveness of enclosures. The procedure involves placing a transmitting antenna and a receiving antenna outside the enclosure. Measurements are made with the door open and with it closed. The difference between the "open" and "closed," expressed in dB, is the shielding effectiveness. Measurements are at ten frequency points ranging from 0.01 to 1000 Mhz. Depending on the enclosure design and frequency of the EMI/RFI, the attenuation of a standard steel enclosure without modification will vary between 0 and 20 dB. With (special conductive gaskets, metal screens for windows, etc.) to incorporate EMI/RFI protection, the enclosure can provide attenuation between 40 and 80 dB.

While an ideal choice for harsh corrosive environments, non-metallic enclosures provide no EMI/RFI shielding. Fiberglass enclosures coated internally with a thin coating of highly conductive nickel provide excellent EMI/RFI shielding. The av-

NEMA TYPE	LOCATION	DESCRIPTION
1	Non-Hazardous	Indoor Use: Falling Dirt.
2	Non-Hazardous	Indoor Use: Falling Dirt, Dripping & Light Liquid Splashing
3	Non-Hazardous	Indoor or Outdoor Use: Failing Dirt, Rain, Sleet, Snow & Windblown Dust; Undamaged by External Ice Formation
3R	Non-Hazardous	Indoor or Outdoor Use: Falling Dirt, Rain, Sleet, and Snow; Undamaged by External Ice Formation
3S	Non-Hazardous	Indoor or Outdoor Use: Failing Dirt, Rain, Sleet, Snow & Windblown Dust; Ice Laden External Mechanism Remains Operable
4	Non-Hazardous	Indoor or Outdoor Use: Failing Dirt, Rain, Sleet, Snow
4X	Non-Hazardous	Indoor or Outdoor Use: Failing Dirt, Rain, Sleet, Snow
5	Non-Hazardous	Indoor Use: Failing Dirt, Settling Airborne Dust, Lint
6	Non-Hazardous	Indoor or Outdoor Use: Failing Dirt, Hose-Directed Water, Water Entry during Occasional Temporary Sub- metsion at Limited Depth; Undamaged by External Ice
6P	Non-Hazardous	Indoor or Outdoor Use: Failing Dirt, Hose-Directed Water, Water Entry during Prolonged Submersion at Limited Depth; Undamaged by External Ice Formation
7	Hazardous	Indoor: Class I, Div 1, Groups A, B, C or D
8	Hazardous	Indoor or Outdoor: Class I, Div 1, Groups A, B, C or D
9	Hazardous	Indoor: Class II, Div 1, Groups A, B, C or D
10	Hazardous	Constructed to meet req'ts of Mine Safety & Health Admin, 30 C. R., Part 18
12	Non-Hazardous	Indoor (Without Knockouts) Use: Falling Dirt
12K	Non-Hazardous	Indoor (With Knockouts) Use: Falling Dirt, Circulating Dust, Lint, Fibers and Flyings; Dripping and Light Splashing of Liquids
13	Non-Hazardous	Indoor Use: Failing Dirt, Circulating Dust, Lint, Fibers and Flyings; Spraying, Splashing and Seepage of Water

Enclosure Types for All Locations (Courtesy of Robroy Industries, Inc.)

erage attenuation for an enclosure with this protection is 60 dB in the frequency range from 0.01 to 1000 dB.

Even though an enclosure has been listed for an NEMA rating by UL and CSA, performance can be compromised by violation of some simple application and installation guidelines. The following guidelines are recommended during application and installation:

1. The enclosure should be located to provide conduit or cable entrance on the bottom. Bottom entry minimizes the potential ingress of water and other environmental elements that can cause damage to the internal equipment.

2. If a top or side conduit or cable entrance must be used, the enclosure entrance connection should be selected with the same sealing requirements and ratings as the enclosure. While this is equally true for bottom entry connections, water and other elements are less likely to enter an enclosure connection through a bottom connection unless the enclosure is submersed. Proper installation of connections is critical to performance.

3. When an enclosure is unpacked, the enclosure and particularly the gaskets should be checked for damage. Depending on the application environment, a damaged gasket provides an entry path for water and other environmental hazards. Damaged gaskets should be replaced!

4. The enclosure should be mounted according to the manufacturer's specifications.

5. Electrical bonding and grounding requirements are dependent on the application; the design specifications must be observed to prevent safety and operating hazards. The installation must be inspected and approved by the authority having jurisdiction.

6. To prevent deterioration of an enclosure after installation, areas where the protective coating is damaged should be repaired with an appropriate protective material. Most enclosure manufacturers have touchup paints and coatings specifically available for this purpose.

BUSHINGS

Bushings must be used to protect the wire from abrasion wherever a conduit or tubing enters a box or cabinet of any kind unless the design of the box or

**Insulated Bushings
Polypropylene**

**Bushings
Steel/Malleable Iron**

**Insulated Bushings
Steel/Malleable Iron**

**Insulated Ground Bushings – Lay-In Lug
Steel/Malleable Iron**

**Insulated Ground Bushings – Feed-Thru Lug
Steel/Malleable Iron**

Figure 8-2 Five of the most common types of bushings (Courtesy of Raco, Inc. and Crouse-Hinds)

fitting is such as to provide equivalent protection. Their function is to provide a smooth, rounded surface to protect the conductor insulation from abrasion. Some bushings are nonmetallic; while others are a combination of metal and insulated material. Metallic bushings and lock nuts ensure electrical continuity of a metallic raceway system. When non-metallic bushings are used, lock nuts are required

both inside and outside the box for ground continuity. Lock nuts must be drawn up tight.

A number of sections in the Code refer to the requirements for bushings, such as *NEC® 300-15* and exceptions. Raceway cables used as wiring methods must be installed to keep the conductors free from abrasion, cutting, or damage. Most raceway articles refer to bushings too, as in *Section 346-8* for rigid metal conduit, or *Section 347-12* for rigid nonmetallic conduit. These requirements provide safety and ensure that the conductors will not be damaged during and after construction. Grounding bushings or bonding bushings equipped with a bonding jumper or set screw ensure better electrical continuity between steel raceways and a box or cabinet, and are required for services or where the voltage exceeds 150 volts-to-ground and where concentric or eccentric knockouts are encountered. These requirements are found in *Article 250, Part E* on bonding.

NEC® Section 250-92 requires service-equipment bonding and requires that bonding bushings or threaded hubs be used for the bonding of service equipment. For circuits of over 250 volts-to-ground, the electrical continuity of metal raceways and cables must be ensured by one of the methods found in *NEC® 250-97*; where oversize concentric or eccentric knockouts are encountered, bonding bushings or bonding lock nuts must be used.

Insulated bushings must be provided where ungrounded conductors of #4 AWG or larger enter or leave a box (see *NEC® Section 300-4 (f))*. Such bushings are available for all sizes of raceways and wiring methods. It is good practice to require their installation for conductors smaller than #4 where vibration or dampness is likely to be encountered. Other references may be found throughout the *National Electrical Code®* that require protection of the conductors where the wiring method terminates. Good engineering practices require insulating connectors or fittings at these locations, or, where the conduit or raceway enters the box, a bushing to protect these conductors.

COUPLINGS AND CONNECTORS

The *NEC®* recognizes both threaded and threadless couplings and connectors, as well as threaded fittings, for EMT electrical metallic tubing,

Blue ENT (Electrical Nonmetallic Tubing)
Fittings and accessories

One piece quick connect
coupling

One piece quick connect
threaded male adapters

One piece quick connect
snap-in terminator adapters

Carlon's one-piece ENT quick-connect fittings are suitable for damp locations and are concrete-tight when used with Carlon ENT.

All Schedule 40 fittings are compatible with ENT fittings with the use of ENT cement.

This procedure is recommended for use with Carlon 1 1/4"-2" Flex-Plus® Blue® ENT.

When one-piece, quick-connect, snap-in terminator adapters are installed in a concrete application, Carlon's flat-sealing washers must be used on the box connection ends.

Figure 8-3   (Courtesy of Carlon, A Lamson & Sessions Company)

rigid steel conduit, and IMC intermediate steel conduit. Concrete-tight types must be used for embedding in masonry or concrete, and rain-tight types for wet locations. In all cases, couplings and connectors must be made up wrench-tight to ensure electrical continuity for mechanical integrity. Running thread couplings, with or without lock nuts, are not permitted by the *NEC*®. Conduit threads must be free of nonconductive coatings. EMT couplings must be concrete-tight where buried in concrete, or water-

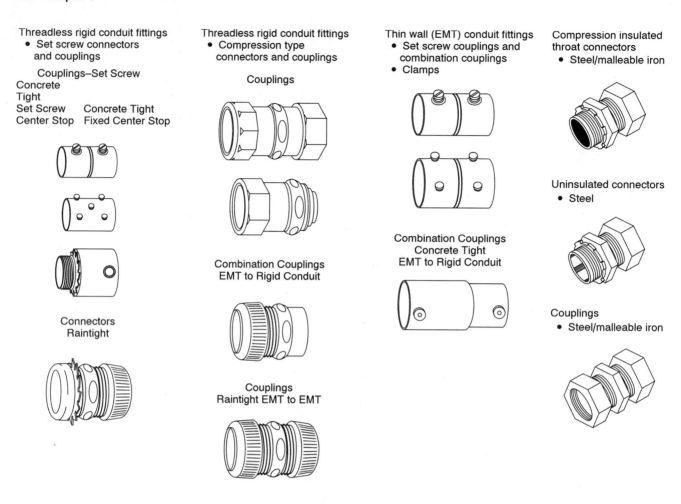

Figure 8-4 Typical metallic raceway fittings (Courtesy of Raco, Inc. & Crouse-Hinds, Division of Cooper Industries)

tight where installed in wet locations. Only thread-less types are used on electrical metallic tubing.

Although these factors are concerned primarily with field workmanship, referring to them in the specifications ensures a good installation. Nonmetallic raceways employ both of the glue-on types, with special installation guidelines for ENT (electrical nonmetallic tubing) on the carton or box. ENT has the solid, snap-on connector and coupling as they are approved concrete-tight with conditions. (See Figures 8-3, 8-4, 8-5, and 8-6.) Connectors for the various other cable wiring methods must be used for the cable type for which they are designed and listed, as required in *NEC® 300-15*. Some connectors may be listed for more than one type of wiring method, such as liquidtight, flexible nonmetallic conduit and

liquidtight, flexible metallic conduit, and types MC and AC cable. The dual listings are marked on the smallest carton for which the connectors are shipped.

CONDUIT BENDS AND ELBOWS

The number of bends or offsets in a conduit run has a direct bearing on the tension needed to pull the conductors into the raceway. Damage to conductor insulation is likely when the number of bends is excessive. For this reason, the number of bends permitted between outlets or pull boxes in a conduit run is limited by the *National Electrical Code®* to the equivalent of four quarter bends, or a total of 360°.

MALE TERMINAL ADAPTERS

For adapting nonmetallic conduits to boxes, threaded fittings, metallic systems. Male threads on one end, socket end on other.

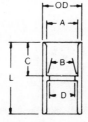

FEMALE ADAPTERS

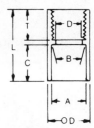

For adapting nonmetallic conduits to threaded fittings, metallic systems. Female threads on one end, socket end on other.

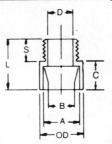

STANDARD COUPLINGS

Socket type for joining nonmetallic conduit.

BOX ADAPTERS FOR ENCLOSURES

Adapts nonmetallic conduit to all electrical enclosures by inserting adapter through knockout and cementing into Carlon couplings.

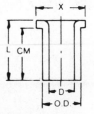

CEMENT JOINTS

Carlon nonmetallic products are joined by means of solvent cement joints. Sizes 1/2″ through 1 1/2″ should be cut square (using a fine tooth handsaw) and deburred. For sizes 2″ through 6″ a miter box or similar saw guide should be utilized to keep the material steady. After cutting and deburring, wipe ends clean of dust, dirt and shavings.

Joining process as follows: Be sure that conduit end is clean and dry. Apply coat of Carlon Solvent Cement (use dauber) to end of conduit, the length of the socket to be attached. Push conduit firmly into fitting while rotating conduit slightly about one-quarter turn to spread cement evenly. Allow joint to set approximately 10 minutes.

Carlon recommends the use of Carlon cement for proper solvent cement joints. Since this cement is prepared particularly for our product compounds and tolerances, we cannot guarantee joints assembled with cement materials supplied by other manufacturers. Regular grade grey solvent cement will accommodate most application situations being of a general purpose nature. In situations requiring an extremely fast-setting joint, (low temperature or difficult installation conditions) Carlon All Weather Quick-Set Cement is recommended. Standard grade clear cement is recommended for noncritical utility applications where gap filling and leak testing are not required.

CEMENT FOR PVC RIGID CONDUIT

STANDARD GRADE CLEAR CEMENT

Recommended installation temperature 40-100°F

MEDIUM GRADE GREY SOLVENT CEMENT

Recommended installation temperature 40-100°F

ALL WEATHER QUICK-SET CLEAR CEMENT

Recommended installation temperature 5-100°F

Figure 8-5   Rigid nonmetallic raceway fittings (Courtesy of Carlon A Lampson Sessions Company.)

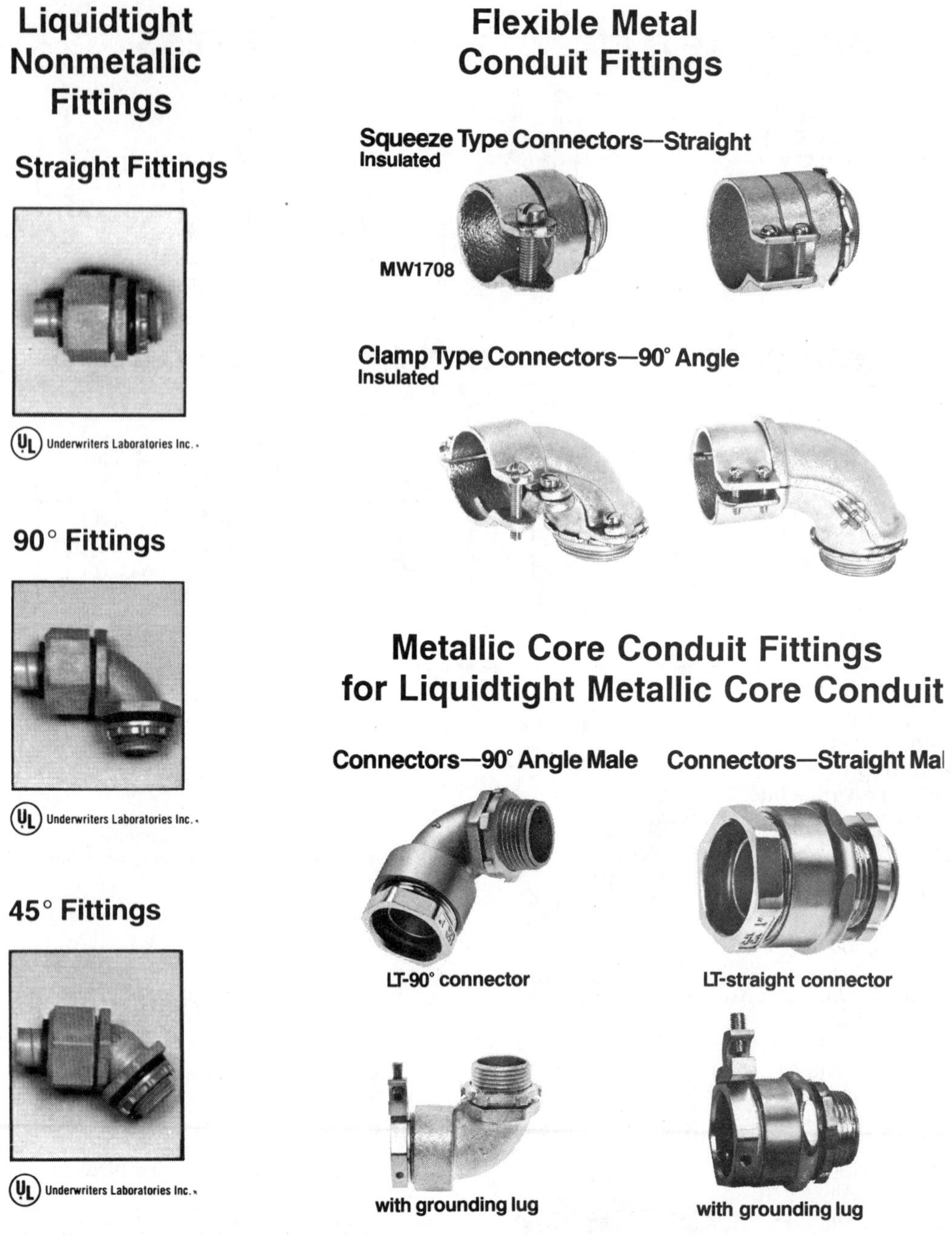

Liquidtight Nonmetallic Fittings

Straight Fittings

(UL) Underwriters Laboratories Inc.·

90° Fittings

(UL) Underwriters Laboratories Inc.·

45° Fittings

(UL) Underwriters Laboratories Inc.·

Flexible Metal Conduit Fittings

Squeeze Type Connectors—Straight
Insulated

MW1708

Clamp Type Connectors—90° Angle
Insulated

Metallic Core Conduit Fittings for Liquidtight Metallic Core Conduit

Connectors—90° Angle Male   **Connectors—Straight Mal**

LT-90° connector   LT-straight connector

with grounding lug   with grounding lug

Figure 8-6   Other types of metallic raceway fittings (Courtesy of Crouse-Hinds, Division of Cooper Industries)

(See Figure 8-8.) The bends may be on more than one plane.

In some cases, it is advantageous to use a greater number of pull boxes than required by the *NEC*® to facilitate the installation of conductors, particularly in long conduit runs. The design engineer may specify where junction boxes are to be placed, or specify maximum lengths between boxes, unless he is willing to

leave such decisions to the contractor. No maximum distance is stipulated in the *NEC*®. However, the location of these junction boxes should be noted on "as-built" drawings at the completion of the job to facilitate maintenance and service of the installation. Factory-made elbows and bends are available in all sizes in all the various raceway methods, and in various radii to fit the conditions of the job. Minimum

Figure 8-7 Factory conduit nipples, elbows, and fittings (Courtesy of Picoma Industries, Inc.)

radii, however, are established for both factory-made and field bends. Field bends include those made in a contractor's shop.

CONDUIT BODIES

A conduit body is defined in *NEC® Article 100* as a separate portion of a conduit or tubing system that provides access with a removable cover to the interior of the system at a junction of two or more sections or at a terminal point of the system. Boxes such as FS and FB types or sheet-metal boxes are not classified as conduit bodies (see *Table 370-16 (a)*). Although this device has been used in raceway systems for many years, the 1975 *NEC®* was the first edition to include regulations for its use. Following the standardization of the dimensions for this type of fitting, *Section 370-16(a)* requires that conduit bod-

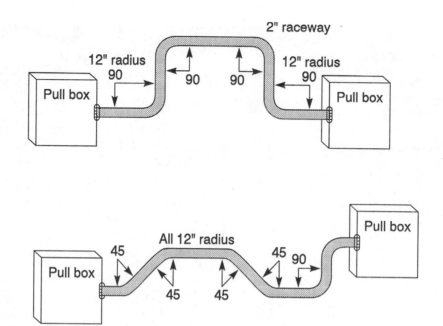

Equivalent of four quarter bends—eight 45° bends,
six 60° bends, or two 90° bends and four 45° bends.

Figure 8-8   Electrical raceway bends between boxes (Courtesy of American Iron & Steel Institute)

ies have a cross-sectional area of not less than twice the cross-sectional area of the largest conduit or tubing to which they are attached, when enclosing #6 AWG conductors or smaller. The maximum number of conductors permitted shall be the maximum number permitted in *Table 1, Chapter 9* for the conduit to which it is attached.

Conduits having provision for fewer than three conduit entries are not permitted to contain splices, taps, or devices unless they comply with the provis-

ions of *NEC® Section 370-16(b)*, which limits the wire fill per cubic inch of space required. In addition, they must be supported in a rigid, secure manner. *NEC® 370-17* requires that all conductors entering a conduit body be protected from abrasion. The manufacturers of conduit bodies generally design these fittings with a smooth, nonabrasive ridge where the conduit enters the body. Conduit bodies are available for all wiring methods and in many configurations to afford neat, easy installation.

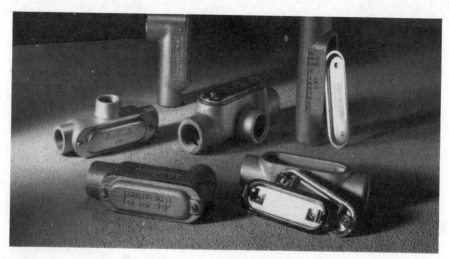

Figure 8-9   Conduit bodies (condulets) shall be installed in accordance with *NEC® Article 370*. (Courtesy of Crouse-Hinds, Division of Cooper Industries)

Figure 8-10   Explosion-proof outlet box GUAX (Courtesy of Crouse-Hinds, Division of Cooper Industries)

OUTLET DEVICE, PULL, AND JUNCTION BOXES

Outlet device boxes are intended to be installed in accordance with UL 514. Outlet device, pull, and junction boxes provide access points for pulling and feeding conductors into either raceways or terminating cable systems. They can be used effectively as T, cross, offset and right-angle junctions for single or parallel groups of the selected wiring method. They are required in electrical wiring methods where there would otherwise be a greater number of bends or offsets than permitted, or for convenience and ease in making the installation. The larger the box, the easier it is to install or position conductors, make taps, or install cable supports.

Field experience has proven that it is inadvisable to skimp on the size or number of boxes. However, an excessive number introduces unnecessary

Figure 8-11   (Courtesy of Carlon, a Lamson & Sessions Company)

Nonmetallic products
For nonmetallic sheathed cable
thermoset set-up & 4" square boxes

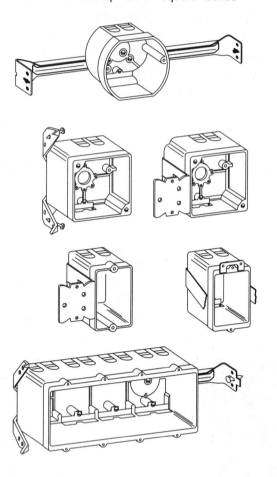

Figure 8-12   Common types of lighting and device outlet boxes (Courtesy of Raco, Inc.)

handling and splicing of conductors. For large-size raceway installations, pull boxes generally are of special design, custom built to fit the job. To ensure sufficient work spaces, such boxes are sized by *NEC® Section 370-28,* which specifies the minimum dimensions for boxes connected to raceways of 3/4-inch trade size or larger and containing conductors or cables made up of conductors size #4 or larger. Boxes of lesser dimensions than those described in *Section 370-28* may be used as pull or junction boxes under some conditions. The attached wiring method must contain less than the maximum allowable number of conductors, and the box must carry a permanent marking to indicate the maximum allowable number and size of conductors that may enter or

terminate in the box. Give careful attention to design to ensure that both these requirements are met.

Many standard boxes are listed in *NEC® Table 370-16 (a).* This table lists the dimensions of these boxes and the maximum number of conductors permitted in each size box if all the conductors are the same size. *Table 370-16 (b)* lists the volume required per conductor where more than one size of wire enters a box. These figures must also be used in determining the allowable number of conductors in boxes of less than 100-cubic-inch capacity other than those listed in *Table 370-16 (a).* This section includes many restrictions to be considered in actual application of these tables. Among them are reductions in the allowable number of conductors when straps (yoke), fixtures, cable clamps, or grounding conductors are contained in the box.

In straight pulls where the conduit enters or leaves on the opposite side of the box, the box length must be at least eight times the nominal diameter of the largest raceway. Good practice, however, suggests a minimum length of twelve times the diameter for ease of handling conductors.

Angle pulls are those where conductors enter one side of the box and leave from other than the opposite side. Application of these rules to a specific condition are given in Figure 8-15 and the accompanying example. The distance between the conduit entry inside the box and the opposite wall of the box must be at least six times the diameter of the largest conduit, plus the sum of the diameter of parallel conduits entering the box. Thus, the box containing only angle pulls may be square if the same number and size of conduits enter and leave the box (dimensions A and B of Figure 8-14). The distance between conduit entries enclosing the same conductor (dimension C of Figure 8-14 must be at least six times the trade diameter of the larger raceway, but it is recommended that ten times the diameter of the largest conduit be used). The depth of the box should be governed by the lock nut dimensions of the conduit involved as given in the *NEC® Appendix, Figure A-8.* Also, ample allowance for installation of any conductor racks or supporting facilities.

Combination boxes involving straight and angle pulls are sized by combining the preceding rules and as illustrated in Figure 8-15. The required minimum box length shown is governed by the size of the

Figure 8-13   Common types of lighting and device outlet boxes (Courtesy of Raco, Inc.)

straight-through conduits (8 × 3 = 24). The required minimum box width is determined by the formula for angle pull (6 × 2 + 2 + 2 = 16). The depth of the unit is governed by the lock nut spacing for the four 3-inch, straight-through conduits. In all cases the junction or pull-box dimensions must be sufficient to provide room for the fittings, lock nuts, and bushings. See Appendix, Figure A-9 for recommended minimum spacing for different sizes of raceways at junction or pull boxes.

To determine the minimum size of a pull box permitted by *NEC®* for cables, assume that the min-

imum size of conduit is permitted for the number and size of conductors in each cable. For example, a Type MC cable containing three #3 type THW conductors is considered to be the same size as a 1 1/4-inch conduit, because that size of conduit is required for this size and type and number of conductors (*NEC® Appendix C).*

Where the number of conductors or cables is sufficient for boxes with a dimension in excess of 6 feet, *NEC® Section 370-28(b)* requires that the box contain facilities for cable and racking conductors in an approved manner. In general, these facilities ne-

a. Straight pull requires at least 8 times the trade diameter of the largest raceway entering the enclosure.

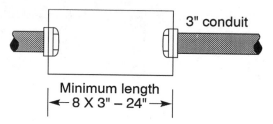

The width and depth of the box is governed by the size and number of the conduit(s) involved. Dimensions of lock nuts and bushings also must be considered (Appendix Fig. A-8)

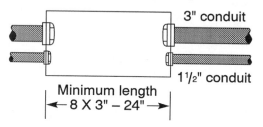

b. Angle or U-pulls require at least 6 times the trade diameter of the largest raceway in a row plus the sum of the diameters of all other raceways entries in the same row or the same wall of the box. Each row shall be calculated individually and the single row that provides the maximum distance shall be used.

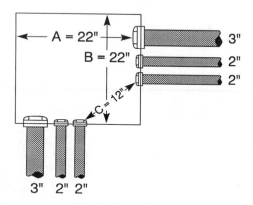

Dimension C = the distance between raceway entries enclosing the same conduits and shall not be less than 6 times the trade diameter of the larger raceway.
All configurations must be calculated and the largest dimension must be used.

Figure 8-14 Pull and junction boxes—minimum size as required in *NEC® Section 370-28* (Courtesy of American Iron & Steel Institute)

cessitate a larger box than is otherwise necessary, but the ease of making up the necessary connections and accomplishing a workmanlike job more than compensates for this difference.

FIXTURE AND FAN SUPPORT

In recent times ceiling fans have again become popular in homes, offices, and warehouses. Because of that popularity, the *National Electrical Code®* has recognized special requirements for hanging and mounting these fans and fan-light combinations. *Section 422-18* requires ceiling fans that do not exceed 35 pounds with or without accessories to be supported by outlet boxes identified for such use and in accordance with *Sections 370-23* and *370-27*. *Section 370-27(c)* states that outlet boxes shall not be used as the sole support for ceiling (paddle) fans except boxes listed for this application. Because of this increased demand, a number of manufacturers have developed installation kits with boxes designed for this support, or brackets designed to carry the weight with boxes. *NEC® 410-16* has similar requirements for light fixtures weighing over 50 pounds. These newly developed box assemblies meet those requirements ideally. (See Figure 8-16.)

CABLES AND CONDUCTORS IN RACEWAY REQUIRED SUPPORTS

Boxes must be included in vertical raceways for installation of suitable conductor supports where the weight of the cable places an excessive strain on the conductor terminals. This is particularly important in high-rise buildings where heavy conductors are involved. *NEC® Section 300-19* requires a support at or near the top of a vertical raceway, plus additional supports at intervals not greater than stated in *Table 300-19 (a)*.

An exception is permitted where the total length of the vertical raceway is less than 25% of the spacing given in the table for the size and kind of conductor involved. In such cases, no support is required. Conduit and conductor supports with split-taper, hard-fiber bushings are available that fit over the end of the conduit riser. The size of the support box is determined by the rules governing pull boxes. Boxes

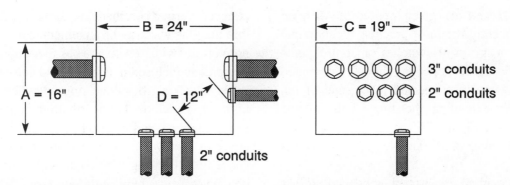

Four 3" rigid steel conduits or EMT, straight pull.
Three 2" rigid steel conduits or EMT, angle pull.
Recommended minimum spacing (Table G) for 3" conduit, 4 3/4"; for 2" conduit, 3 38"; and for 3" to 2" conduits, 4".

For depth of box (A)
Three 2" conduits, angle pull:
A = (6 × 2") + 2" = 2" = 16"
Spacing for two rows of conduit:
Edge of box to center of 3" conduit = 2 3/8"
Center to center, 3" to 2" conduit = 4
Minimum distance, D = 8 1/2
Total = 14 7/8"
Required depth: Use 16"

For length of box (B)
Four 3" conduits, straight pull: B = 8 × 3" = 24"
2" conduit, angle pull: B = A = 16"
Required length: Use 24"

For width of box (C)
Recommended spacing for 3" conduits = 4 3/4"
C = 4 × 4 3/4" = 19"
Required width: Use 19"

Angle pull distance (D)
D = 6 × 2" = 12"
Required distance: Use 12"

Figure 8-15   Pull and junction boxes—minimum size as required in *NEC® Section 370-28* (Courtesy of American Iron & Steel Institute)

used as combined splice and support boxes should be placed with ample space left above the support units to make the splices. Vertically installed raceways can be properly supported by the use of clamp-type floor brackets at suitable intervals, depending on their size. The support requirements for the various wiring methods are contained within the article covering each wiring method. These articles must be consulted when installing.

CABINETS AND CUTOUT BOXES— UL 50

Cabinets and cutout boxes shall be installed in accordance with *NEC® Article 373* and UL 50. Cabinet and cutout boxes are designed for the enclosure of switches and overcurrent devices. They may be surface mounted or recessed. They must have ample strength and rigidity, and if made of sheet steel, to be of not less than .053 inch thick. The size of a cabinet or cutout box must be sufficient for clearance between devices and conductors in the box to allow the box door to be closed when the switches or other devices in the box are in any position. They are also required to have a back-wiring space or one or more side-wiring spaces that contain equipment such as overcurrent protective and switching devices connected to more than eight conductors for any use, including branch-circuit, feeder-type conductors, but not including supply conductors or feeder continue-through conductors. These provisions are included in *Section 373-11*.

Too often these wiring spaces become overcrowded with branch-circuit conductors, feeders, taps, and continuation of feeder conductors to the next cabinet or cutout box when closing a panelboard or switch. The cabinet should not be used as a feeder raceway. The feeder conduit of a pull-box-to-pull-box connection adjacent to the panelboard should be extended for short taps to the panelboard buses. Such taps are permitted in *NEC®*

Section 240-21 and are given in Figure 4-17, on page 70. Frequently the most practical, economical way to serve a group of switches or control units is by a single feeder circuit. The recommended method is a wireway connected to the end of the feeder conduit extending the full length of the switch group (see Figure 4-19, page 72). Short taps from feeder conductors supply the individual switches.

Another method consists of is a junction box with conduit extensions to each switch or circuit-breaker enclosure. Although this satisfies the *NEC®* requirements, it involves a multiplicity of conductor connections in a confined area and is much less flexible than the wireway arrangement. The *NEC®* does not permit switch enclosures as raceways. Common equipment enclosed in cabinets and cutout boxes are meters, sockets, disconnecting switches fused and unfused, panelboards, distribution panels, and so forth. Cabinets and cutout boxes may have a hinge cover or removable cover and must have sufficient space for all wiring they contain. The deflection of conductors must have sufficient space where entering or leaving the cabinet or cutout box and where terminating on the enclosed electrical equipment (circuit breaker or switch).

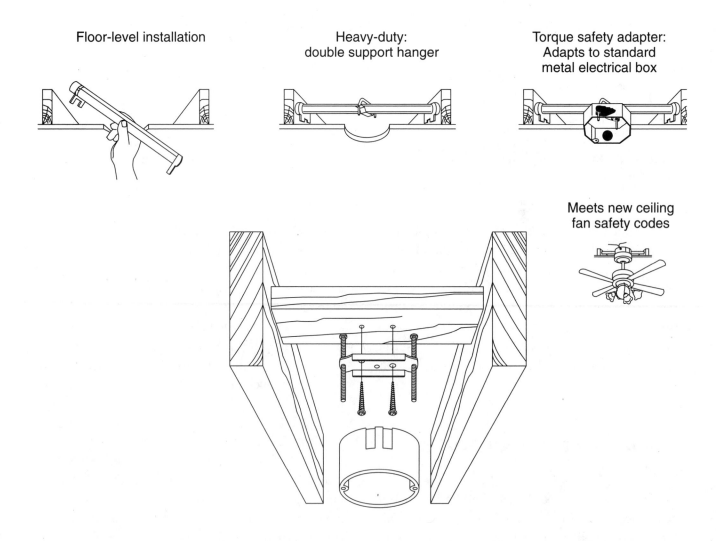

Figure 8-16   Complies with *NEC®* *Sections 370-23, 370-27,* and *422-18* as acceptable support methods for paddle fans and lighting fixtures (Courtesy of Reiker Enterprises Inc.)

Table 300-19(a). Spacings for Conductor Supports

		Conductors	
Size of Wire	Support of Conductors in Vertical Raceways	Aluminum or Copper-Clad Aluminum	Copper
18 AWG through 8 AWG	Not greater than	100 ft	100 ft
6 AWG through 1/0 AWG	Not greater than	200 ft	100 ft
2/0 AWG through 4/0 AWG	Not greater than	180 ft	80 ft
Over 4/0 AWG through 350 kcmil	Not greater than	135 ft	60 ft
Over 350 kcmil through 500 kcmil	Not greater than	120 ft	50 ft
Over 500 kcmil through 750 kcmil	Not greater than	95 ft	40 ft
Over 750 kcmil	Not greater than	85 ft	35 ft

Note: For SI units, 1 ft = 0.3048 m.

Figure 8-17 Vertical conductor support requirements (Reprinted with permission from NFPA 70-1999)

HARDWARE, SUPPORTS, CLIPS, AND STRAPS

These hardware items, although required methods in the *NEC®* for each specific wiring method, are not evaluated by recognized testing laboratories but are evaluated individually by manufacturers as to load

Figure 8-18 Typical cabinet manufactured in accordance with *NEC®* Article 373 (Courtesy of Unity Manufacturing)

testing and use. Generally, catalog data provide the intended use and weight loads permitted. *NEC® Article 300, Section 300-11* provides specific requirements for all the general wiring methods. However, the strap or support method selected must provide secure support as required in each individual article of the *NEC®*. For examples see Figures 8-19 and 8-20.

SWITCHBOARDS AND PANELBOARDS

Switchboards shall be installed in accordance with *NEC® Article 384* and UL 891, and panelboards shall be installed in accordance with *Article 384* and UL 891. A switchboard is a component of an electrical distribution system that is essentially a panel supporting switches for buses and overcurrent devices, accessible from front and/or rear, and not intended to be installed in a cabinet. A panelboard is similar, except that it may or may not contain switches, is designed to be placed in a cabinet or cutout box as per *Article 373*, and is accessible only from the front. Because they are enclosed, panelboards are subject to more capacity restrictions than switchboards.

NEC® requirements for switchboards are for the most part related to clearances to allow access and to prevent overheating of adjacent material or constructions. A special type of panelboard is designated in *Section 384-14* as a lighting and appliance branch-circuit panelboard. More than 10% of this type's overcurrent devices are rated at 30 amperes or less, for which neutral connections are provided. Panelboards are not permitted to contain more than 42 overcurrent devices in addition to any main overcurrent device. A two-pole circuit breaker is considered as two devices; a three-pole circuit breaker as three devices. It is also required to have a physical means to prevent the installation of a greater number of overcurrent devices than that for which it is designed and approved. Panelboards may be equipped with circuit breakers or fuses, with or without snap switches.

The snap switches are required to have a rating of not less than that of the feeder ampacity, as determined by the connected load. Panelboard overcurrent devices must not be loaded to more than 80% of their rating when loads are continuous for more than

Snap Strap Conduit Wall Hangers

Clamp solves bowing problems resulting from the expansion and contraction of conduit caused by varying temperature changes.

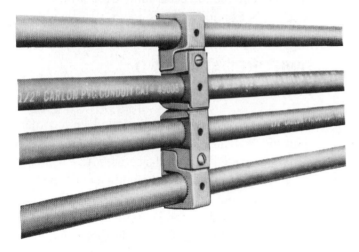

Two Hole Snap Strap Clamp

To be used in accordance with conduit spacing requirements per NEC®, Article 347-8.

To be used in accordance with conduit spacing requirements per NEC®, Article 347-8.

One Hole Snap Strap Clamp

Nonmetallic clamp

Figure 8-19   Common types of supports for wiring methods (Courtesy of Carlon, A Lamson & Sessions Company)

three hours, unless the panelboard and all the overcurrent devices are approved for continuous duty at 100% loading. Overcurrent protection for panelboards having snap switches rated at 30 amperes or fewer is limited to 200 amperes. Circuit breakers are not considered snap switches. Each lighting and appliance branch-circuit panelboard is required to have individual overcurrent protection, which may consist of not more than two main circuit breakers or sets of fuses, with a combined rating not in excess of the rating of the panelboard except (1) when the feeder overcurrent device has a rating not greater than that of the panelboard; or (2) when a lighting and appliance branch circuit is used as the service equipment for an individual residential occupancy, such as an apartment, provided each set of fuses supplying 15- or 20-ampere branch circuits is protected on the supply side with a main overcurrent device. *NEC® Sections 230-71* and *230-90* limit the total number of main disconnecting means and sets of overcurrent devices to six where they constitute the service equipment. *Section 225-33* requires feeders to the second building terminate in a maximum of six disconnects as per *Section 225-33(a)*.

When panelboard is supplied through a trans-

former, the overcurrent protection shall be located on the secondary side of the transformer; however, the secondary side of a single transformer having a two-wire, single-voltage secondary shall be considered the overcurrent protection. A three-phase disconnect or overcurrent device shall not be connected to the bus of any panelboard that has less than three-phase buses.

Backfed devices permitted as the main disconnecting means for a plug-in-type overcurrent protection shall be secured by an additional fastener that requires other than a pull to release the device from the mounting on the panel. Panelboards in wet or damp locations shall be installed to comply with *NEC® Section 373-2(a)*. Panelboards, cabinets, and panelboard frames, if made of metal, shall be in physical contact with each other and grounded in accordance with *Article 250* or *Section 384-3(c)*. When the panelboard is used with a nonmetallic raceway or cable, or where separate grounding conductors are provided, a terminal bar for the grounding conductor shall be secured inside the cabinet. The terminal bar shall be bonded to the cabinet and panelboard frame, if made of metal. Otherwise it shall be connected to the grounding

One hole push-on
straps 1/2" through 4"
Stamped steel

One hole straps
1/2" through 4"
Malleable iron

One hole straps
3/8" through 6"
Malleable iron

Heavy duty two hole
straps 1/2" through 4"
Stamped steel

Heavy duty two hole
straps 3/8" through 6"
Stamped steel

Two hole straps
1/2" through 4"
Stamped steel

Nail-up straps
1/2" through 1"
Stamped steel

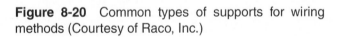

Figure 8-20 Common types of supports for wiring methods (Courtesy of Raco, Inc.)

Figure 8-21 Panelboards (Courtesy of Square D Company)

conductor that is run with the conductors feeding the panelboard.

Grounding conductors shall not be connected to a terminal bar provided for grounded conductors, unless the bar is identified for the purpose and is located where connection is made from grounded conductor to grounding electrode as permitted or required in *NEC® Article 250*. Wires entering and terminating inside panelboards are required to have bending space sized in accordance with *Table 373-6 (b)* for the largest conductor entering or leaving the enclosure. Side-wiring bending space shall be in accordance with *Table 376(a)* for the largest conductor to be terminated in that space. Four exceptions can be found in *Section 384-35*, which apply to panelboards and modify the rules in *Article 373*.

REVIEW QUESTIONS

1. What size outlet box is required to accommodate a duplex receptacle with two no. 12/2 w grd nonmetallic-sheathed conductors, one entering the outlet box to feed the receptacle and one leaving the outlet box to feed receptacles around the wall?

2. Are connectors and couplings for all wiring methods required to be "listed"?

3. What is the maximum number of circuit breakers (overcurrent devices) permitted in a panelboard?

4. What maximum fixture weight is a standard lighting outlet box permitted to support?

5. What minimum size of junction box may be used to accommodate two or three rigid conduits entering the side of the box and leaving through the bottom opposite the side?

6. What is the minimum size angle/straight pull box for the installation shown in Figure 8-22?

7. Is it permissible to splice conductors in a conduit body (condulet) with two openings?

8. Is it permissible to support a ceiling (paddle) fan from a standard outlet box?

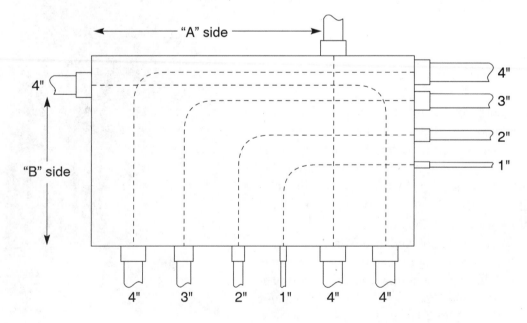

Figure 8-22

9. What is the minimum vertical rise in a raceway system before conductor support is required?

10. What constitutes a lighting and branch-circuit panelboard?

11. What maximum number of disconnects is permitted to protect individually a lighting and branch-circuit panelboard?

Chapter 9

Hazardous Locations

Today, engineers and contractors design and install electrical systems with a high level of confidence. However, the subject of hazardous (classified) locations is still viewed with uncertainty by many in the industry.

Actually, hazardous location installations are more exacting than difficult or mysterious. *Articles 500* through *505* of the *National Electrical Code®* define various types of hazardous locations and provide performance and installation standards.

There are many rules to guide you. But safety starts with a good understanding of two basics: proper classification of locations and selection of the correct equipment.

When an electrical system, or any portion of it, is installed in an area containing a hazardous atmosphere, protection against possible explosions is a major design factor. All the basic design considerations still apply to determination of conductor sizes and general layout of the overall system. In addition, the portion of the system passing through an area where flammable gases, combustible dust, or ignitable fibers may exist must have equipment enclosures especially designed to prevent electrical sparks and arcs from igniting the surrounding atmospheres and causing serious fires or explosions. For this portion of the design the electrical engineers must be doubly careful to select equipment specifically approved for the conditions encountered.

The *National Electrical Code®* describes three classifications of hazardous locations where an unconfined spark could result in explosion or fire. Each classification identifies conditions that may exist under normal operations; under occasional but frequent conditions resulting from maintenance operations; or under conditions of accidental breakdown that may cause the release of flammable or readily combustible materials. *NEC® Articles 500-505, 510, 511, 513, 514, 515, 516*, and *517* contain rules covering the type of wiring and apparatus to be installed in hazardous locations. Because the actual degree of hazard is frequently evaluated by local inspection and fire department authorities, they should be consulted during design of the wiring system that may be in the hazardous area.

A new *Article 505* was added to the 1996 *NEC®* introducing the European zone concept as a new optional method for classifying Class I hazardous (Classified) locations. In the 1999 *NEC®* there was a substantial effort to make the zone concept article, *505*, a stand-alone article. This has not been totally accomplished; however, engineers now have a clear choice of using either the class-division American system or the class-zone European system for designing and installing Class I hazardous electrical installations, for all voltages in locations where fire or explosive hazards may exist due to flammable gases or vapors and flammable liquids. Refer to Appendix A-21A and A-21B for Class I Division-Zone comparison charts

CODE CLASSIFICATIONS

The code divides hazardous locations into three classes, plus divisions and groups.

179

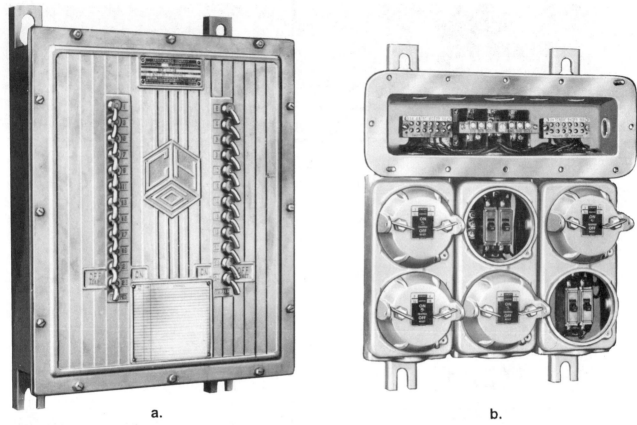

Figure 9-1   Electrical panelboards for hazardous locations. (Courtesy of Crouse-Hinds, Division of Cooper Industries) (a) Suitable for Classes II & III Groups E,F, G; (b) Suitable for Classes I, II, & III Groups C,D,E,F,G

Class I (*NEC® 500-7*) covers areas where gases or vapors are present in sufficient quantities to produce an explosive or ignitable environment. Class II (*NEC® 500-8*) locations are hazardous because of the presence of combustible dust. Class III (*NEC®*

WHAT ARE HAZARDOUS (CLASSIFIED) LOCATIONS?

Locations are classified depending on the properties of the flammable vapors, liquids, or gases, or combustible dusts of fibers which may be present and the likelihood that a flammable or combustible concentration or quantity is present.

Each room, section or area must be individually classified by indicating:

— Class

— Division

— Group

Figure 9-2

WHAT ARE THE CLASSES, DIVISIONS, (ZONES) AND GROUPS?

Class I

Zone 0	Division 1	Zone 1	Dvision 2	Zone 2
Group 11C	Group A	Group 11C	Group A	Group 11C
C	B	C	B	C
B	C	B	C	B
A	D	A	D	A

Class II

Division 1	*Division 2*
Group E	Group F
F	G
G	

Class III

Division 1	*Division 2*
No Groups	No Groups

Figure 9-3

500-9) locations are hazardous because of the presence of easily ignitable fibers or flyings.

Under all classes, Division 1 identifies situations where the hazard is "normally" expected to be present. Division 2 is where hazardous location substances are *not* normally present in the atmosphere, except under "abnormal" conditions. Class I can be divided either by the division or the zone concepts. When designing by *Article 505*, Class I is divided into three levels: Zone 0, Zone 1, Zone 2. One example of a Class I, Division 2 area is where flammable liquids or gases are confined in closed containers or systems from which they can't escape except through an accident.

A Class I, Zone 0 location is a location (1) in which ignitable concentrations of flammable gases or vapors are present continuously or (2) in which ignitable concentrations of flammable gases or vapors are present for long periods of time.

A Class I, Zone 1 location is a location (1) in which ignitable concentrations of flammable gases or vapors are likely to exist under normal operating conditions or (2) in which ignitable concentrations of flammable gases or vapors may exist frequently because of repair or maintenance operations or because of leakage or (3) in which equipment is operated or processes are carried on, of such a nature that equipment breakdown or faulty operations could result in the release of ignitable concentrations of flammable gases or vapors and also cause simultaneous failure of electrical equipment in a mode to cause the electrical equipment to become a source of ignition; or (4) that is adjacent to a Class I, Zone 0 location from which ignitable concentrations of vapors could be communicated, unless communication is prevented by adequate positive-pressure ventilation from a source of clean air and effective safeguards against ventilation failure are provided.

A Class I, Zone 2 location is a location (1) in which ignitable concentrations of flammable gases or vapors are not likely to occur in normal operation and if they do occur will exist only for a short period or (2) in which volatile flammable liquids, flammable gases, or flammable vapors are handled, processed, or used, but in which the liquids, gases, or vapors normally are contained with in closed containers or closed systems from which they can escape only as a result of accidental rupture or breakdown of the containers or system, or as the result of the abnormal operation of the equipment with which the liquids or gases are handled, processed, or used or (3) in which ignitable concentrations of flammable gases or vapors normally are prevented by positive mechanical ventilation, but which may become hazardous as the result of failure or abnormal operation of the ventilation equipment or (4) that is adjacent to a Class I, Zone 1 location, from which ignitable concentrations of flammable gases or vapors could be communicated, unless such communication is prevented by adequate positive-pressure ventilation from a source of clean air and effective safeguards against ventilation failure are provided.

In Class I, Groups A, B, C, and D divide materials by flame propagation characteristics, explosion pressures, and other factors. Class II, Groups E, F, and G define ignition temperature and conductivity of combustible dusts. For the hazardous group atmospheres identifications see *NEC® Section 500-3 (a)* and *(b)*. For example, Group A comprises atmospheres containing acetylene.

Evaluation of Hazardous Areas

Each area that contains gases or dusts that are considered hazardous must be carefully evaluated to make certain the correct electrical equipment is selected. Many hazardous atmospheres are Class I, Group D, or Class II, Group G. However, certain areas may involve other groups, particularly Class I, Groups B and C. Conformity with the *National Electrical Code®* requires enclosures and fittings approved for the hazardous gas or dust involved.

For the requirements for electrical and electronic equipment and wiring for all voltages where fire or explosion hazards may exist due to flammable gases or vapors, flammable liquids, or combustible dusts or fibers. Refer to *Articles 500* through *504* in Class I, Division 1 or Division 2, Class II, Division 1 or Division 2, and Class III, Division 1 or Division 2 hazardous (classified) locations. Refer to *Article 505* for an alternative to the division classification system covered in *Article 500. Article 505* covers the requirements for the zone classification system for electrical and electronic equipment and wiring for all voltages in Class I, Zone 0, Zone 1, and Zone 2 hazardous classified locations.

Classification of Hazardous Areas

Classification of specific areas is determined by the "code-enforcing authority," who may be a representative of the insurance underwriters, municipal electrical inspector, fire marshal, or member of the corporate safety organization.

Even though the professionals who classify locations "live by the Code," many factors may not be obvious. The distance from the hazardous source may affect the classification of the area. Ventilation and air currents must be considered. Temperatures, topography, walls, and special safeguards may have an effect.

Incorrect classification could lead to the use of unsafe equipment. This may result in fires, explosions, even loss of life and property.

With so many code requirements and location conditions to evaluate, experience, technical knowledge, and judgment are essential for correct classification of hazardous areas.

Hazardous locations are areas with a potential for explosion and fire due to flammable gases, vapors, or finely pulverized dusts in the atmosphere, or because of the presence of easily ignitable fibers or flyings. Hazardous locations may result from the normal processing of volatile chemicals, gases, grains, or they may result from accidental failure of storage systems for these materials. A hazardous location may also be created when volatile solvents or fluids, used in a normal maintenance routine, vaporize to form an explosive atmosphere.

Regardless of the cause of a hazardous location, every precaution must be taken to prevent ignition of the atmosphere. Certainly no open flames are permitted in these locations, but what about other sources of ignition?

In addition to all the requirements for classifying the class-division system, the Class I zone concept must also have the supervision of work, the classification of areas, the selection of equipment and wiring methods, and the installation made under the supervision of a "registered qualified professional engineer".

Electrical Sources of Ignition

A source of ignition is simply the energy required to touch off an explosion in a hazardous atmosphere.

Electrical equipment can be a source of this ignition energy. The normal operation of switches, circuit breakers, motor starters, contactors, plugs, and receptacles releases this energy in the form of arcs and sparks as contacts open and close, which make and break circuits.

Electrical equipment such as lighting fixtures and motors are classified as "heat producing"; they become sources of ignition if they reach a surface temperature that exceeds the ignition temperature of the particular gas, vapor, or dust in the atmosphere.

It is also possible that an abnormality or failure in an electrical system can cause an ignition. A loose termination in a splice box or a loose lamp in a socket can be the source of both arcing and heat. The failure of insulation from cuts, nicks, or aging can also be an ignition source from sparking, arcing, and heat.

Applications for Hazardous Location Equipment

Hazardous location equipment may be required in any area where flammable gases, vapors, or finely pulverized dusts in the atmosphere are sufficient to create a threat of explosion or fire. It may also be required where easily ignitable fibers or flyings are present.

Look for the Label

Equipment for hazardous (classified) locations must be clearly marked with the class, group, and temperature for which it is designed. Temperature is marked as a maximum safe operating temperature or as an identification number, which is listed in *Table 500-5(d)* of the Code. There are exceptions, so check the Code.

In most cases there is a trademark of the testing organization on the label—Underwriters Laboratory (UL), Factory Mutual (FM), or Canadian Standards Association (CSA). These organizations are certifying agencies. Their trademarks indicate that the

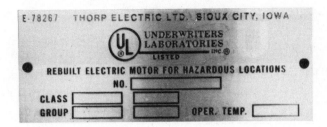

Figure 9-4 Listing label must include class, group, and operating temperature (This material extracted from Underwriters Laboratories Inc. 1998 *General Information for Electrical Construction, Hazardous Location, and Electrical Heating and Air Conditioning Equipment Directory*.)

equipment has been properly tested and has met the standards for the specified application.

Hazardous Locations and the *National Electrical Code*®

The *National Electrical Code*® treats installations in hazardous locations in *Articles 500* through *517*. Each hazardous location can be classified by the definitions in the *NEC*®. Interpretations of these classifications and applications follow.

Explosionproof apparatus. Enclosed apparatus capable of withstanding an explosion of a specified gas or vapor that may occur within it and of preventing the ignition of a specified gas or vapor surrounding the enclosure by sparks, flashes, or explosion of the gas or vapor within, and that operates at such an external temperature that a surrounding flammable atmosphere will not be ignited thereby.

Dust-ignitionproof. NEC® *502-1* states: "as used in this Article, shall mean enclosed in a manner that will exclude dusts and, where installed and protected in accordance with this code, will not permit arcs, sparks, or heat otherwise generated or liberated inside the enclosure to cause ignition of exterior accumulations or atmospheric suspensions of a specified dust on or in the vicinity of the enclosure."*

Hazardous Location. An area where the possibility of explosion and fire is created by the presence of flammable gases, vapors, dusts, fibers, or flyings.

Class I (NEC® *Section 500-7).* Those areas in which flammable gases or vapors may be present in sufficient quantities to be explosive or ignitable.

Class II (NEC® *Section 500-8).* Those areas made hazardous by the presence of combustible dust.

Class III (NEC® *Section 500-9).* Those areas in which there are easily ignitable fibers or flyings owing to type of material being handled, stored, or processed.

Division 1 (NEC® *Section 500-7, 8, 9).* Division 1 in the normal situation of hazard is expected to be present in everyday production operations or during frequent repair and maintenance.

Division 2 (NEC® *Section 500-7, 8, 9).* Division

Figure 9-5 Explosionproof switch and receptacle (Courtesy of Crouse-Hinds, Division of Cooper Industries)

*(Reprinted with permission from NFPA 70-1999.)

Hazardous(Classified) Locations in Accordance with *Article 500, National Electrical Code*® – 1999

Class I
Flammable Gases
or Vapors

Division 1

• Exists under normal conditions
• May exist because of:
    - repair operations
    - maintenance operations
    - leakage
• Released concentration because of:
    - breakdown of equipment
    - breakdown of process
    - faulty operation of equipment
    - faulty operation of process that
      causes simultaneous failure of
      electrical equipment

Division 2

• Liquids and gases are in closed containers or the systems are:
    - handled
    - processed
    - used
• Concentrations are normally prevented
  by positive mechanical ventilation.
• Adjacent to a Class I, Division 1 location

Group A: Atmospheres containing acetylene.

Group B: Atmospheres such as butadiene, ethylene oxide, propylene oxide,
            acrolein, or hydrogen (or gases or vapors equivalent in hazard to
            hydrogen, such as manufactured gas).

Group C: Atmospheres such a cyclopropane, ethyl ether, ethylene, or gases or
            vapors equivalent in hazard.

Group D: Atmospheres such as acetone, alcohol, ammonia, benzine, benzol
            butane, gasoline, hexane, lacquer solvent vapors, naphtha, natural
            gas, propane, or gases or vapors equivalent in hazard.

Figure 9-6

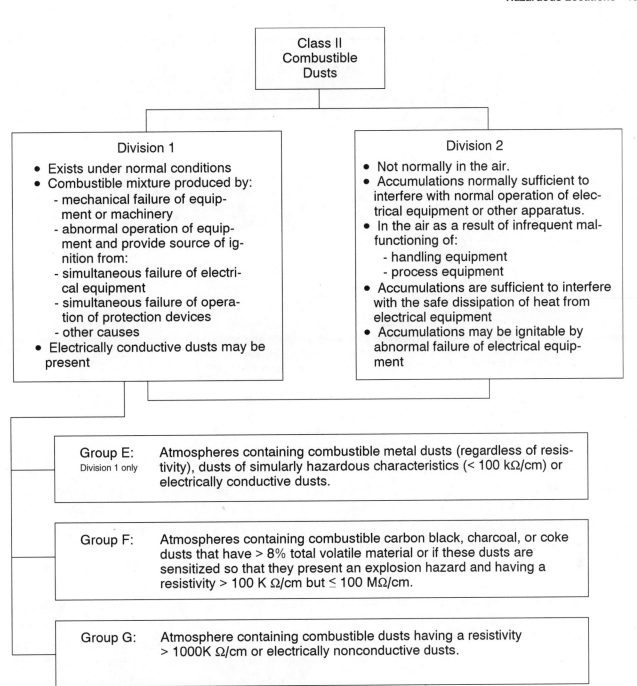

Figure 9-7

2 in the abnormal situation is expected to be confined within closed containers or closed systems; danger is present only through accidental rupture, breakage, or unusually faulty operation.

Groups (NEC® Section 500-5). The gases and vapors of Class I locations are broken into groups A, B, C, and D by the Code. These materials are grouped according to their explosion pressures and other flammable characteristics. Class II, which covers dust lo-

cations, comprises Groups E, F, and G. These groups are classified according to the ignition temperature and the conductivity of the hazardous substance.

Seals (NEC® Section 501-5 and *502-5).* Special fittings are required in Class I locations to minimize the passage of gases and vapors or, in the case of an explosion, the passage of flames. They are also required in Class II locations to prevent the passage of combustible dust.

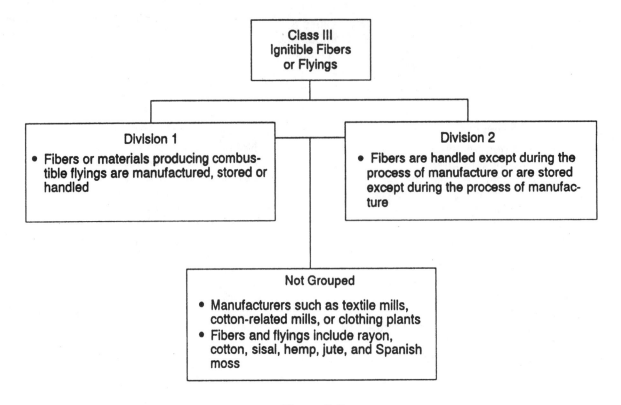

Figure 9-8

Articles 500 through 503 (1987 NEC®). These articles detail the requirements for the installation of wiring or electrical equipment in hazardous locations. These articles, along with other applicable regulations, local governing inspection authorities, insurance representatives, and qualified engineering/technical assistance should guide you in the installation of wiring or electrical equipment in any hazardous or potentially hazardous location.

CLASSIFICATION OF AREA

Area classification requires experience and sound judgment for the basic application of the installation and equipment required for the operations of that area. Electrical area classification is performed to prevent explosions and fires caused by electrical equipment, arcing and sparking, high temperatures, or electrical system faults, which can result in personal injury, loss of life, and destruction of plant equipment. Installation of correct electrical equipment and materials and the proper installation techniques can contribute significantly toward reducing the risk of such occurrences.

No one individual should be totally responsible for the ultimate classification of an area. It should be the result of group meetings and agreement between the customer, who can provide experience and history related to the proposed area or similar areas using similar hazardous materials, his or her personnel, plant operations, plant maintenance, plant safety department on the one hand, and in-house engineering staff, electrical engineers involved with the design, architectural designers involved with the architectural appurtenances, the authority having jurisdiction (which may consist of local and state authorities, electrical and fire inspectors, and building officials) and the operating company's underwriters (insurers) on the other hand. Careful coordination among these groups not only ensures the correct classification, but also provides alternatives to reduce the hazardous areas to a minimum.

Many industry codes are required in making these decisions. You must first have an industry code that is related specifically to the industry chemical or hazardous material for which this area is designed, such as API (American Petroleum Institute). The API has unique recommendations for the classification of areas containing petroleum facilities, such as APIRP 500. Other industries have similar application codes or guides. The representative for the operating company should be able to provide that information.

The National Fire Protection Association (NFPA) has a number of standards that can assist in making this determination. NFPA 497 concerns the classification of Class I hazardous locations for electrical installations in chemical plants. NFPA 499 is a classification of Class II hazardous locations for electrical installations in chemical process areas. NFPA 497 is a classification standard for gases and vapors for electrical equipment in Class I hazardous locations. NFPA 499 is a classification for combustible dusts for electrical equipment in hazardous locations. The *National Electrical Code®, Chapter 5* contains mandatory requirements as adopted by the jurisdiction in which the facility is being built.

You must check with the authority having jurisdiction in that area, state, county, or city to verify which edition of *NEC®* has been adopted and if there are amendments. NFPA-321 covers the classification of flammable and combustible liquids. NFPA-325-M-1994 describes the fire hazard for those flammable liquids, gases, and volatile solids. This publication provides additional information to help determine the gas groups not described in NFPA-497 or elsewhere. NFPA-496-1993 deals with purged and pressurized enclosures. NFPA-493 concerns intrinsically safe apparatus and associated apparatus for use in Class I, II, and III, Division 1 hazardous locations. NFPA-30-1996 is useful for flammable and combustible liquids and covers the ventilation, handling, and storage of such combustible liquids.

Other standards may be useful in making the proper classification, but class must first be established. The *National Electrical Code®* classifies these areas in *Chapter 5, Article 500*. Class I areas are those in which flammable gases or vapors are present. Class II deals with combustible dust. Class III covers ignitable fibers or flyings, in *Articles 501, 502, and 503*. *NEC® Article 504* covers intrinsically safe systems and is discussed later in this article.

Next, the group must be established. The group is listed in alpha designations in *NEC®*, which gives a partial list of the most commonly encountered materials that have been tested. They are grouped on the basis of flammability. One should utilize the expertise of chemical process engineers, safety-loss prevention engineers, or other experts in developing and establishing the group and class for the area being designed.

Then the division must be established. *NEC®* recognizes two divisions. The Division 1 criterion is that the location is likely to have volatile liquids, vapors, or dust present under normal conditions. The Division 2 criterion is that the area is likely to have flammable or volatile gases, vapors, or dust present only under abnormal conditions, such as accidental spills or releases of those volatile materials. One should classify the division very carefully, because incorrect area classification of Division 2 permits a much more liberal wiring method with fewer built-in safety guards. Misclassification of these areas can greatly reduce the safety factor desired under future operating conditions. Too often, operations develop poor janitorial maintenance after only a few years, creating a much more hazardous condition than was originally designed to protect against at the time the facility was built.

Additional references for the application and use of hazardous materials may be found in *Appendix A* of the *National Electrical Code®*.

CLASS I

Class I Locations

Class I locations are those in which flammable gases or vapors are or may be present in quantities sufficient to produce explosive or ignitable mixtures.

Class I, Division 1

These are Class I locations where a hazardous atmosphere is expected during normal operations. It may be present continuously, intermittently, periodically, or during normal repair or maintenance operations. Division 1 locations also result from a breakdown in the operation of processing equipment, with the release of hazardous vapors and the simultaneous failure of electrical equipment.

Class I, Zone 0

A Class I, Zone 0 location is a location in which ignitable concentrations of flammable gases or vapors are present continuously; or in which ignitable concentrations of flammable gases or vapors are present for long periods of time. See Figure A-21 for additional information on the European zone system covered in the 1999 *NEC®Article 505.*

Class I, Zone 1

A Class I, Zone 1 location is a location:

1. In which ignitable concentrations of flammable gases or vapors are likely to exist under normal operating conditions.
2. An area adjacent to a Class I, Zone 0 location from which ignitable concentrations of vapors could be communicated, unless communication is prevented by adequate positive pressure ventilation from a source of clean air and effective safeguards against ventilation failure are provided.
3. An area in which ignitable concentrations of flammable gases or vapors may exist frequently because of repair or maintenance operations, because of leakage
4. Where equipment is operated or processes are carried on, of such a nature that equipment breakdown or faulty operations could result in the release of ignitable concentrations of flammable gases or vapors. And may cause simultaneous failure of electrical equipment in a mode to cause the electrical equipment to become a source of ignition.

Class I, Division 2

These are Class I locations in which volatile flammable liquids or gases are handled, processed, or used, but in which they are normally confined within closed containers or closed systems from which they can escape only through accidental rupture or breakdown of the containers or systems. Hazardous conditions occur only under abnormal conditions.

Class I, Zone 2

A Class I, Zone 2 location is a location:

1. In which ignitable concentrations of flammable gases or vapors are not likely to occur in normal operation and if they do occur will exist only for a short period.
2. An area that is adjacent to a Class I, Zone 1 location, from which ignitable concentrations of flammable gases or vapors could be communicated, unless such communication is prevented by adequate positive-pressure ventilation from a source of clean air, and effective safeguards against ventilation failure are provided.
3. An area where volatile flammable liquids, flammable gases, or flammable vapors are handled, processed, or used, but in which the liquids, gases, or vapors normally are confined within closed containers or closed systems from which they can escape;
    (a) As a result of accidental rupture or breakdown of the containers or system, or
    (b) As a result of the abnormal operation of the equipment with which the liquids or gases are handled, processed, or used.
4. An area where ignitable concentrations of flammable gases or vapors normally are prevented by positive mechanical ventilation, but which may become hazardous as a result of failure or abnormal operation of the ventilation equipment.

EQUIPMENT FOR CLASS I LOCATIONS—DIVISIONS 1 AND 2

Devices for Class I locations are housed in enclosures strong enough to contain an explosion if the hazardous vapors enter the enclosure and are ignited. These enclosures then cool and vent the products of combustion in such a way that the surrounding atmosphere is not ignited.

Heat-producing equipment such as lighting fixtures, for hazardous locations, must not only contain the explosion and vent the cooled products of combustion, but must also be designed to operate

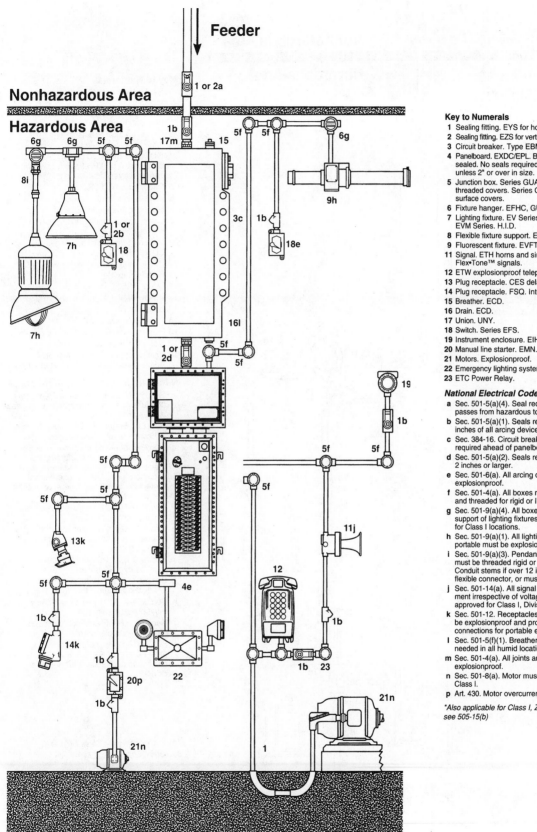

Feeder

Nonhazardous Area

Hazardous Area

Key to Numerals

1 Sealing fitting. EYS for horizontal or vertical.
2 Sealing fitting. EZS for vertical or horizontal conduits.
3 Circuit breaker. Type EBM.
4 Panelboard. EXDC/EPL. Branch circuits are factory sealed. No seals required in mains or branches unless 2" or over in size.
5 Junction box. Series GUA, GUB, GUJ, have threaded covers. Series CPS has ground flat surface covers.
6 Fixture hanger. EFHC, GUAC, or EFH.
7 Lighting fixture. EV Series incandescent and EVM Series. H.I.D.
8 Flexible fixture support. ECHF.
9 Fluorescent fixture. EVFT.
11 Signal. ETH horns and sirens. ESR bells, Flex•Tone™ signals.
12 ETW explosionproof telephone.
13 Plug receptacle. CES delayed action.
14 Plug receptacle. FSQ. Interlocked with switch.
15 Breather. ECD.
16 Drain. ECD.
17 Union. UNY.
18 Switch. Series EFS.
19 Instrument enclosure. EIH.
20 Manual line starter. EMN.
21 Motors. Explosionproof.
22 Emergency lighting system. ELPS.
23 ETC Power Relay.

National Electrical Code **References**

a Sec. 501-5(a)(4). Seal required where conduit passes from hazardous to nonhazardous area.
b Sec. 501-5(a)(1). Seals required within 18 inches of all arcing devices.
c Sec. 384-16. Circuit breaker protection required ahead of panelboard.
d Sec. 501-5(a)(2). Seals required if conduit is 2 inches or larger.
e Sec. 501-6(a). All arcing devices must be explosionproof.
f Sec. 501-4(a). All boxes must be explosionproof and threaded for rigid or IMC conduit.
g Sec. 501-9(a)(4). All boxes and fittings for support of lighting fixtures must be approved for Class I locations.
h Sec. 501-9(a)(1). All lighting fixtures fixed or portable must be explosionproof.
i Sec. 501-9(a)(3). Pendant fixture stems must be threaded rigid or IMC conduit. Conduit stems if over 12 inches must have flexible connector, or must be braced.
j Sec. 501-14(a). All signal and alarm equipment irrespective of voltage must be approved for Class I, Division 1 locations.
k Sec. 501-12. Receptacles and plugs must be explosionproof and provide grounding connections for portable equipment.
l Sec. 501-5(f)(1). Breathers and drains needed in all humid locations.
m Sec. 501-4(a). All joints and fittings must be explosionproof.
n Sec. 501-8(a). Motor must be suitable for Class I.
p Art. 430. Motor overcurrent protection.

Also applicable for Class I, Zone 1, see 505-15(b)

Figure 9-9 Diagram for Class I Division 1 power installation (Courtesy of Crouse-Hinds, Division of Cooper Industries)

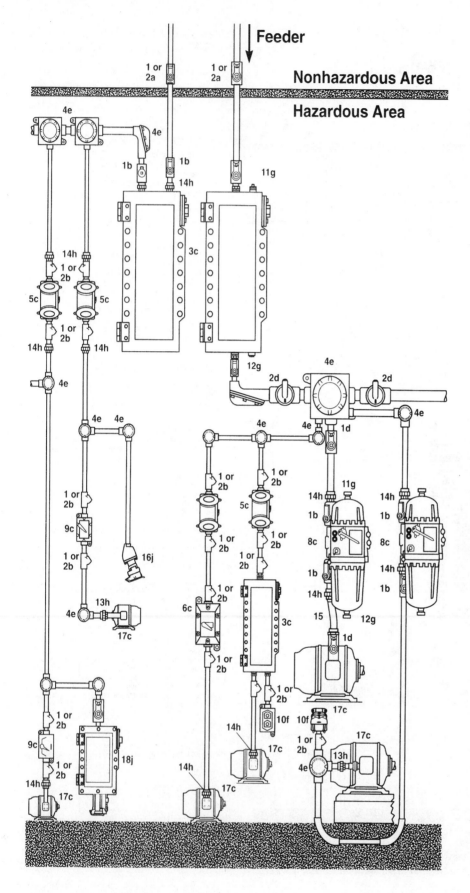

Feeder

Nonhazardous Area

Hazardous Area

Key to Numerals

1 Sealing fitting. EYS for horizontal or vertical.
2 Sealing fitting. EZS for vertical or horizontal conduits.
3 Circuit breaker EBMB.
4 Junction box Series GUA, GUB, and GUJ have threaded covers. Series CPS and Type LBH have ground flat surface covers.
5 Circuit breaker FLB.
6 Manual line starter EMN.
7 Magnetic line starter EBMS.
8 Combination circuit breaker and line starter EPC.
9 Switch or motor starter. Series EFS, EDS, or EMN.
10 Pushbutton station. Series EFS or OAC.
11 Breather. ECD.
12 Drain. ECD.
13 Union. UNF.
14 Union. UNY.
15 Flexible coupling. EC.
16 Plug receptacle. CES. Factory sealed.
17 Motor for hazardous location.
18 Plug receptacle. EBBR Interlocked Arktite® receptacle with circuit breaker.

National Electrical Code **References**

a Sec. 501-5(a)(4). Seals required where conduits pass from hazardous to nonhazardous area.
b Sec. 501-5(a)(1). Seals required within 18 inches of all arcing devices.
c Art. 430 should be studied for detailed requirements for conductors, motor feeders, motor feeder and motor branch circuit protection, motor overcurrent protection, motor controllers, and motor disconnecting means.
d Sec. 501-5(a)(2). Seals required if conduit is 2 inches or larger.
e Sec. 501-4(a). All boxes must be explosionproof and threaded for rigid or IMC conduit.
f Sec. 501-6(a). Pushbutton stations must be explosionproof.
g Sec. 501-5(f)(1). Breathers and drains needed in all humid locations.
h Sec. 501-4(a). All joints and fittings must be explosionproof.
i Sec. 501-4(a). Flexible connections must be explosionproof.
j Sec. 501-12. Receptacles and plugs must be explosionproof, and provide grounding connections for portable devices.

Also applicable for Class I, Zone 1, see 505-15(b)

Figure 9-9 (continued)

with surface temperatures below the ignition temperatures of the hazardous atmosphere.

Because the different vapors and gases making up hazardous atmospheres have varying properties, they have been placed in groups based on common flame-propagation characteristics and explosion pressures. These groups are designated A, B, C, and D, and the equipment selected must be suitable for the respective group of hazardous gas or vapor with regard to flame propagation, explosion pressures, and operating temperatures. See Figure A-14 (Appendix).

Reference to the *National Electrical Code®* indicates that most of the equipment for Class I, Division 2 applications is the same as that for Division 1 applications. However, in certain instances, standard location equipment may be used for some of the Class I, Division 2 applications if the appropriate restrictions are followed. The Code contains the specific rules.

Zones 0, 1 & 2

Electrical equipment installed in hazardous (classified) locations shall be installed in accordance with the instructions (if any) provided by the manufacturer.

In Class I, Zone 0 locations, only equipment specifically listed and marked as suitable for the location shall be permitted.

In Class I, Zone 1 locations, equipment specifically listed and marked as suitable for the location or Intrinsically safe equipment listed for use in Class I, Division 1 locations for the same gas, or as permitted by Section 505-5(d), and with a suitable temperature rating are permitted.

In Class I, Zone 2 locations:

1. Equipment specifically listed and marked as suitable for the location,
2. Equipment listed for use in Class I, Zone 0, or Zone 1 or Class I, Division 1, Division 2 locations for the same gas, or as permitted by Section 505-5(d), and with a suitable temperature rating.

In Class I, Zone 2 locations, the installation of open or nonexplosionproof or nonflameproof enclosed motors, such as squirrel-cage induction motors without brushes, switching mechanisms, or similar arc-producing devices, shall not be permitted.

Class I Locations

- Petroleum refining facilities
- Dip tanks containing flammable or combustible liquids
- Dry cleaning plants
- Plants manufacturing organic coatings
- Spray finishing areas (residue must be considered)
- Petroleum dispensing areas
- Solvent extraction plants
- Plants manufacturing or using pyroxylin (nitro-cellulose) type and other plastics (Class II also)
- Locations where inhalation anesthetics are used
- Utility gas plants and operations involving storage and handling of liquified petroleum and natural gas
- Aircraft hangars and fuel servicing areas
- Anesthetizing locations

WIRING METHODS—
DIVISIONS 1 AND 2

NEC® Article 501 covers the requirements for Class I locations. Wiring methods are covered in *Section 501-4(a)* and *(b)* for Division 1 and Division 2 locations. *Section 501-4(a)* requires wiring methods of threaded rigid metal conduit, threaded steel intermediate metal conduit (IMC), or Type MI cable with termination fittings approved for the location.

In industrial establishments with restricted public access, where the conditions of maintenance and supervision ensure that only qualified persons will service the installation the following wiring methods shall also be acceptable.

1. Type MC cable, listed for use in Class I, Division 1 locations, with a gas/vaportight continuous corrugated aluminum sheath, an overall jacket of suitable polymeric material, separate grounding conductors in accordance with Section 250-122, and provided with termination fittings listed for the appli-

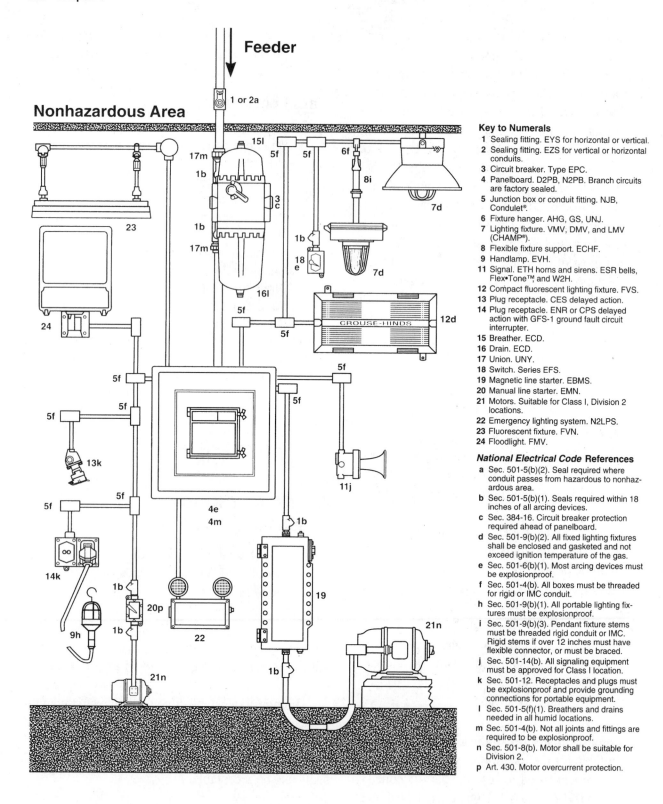

Feeder

Nonhazardous Area

1 or 2a

Key to Numerals

1 Sealing fitting. EYS for horizontal or vertical.
2 Sealing fitting. EZS for vertical or horizontal conduits.
3 Circuit breaker. Type EPC.
4 Panelboard. D2PB, N2PB. Branch circuits are factory sealed.
5 Junction box or conduit fitting. NJB, Condulet®.
6 Fixture hanger. AHG, GS, UNJ.
7 Lighting fixture. VMV, DMV, and LMV (CHAMP®).
8 Flexible fixture support. ECHF.
9 Handlamp. EVH.
11 Signal. ETH horns and sirens. ESR bells, Flex•Tone™ and W2H.
12 Compact fluorescent lighting fixture. FVS.
13 Plug receptacle. CES delayed action.
14 Plug receptacle. ENR or CPS delayed action with GFS-1 ground fault circuit interrupter.
15 Breather. ECD.
16 Drain. ECD.
17 Union. UNY.
18 Switch. Series EFS.
19 Magnetic line starter. EBMS.
20 Manual line starter. EMN.
21 Motors. Suitable for Class I, Division 2 locations.
22 Emergency lighting system. N2LPS.
23 Fluorescent fixture. FVN.
24 Floodlight. FMV.

National Electrical Code References

a Sec. 501-5(b)(2). Seal required where conduit passes from hazardous to nonhazardous area.
b Sec. 501-5(b)(1). Seals required within 18 inches of all arcing devices.
c Sec. 384-16. Circuit breaker protection required ahead of panelboard.
d Sec. 501-9(b)(2). All fixed lighting fixtures shall be enclosed and gasketed and not exceed ignition temperature of the gas.
e Sec. 501-6(b)(1). Most arcing devices must be explosionproof.
f Sec. 501-4(b). All boxes must be threaded for rigid or IMC conduit.
h Sec. 501-9(b)(1). All portable lighting fixtures must be explosionproof.
i Sec. 501-9(b)(3). Pendant fixture stems must be threaded rigid conduit or IMC. Rigid stems if over 12 inches must have flexible connector, or must be braced.
j Sec. 501-14(b). All signaling equipment must be approved for Class I location.
k Sec. 501-12. Receptacles and plugs must be explosionproof and provide grounding connections for portable equipment.
l Sec. 501-5(f)(1). Breathers and drains needed in all humid locations.
m Sec. 501-4(b). Not all joints and fittings are required to be explosionproof.
n Sec. 501-8(b). Motor shall be suitable for Division 2.
p Art. 430. Motor overcurrent protection.

*Also applicable for Class I, Zone 2, see 505-15(c)

Figure 9-10   Diagram for Class I Division 2 power and lighting installation (Courtesy of Crouse-Hinds, Division of Cooper Industries)

Figure 9-13 Explosionproof flexible conduit is approved for limited use, where necessary for flexibility. Use only approved wiring methods, such as pictured. (Courtesy of Crouse-Hinds, Division of Cooper Industries)

Figure 9-11 Explosionproof light fixture (Courtesy of Crouse-Hinds, Division of Cooper Industries)

Figure 9-12 Only approved thread lubricants should be used (Courtesy of Crouse-Hinds, Division of Cooper Industries)

cation. For general restrictions on use of Type MC cable see NEC article 334.

2. Type ITC cable, listed for use in Class I, Division 1 locations, with a gas/vaportight continuous corrugated aluminum sheath, an overall jacket of suitable polymeric material, separate grounding conductor(s), and provided with termination fittings listed for the application.

All boxes, fittings, and joints must be threaded for a connection to conduit or cable terminations and must be explosionproof. Threaded joints must be made up with at least five threads fully engaged. MI cable must be installed and supported to avoid tensile stress at the terminations. When flexible connections are necessary, such as at motor terminals, flexible fittings approved for Class I locations must be used.

NEC® Section 501-11, however, permits flexible cord or connection between portable lighting equipment or other portable utilization equipment and a fixed portion of its supply circuit with rigid requirements: extra-hard usage cord must be employed in addition to the conductors in the circuit and a grounding conductor complying with *Section 400-23*. It must be connected to terminals or supply conductors in an approved manner and be supported by clamps or other suitable means so that there will be no tension on the terminal connections. The connection must be provided with seals where the cord enters boxes, fittings, or enclosures of explosionproof type.

Areas classified as Class I Division 2 have somewhat less stringent requirements. In these areas the following wiring methods are permitted.

1. Threaded rigid metal conduit
2. Threaded steel intermediate metal conduit

3. Enclosed gasketed busways and wireways
4. Type PLTC cable in accordance with the provisions of *Article 725*
5. Type MI, MC, MV, or TC cable with approved termination fittings
6. Type ITC cable in cable trays, in raceways, supported by messenger wire, or directly buried where the cable is listed for this use
7. Types ITC, PLTC, MI, MC, MV, or TC cable and shall be permitted to be installed in cable tray systems and shall be installed in a manner to avoid tensile stress at the termination fittings.
8. Boxes, fittings, and joints shall not be required to be explosionproof except as required by Sections *501-3(b)(1), 501-6(b)(1),* and *501-14(b)(1).*

When provisions must be made for limited flexibility, such as at motor terminals, flexible metal conduit with approved fittings, armored cable with approved fittings, liquidtight flexible metal conduit with approved fittings, liquidtight flexible nonmetallic conduit with approved fittings, or flexible cord approved for hard uses provided with approved bushed fittings may be used. The additional conductor for grounding shall be included in the flexible cord. Wiring that cannot release sufficient energy to ignite a specific ignitable atmospheric mixture by opening, shorting, or grounding may be permitted in this area using any of the methods suitable for wiring in ordinary locations.

Zone 0, 1 and 2 Wiring Methods

Class I, Zone 0. Only intrinsically safe wiring in accordance with *Article 504*, or nonconductive optical fiber cable or systems with an approved energy-limited supply are acceptable.

Class I, Zone 1. All wiring methods permitted for Class I, Division 1 locations which maintain the integrity of protection techniques are acceptable. For example, equipment with type of protection "e" requires that all cable and conduit fittings incorporate suitable methods to maintain the ingress protection of the enclosure and different electrical enclosures provide different degress of ingress protoection.

Class I, Zone 2. All wiring methods permitted

for Class I, Division 1 or Class I, Division 2 locations which maintain the integrity of protection techniques are acceptable. For example, equipment with type of protection "nR" requires that cable and conduit fittings incorporate suitable sealing methods to maintain the restricted breathing technique.

SEALING REQUIREMENTS— DIVISIONS 1 AND 2

All wiring methods in Class I locations are required to be sealed in accordance with *Section 501-5(a)* and *(b)*. In each conduit run passing from a Class I location into an unclassified area, a sealing fitting is permitted on either side of the boundary of such location but must be so designed and installed as to minimize the amount of gas or vapor that may enter the conduit system within the Class II location from being communicated to the conduit beyond the seal. Rigid metal conduit or threaded steel intermediate conduit shall be used between the sealing fitting and the point at which the conduit leaves the classified location, and a threaded connection shall be used at the sealing fitting. No fitting or coupling is permitted between the sealing fitting and the point at which the conduit leaves the classified location.

Conduit systems terminating in an open raceway in an outdoor, unclassified area are required to be sealed between the point at which the conduit leaves the classified location and the point at which it enters the open raceway. The specific requirements for installing the seals are found in *NEC® Section 501-5(c)* and the manufacturer's instructions. Installation of these seals is critical; improperly made, they will not provide the protection needed to isolate accidental explosions. Raceways in Class I hazardous locations must be provided with sealing fittings to prevent the passage of gases, vapors, or flames through the raceway from one portion of an electrical installation to another, and to prevent pressure piling or cascading. Pressure piling may occur in these locations when an explosion in some part of the system, such as a switch enclosure, is not effectively blocked or sealed off. The first explosion may be rendered harmless if properly confined by a seal to the switch enclosure. If allowed to enter the raceway, it builds up great pressure and another, more power-

Figure 9-14 Seal-off fitting approved for horizontal or vertical installation. (Crouse-Hinds, Division of Cooper Industries)

ful explosion may follow. The second explosion builds up additional pressures in the gas-filled conduit and a greater explosion results. This may continue in very long raceway runs not provided with intermittent seals until some part of the system ruptures.

The hot or burning gases could then escape and ignite flammable gases in the surrounding atmosphere, causing a general explosion. Proper sealing is one of the most important requirements for safety in a Class I hazardous location. Explosive vapors are prevented from passing to a nonhazardous area where they could be ignited and carry the flame back to the explosive vapor location. In the manner of a pilot light on a gas range, arcs, sparks, and hot particles from an arcing device are also prevented from entering the raceway by the seal. Sealing fittings permit the pouring of a water-base sealing compound into the fitting around and between the conductors.

The raceway should first be blocked at the bottom of the sealing fitting with a packed material to prevent the sealing mixture from draining out. The liquid mixture solidifies after a few hours and becomes very hard. Proper mixing of the compound and liquid is essential; otherwise the seal will not prevent the pressure piling. Even though it may prevent the passage of vapors, arcs, and hot particles, all voids under, between, and over conductors must

be filled. If the sealing mixture is to perform its function of preventing the passage of gases, vapors, or flames through the raceway, complete elimination of the pressure piling is necessary and can be achieved by installing effective seals in every raceway regardless of size where it enters or leaves any junction box at regular intervals, in all raceway runs, and at points at which they presently are required by the *NEC*®.

In the absence of tests for various gases, determine the proper spacing of seals. In long conduit runs it is not unreasonable to specify that they be installed at 50-foot intervals in all conduit runs within a Class I, Division 1 location. If installed in accordance with the *NEC*® requirements, a rigid steel conduit system with suitable enclosure and seals effectively contains an internal explosion and prevents the escape of burning gases until they are cooled sufficiently to prevent the ignition of flammable vapors in the surrounding atmosphere. The gases are cooled by expansion as they escape through the machine-grounded joints of predetermined width and through threaded joints. Ground joints, unless properly maintained, are not as reliable as threaded joints. A single loose screw in a cover of a ground joint or scratches in ground surfaces can render an otherwise explosionproof enclosure ineffective. Threaded joints, on the other hand, are reliable because the gases escape only by following the threads. The *NEC*® requires at least five full threads to be engaged at each coupling.

Effective seals must be placed in a raceway as required by the *NEC*® within 18 inches of an arcing device or at the point where the raceway leaves or enters a hazardous area. For a raceway 2 inches or larger in diameter, seals are required within 18 inches of any junction box containing splices, taps, or terminal connections. No coupling or fitting is permitted between the seal and the boundary between the hazardous and nonhazardous areas. In many cases, special panelboards and controls are designed with factory seals that are an integral part of the equipment for Class I locations. Because such seals effectively isolate the arc-producing segments of the assembly, additional seals are not generally required in the conduit leaving these panelboards. Incandescent and fluorescent fixtures in Class I locations are also available with integral seals.

Zone 0, 1 and 2 Seals

Class I, Zone 0. Seals are required within 10 feet of where a conduit leaves a Zone 0 location. Unions, couplings, boxes, or fittings are not permitted except reducers at the seal, in the conduit run between the seal and the point at which the conduit leaves the location. Rigid unbroken conduit that passes completely through the Zone 0 location with no fittings less than 12 inches beyond each boundary, shall not be required to be sealed, if the termination points of the unbroken conduit are in unclassified locations. Seals shall be provided on cables at the first point of termination after entry into the Zone 0 location. Seals are not be required to be explosionproof or flameproof.

Class l, Zone 1. Sealing and drainage shall be provided in accordance with *Section 501-5(a), (c), (d),* and *(f)*, except where the term "Division 1" is used "Zone 1" shall be substituted.

Class l Zone 2. Sealing and drainage shall be provided in accordance with *Section 501-5(b), (c), (e),* and *(f)*, except where the term "Division 2" is used, "Zone 2" shall be substituted.

CLASS II

Class II Locations

Class II locations are hazardous because of the presence of combustible dust.

Class II, Division 1. These are Class II locations where combustible dust may be in suspension in the air under normal conditions in sufficient quantities to produce explosive or ignitable mixtures. This may occur continuously, intermittently, or periodically. Division 1 locations also exist where failure or malfunction of machinery or equipment might cause a hazardous location while providing a source of ignition with the simultaneous failure of electrical equipment. Included also are locations in which combustible dust of an electrically conductive nature may be present.

Class II, Division 2. A Class II, Division 2 location is one in which combustible dust is not normally in suspension in the air and normal operations will not

Figure 9-15 An example of a Class II panelboard (Courtesy of Crouse-Hinds, Division of Cooper Industries)

put the dust in suspension, but where accumulation of the dust may interfere with the safe dissipation of heat from electrical equipment or where accumulations near electrical equipment may be ignited by arcs, sparks, or burning material from the equipment.

Equipment for Class II Locations

The enclosures that house devices in Class II locations are designed to seal out dust. Contact between the hazardous atmosphere and the source of ignition has been eliminated and no explosion can occur within the enclosure.

As in Class I equipment, heat-producing equipment must be designed to operate below the ignition temperature of the hazardous atmosphere. However, in Class II equipment, additional consideration must be given to the heat buildup that may result from the layer of dust that settles on the equipment.

Dusts have also been placed in groups E, F, and G based on their particular hazardous characteristics and the dusts' electrical resistivity. It is important to select equipment suitable for the specific hazardous group. See Figure A-14 (Appendix.)

NOTE: When the division is not specified, the equipment is suitable for both Division 1 and Division 2 applications.

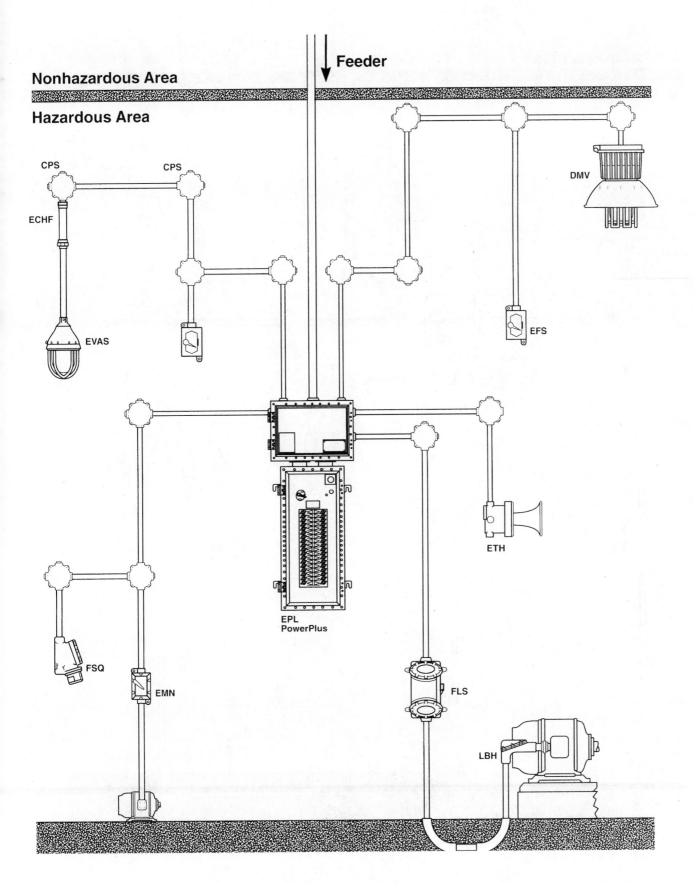

Feeder

Nonhazardous Area

Hazardous Area

CPS

CPS

ECHF

EVAS

DMV

EFS

EPL
PowerPlus

ETH

FSQ

EMN

FLS

LBH

Figure 9-16 Class II lighting installation (Courtesy of Crouse-Hinds, Division of Cooper Industries)

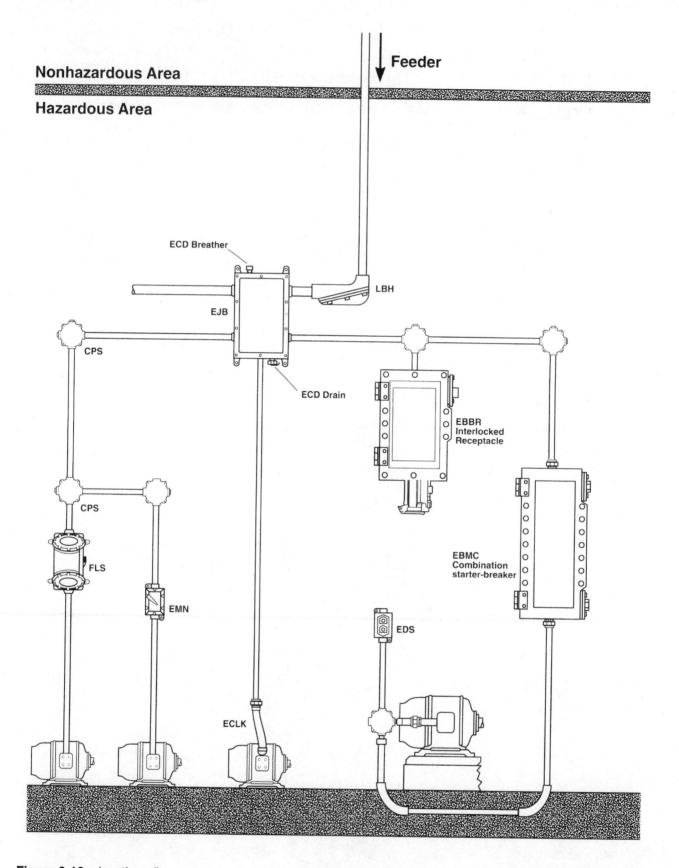

Figure 9-16 (continued)

Figure 9-17 Class II switched plug receptacle (Courtesy of Crouse-Hinds, Division of Cooper Industries)

Class II Locations

- Grain elevators and bulk handling facilities
- Manufacture and storage of magnesium
- Manufacture and storage of starch
- Fireworks manufacture and storage
- Flour and feed mills
- Areas for packaging and handling of pulverized sugar and cocoa
- Facilities for the manufacture of magnesium and aluminum powder
- Spice grinding plants
- Confectionary manufacturing plants

WIRING METHODS

Wiring methods for Class II, Division 1 locations, although similar to these for Class I, Division 1 locations, have a very different purpose. In Class II, Division 1 locations combustible dust is in the air under normal operating conditions in quantities sufficient to produce explosive and ignitable mixtures. The wiring methods and equipment are designed to keep these flyings and

combustible dusts out of the electrical system, whereas the Class I, Division 1 location electrical wiring methods and equipment are designed to contain an explosion.

NEC® 502-4 covers the wiring methods for Class I, Division 1 locations; they shall be made in threaded rigid metal conduit, threaded steel intermediate conduit, or Type MI cable with termination fittings approved for the location. MI cable when installed also must be supported to avoid tensile stress on the termination fittings.

In industrial establishments with restricted public access, where the conditions of maintenance and supervision ensure that only qualified persons will service the installation Type MC cable, listed for use in Class I, Division 1 locations, with a gas/vaportight continuous corrugated aluminum sheath, an overall jacket of suitable polymeric material, separate grounding conductors in accordance with *Section 250-122,* and provided with termination fittings listed for the application is permitted. For general restrictions on use of Type MC cable see *NEC Article 334.* All fittings and boxes in these locations must be provided with threaded bosses for connection to the conduit and cable terminations. They must have close-fitting covers and shall have no openings such as holes for attachment screws through which dust might enter or through which sparks or burning material might escape. Fittings and boxes in which taps, joints, or termination connections are made, or that are used in locations where dusts of a combustible, electrically conductive nature are present, shall be of the approved type for Class II locations.

When it is necessary to use flexible connections, dust-tight flexible connectors and liquidtight flexible metal conduit with approved fittings, liquidtight flexible nonmetallic conduit with approved fittings, or flexible cord approved for extra-hard usage and provided with bushed fittings shall be used. The flexible cords, like MI cable, must be dust-tight at both ends, with tension permitted on the connections. They shall be made of a type of insulation approved for the condition present or protected by a suitable sheath.

Class II, Division 2 locations permit the use of rigid metal conduit, intermediate metal conduit, electrical metallic tubing, dust-tight wireways, Types MI, MC, or SNN cable with approved termination fittings, or Type MC, PLTC, ITC or TC cable installed in ventilated channel-type cable trays in a single layer with a space of not fewer than the larger cable diameter between the two adjacent cables. Wiring in nonincendive circuits that under normal conditions cannot release sufficient energy to ignite a specific combustible dust mixture by opening, shorting, or grounding shall be permitted using any of the methods suitable for wiring in ordinary locations.

SEALING REQUIREMENTS

Combustible and explosive dusts constitute a hazard in Class II locations. Such dusts do not travel for great distance within a conduit. Therefore, seals are not required if the conduit running between a dust-ignition-proof enclosure and one that is not dust-ignition-proof consists of either a horizontal section not fewer than 10 feet in length or a vertical section not fewer than 5 feet in length extending downward from the dust-ignition-proof enclosure through the raceway. Where such conduit runs are not provided, seals must be used to prevent the entrance of dust into the ignitionproof enclosure through the raceway. Special sealing fittings and poured sealing compound are not required, although they may be used. The term *dust-ignition-proof* is defined in *NEC® Section 502-1*.

CLASS III LOCATIONS

Class III locations are hazardous because of the presence of easily ignitable fibers or flyings, but the fibers or flyings are not likely to be in suspension in the air in quantities sufficient to produce ignitable mixtures.

Equipment for Class III Locations

Class III locations require equipment that is designed to prevent the entrance of fibers and flyings, prevent the escape of sparks or burning material, and operate at a temperature below the point of combustion.

Class III

- Woodworking plants
- Textile mills

- Cotton gins and cotton seed mills
- Flax producing plants

WIRING METHODS

Wiring methods in Class III, Division 1 or Division 2 locations shall be made with rigid metal conduit, rigid nonmetallic conduit, intermediate metal conduit, electrical metallic tubing, dust-tight wireways, or Type MI, MC, or SNM cable with approved termination fittings. All boxes and fittings shall be dust-tight. When necessary to employ flexible connections, dust-tight flexible connectors, liquidtight flexible metal conduit with approved fittings, liquidtight flexible nonmetallic conduit with approved fittings, or flexible cord approved for extra-hard usage and provided with bushed fittings shall be used. In Class III, Division 2 locations, in sections, compartments, or areas solely for storage and containing no machinery, open wiring on insulators is permitted when installed in accordance with *NEC® Article 320*, but only on the condition that protection as required by *NEC® 320-14* is provided where the conductors are not run in roof spaces and are well protected from physical damage.

SEALING REQUIREMENTS

Seals are not required in Class III locations.

COMMERCIAL GARAGES, REPAIR, AND STORAGE

Commercial garages are defined by *NEC® Article 511*. Additional information may be obtained in NFPA 88A-1995 for parking structures and NFPA 88B-1997 for repair garages. Commercial garages are defined as locations to service and repair self-propelled vehicles, including, but not limited to, passenger automobiles, buses, trucks, and tractors, in which volatile, flammable liquids are used for fuel or power. Areas in which flammable fuel is transferred to vehicle fuel tanks, however, are required to conform with *Article 514*. Parking garages used for parking or storage and where no repair work is done except the exchange of parts and routine maintenance requiring no use of electrical equipment, open

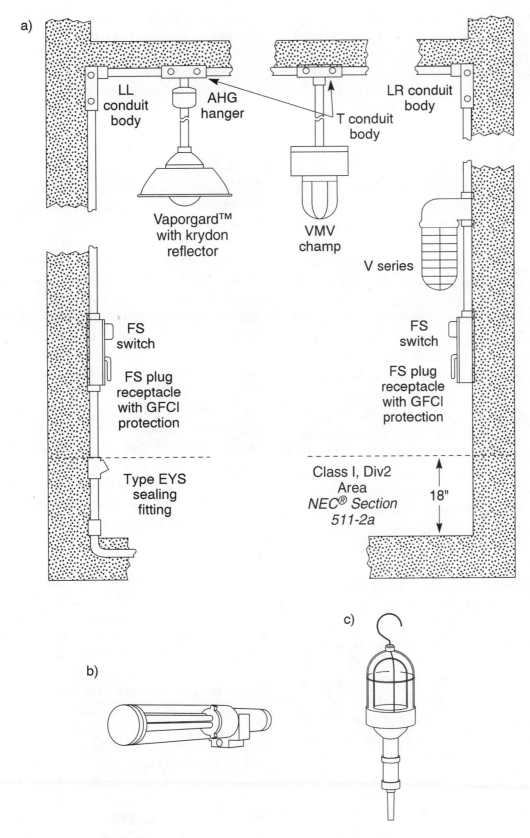

Figure 9-18 (a) Recommended installation for commercial garages. (Courtesy of Crouse-Hinds, Division of Cooper Industries) (b) Illuminator EVFT fluorescent lighting fixture listed for Class I locations (Courtesy of Crouse-Hinds, Division of Cooper Industries) (c) EV106 portable handlamp for use in Class I hazardous locations (Courtesy of Crouse-Hinds, Division of Cooper Industries)

flame, welding, or volatile flammable liquids are not classified, but they are required to be adequately vented to carry the exhaust fumes of the engines.

Areas in commercial garages up to a level of 18 inches above the floor are classified as Class I, Division 2 locations, except where the enforcing agency (AHJ) determines there is mechanical ventilation providing a minimum of four air changes per hour. Any pit or depression below the floor level must be considered a Class I, Division 1 location except where six air changes per hour are exhausted at the floor level of the pit. When it is determined by the AHJ that there is enough mechanical ventilation, lubritoriums, where lubrication is performed, and service rooms without dispensing facilities shall be classified in accordance with *NEC® Article 514, Table 514-2*.

Areas adjacent to the defined locations or with positive pressure ventilation in which flammable vapors are not likely to be released, such as stock rooms, switchboard rooms, offices, and similar locations, need not be classified where mechanically ventilated at a rate of four or more air changes per hour or where effectively cut off by walls or partitions. Adjacent areas that by reason of ventilation, air-pressure differentials, or physical spacings are such that in the opinion of the enforcing agency (AHJ) no ignition hazards exist may be considered nonhazardous and need not be classified. Wiring and equipment located within the Class I locations shall conform to the applicable provisions of *NEC® Article 501*. For raceways imbedded in masonry walls or buried beneath the floor, if any connections or extensions lead into or through the area, seals are required as per *Section 501-5* and *501-5(b)(2)* and apply to both horizontal and vertical boundaries of the defined Class I location.

Wiring above the Class I location must be run in metal raceways, nonmetallic conduit, electrical nonmetallic tubing, flexible metal conduit, liquidtight flexible metal conduit, or liquidtight flexible nonmetallic conduit. This wiring can also be Type MC or MI manufactured wiring systems, SNM or PLTC cable in accordance with *NEC® Article 725*. TC cable, cellular metal floor raceway, or cellular concrete floor raceways are permitted below the floor, but such raceways shall have no connections leading into or through the Class I lo-

cation above the floor. Where a circuit is portable or supplied from pendants and includes a grounded conductor as provided in *Article 200*, receptacles, attachment plugs, connectors, and similar devices shall be of the polarized type, and the grounded conductor of the flexible cord shall be connected to the screw shell of the lampholder or to the grounded terminal of any utilization equipment supplied.

Equipment located above the Class I location shall be at least 12 feet above the floor level, where it may be deemed as arc-producing equipment, such as cutout switches, charging panels, generators, motors or other equipment having make- and break-sliding contacts, which are required to be totally enclosed or so constructed as to prevent the escape of sparks or hot metal particles. This requirement does not include receptacles, lamps, or lampholders. Fixed lighting located over the lanes through which vehicles are commonly driven, or otherwise exposed to physical damage, shall be located not fewer than 12 feet above the floor unless totally enclosed or so constructed as to prevent the escape of arcs, sparks, or hot metal particles. Battery chargers and their control equipment shall not be located within the locations as classified in *NEC® Article 511-3*.

NEC® 511-10 requires that all 125-volt, single-phase, 15- and 20-ampere receptacles installed in areas where electrical diagnostic equipment, electrical hand tools, or portable lights are to be used shall provide ground-fault circuit interrupter protection for personnel. Although this *NEC®* article covers only those areas within the scope of NFPA 88, it may provide good guidelines for similar areas in which other combustible engines are being serviced and repaired. Bear in mind that boat shops, lawnmower shops, and the like, although similar in nature, must comply with *Article 501*.

AIRCRAFT HANGARS

Aircraft hangars are defined by *NEC® Article 513* as locations for storage or servicing of aircraft in which gasoline, jet fuels, or other volatile, flammable liquids or flammable gases are used. They shall not include locations used exclusively for aircraft that have never contained such liquids or gases or that have been drained and properly purged. This

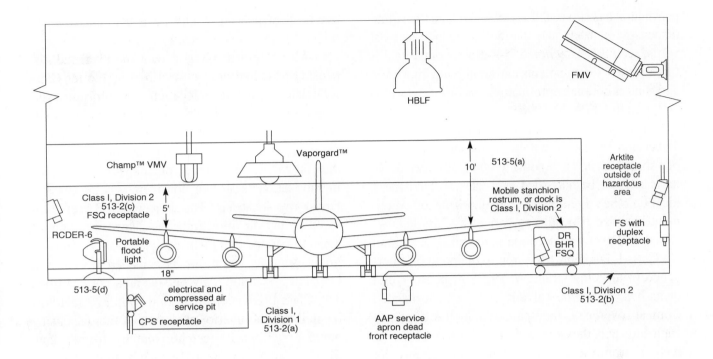

Figure 9-19 Sketch of recommended aircraft hangar installation (Courtesy of Crouse-Hinds, Division of Cooper Industries)

includes areas such as assembly plants and rebuilding plants for aircraft. Although it is increasingly popular for residential developments to be built in which individual aircraft are stored adjacent to the residence, this article has not been revised to cover those areas, which must meet therefore the provisions of *Article 513*. Applicable building codes may also have additional requirements related to residential aircraft hangars.

The entire floor area of aircraft hangers up to 18 inches above the floor and adjacent areas not suitably cut off from the hazardous area, or not elevated at least 18 inches above it, are classified as Class I, Division 2 locations. Pits below grade are classified as Class I, Division 1. Within 5 feet horizontally from the aircraft power plants, fuel tanks, and structures containing fuel, the Class I, Division 1 location extends up to a level that is 5 feet above the upper surface of the wings and engine enclosures. Fixtures and other equipment that produce arcs or sparks are required to be totally enclosed or constructed to prevent the escape of sparks or hot metal particles if less than 10 feet above the aircraft wings or engine enclosures.

GASOLINE DISPENSING AND SERVICE STATIONS AND BULK STORAGE PLANTS

Gasoline dispensing and service stations are defined in *NEC® Article 514* as locations where gasoline or other volatile, flammable liquids or liquified flammable gases are transferred to fuel tanks, including auxiliary fuel tanks of self-propelled vehicles or approved containers. Other areas such as lubritoriums, service rooms, repair rooms, offices, sales rooms, compressor rooms, and similar locations shall comply with *Articles 510* and *511* with respect to the electrical wiring and equipment. Where the AHJ can satisfactorily determine that flammable liquids having a flash point below 38°C (100°F), such as gasoline, are not handled, such locations need not be classified. For additional information regarding safeguards for gasoline dispensing and service stations, see the Automotive and Marine Service Station Code NFPA 30-1996.

For additional information pertaining to LP gas systems other than residential or commercial, see NFPA 58-1998. For storage and handling of liquified petroleum gases, see NFPA 59-1998. For gasoline dispensing stations and marine areas such as boat yards, see *NEC® Section 555-9*. Bulk storage facilities and bulk storage plants must comply with *Article 515*, and additional information for those may also be found in NFPA 30-1996.

The space within a dispenser up to a height of 4 feet and the space within 18 inches horizontally of a dispenser and extending to a height 4 feet above its base must be classified as a Class I, Division 1 location. (See Figure 9-20.) Any wiring or equipment installed beneath any part of a Class I, Division 1 or Division 2 location is classified as within the Class I, Division 1 location to the point of emergence from below grade. Where a dispensing unit or its hose or nozzle valve has not been suspended from an overhead, the space within the dispenser enclosure and the area within 18 inches in all directions from the enclosure that is not suitably cut off by a ceiling or wall is classified as a Class I, Division 1 area. The area within 20 feet in all directions from the Class I, Division 1 area and extending down to grade level is classified as a Class I, Division 2 area. The horizontal area 18 inches above grade and extending 20 feet measured from those points vertically below the outer edges of the

overhead dispensing enclosure is also classified as a Class I, Division 2 area.

All equipment integral with the overhead dispensing hose and nozzle must be suitable for Class I, Division 1 hazardous locations. Lubritoriums, service rooms, repair rooms, and other areas not suitably cut off or elevated at least 18 inches above these areas are classified and considered as commercial garages. Pits or any other space below grade in lubrication areas are considered to be Class I, Division 1 locations. The area within 3 feet measured in any direction from the dispensing point of a hand-operated unit dispensing a Class I flammable liquid, such as white gas, is considered a Class I, Division 2 location. Any outside area not classified as a Class I, Division 1 location within 20 feet horizontally of the interior enclosure of the dispensing pump or within 10 feet horizontally from a tank filled by and fewer than 18 inches above grade is to be classified as a Class I, Division 2 location. (See Figure 9-20.)

A spherical volume within a 3-foot radius of the point of discharge of any tank vent pipe is classified as a Class I, Division 1 location. That part of the spherical volume lying between a 3-foot radius and a 5-foot radius of the point of discharge of any tank

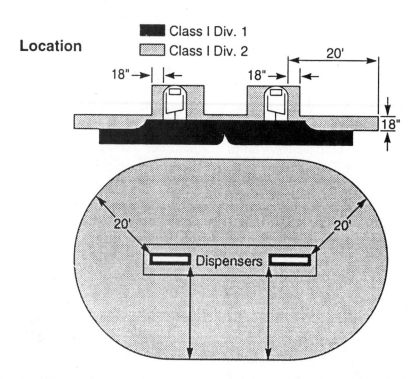

Figure 9-20 Gasoline dispensing and service station installation as per *NEC® Article 514* (Reprinted with permission from NFPA 70-1999.

vent pipe is classified as a Class I, Division 2 location. When the tank vent does not discharge upward, the cylindrical volume below the Class I, Division 1 location and extending to grade level is classified as a Class I, Division 2 location.

Spaces beyond the 5-foot radius from the tank vents that discharge upward, spaces beyond unpierced walls, and areas below grade that lie beneath tank vents are not classified as hazardous. A seal is required in the first fitting after a conduit emerges from the earth or concrete and enters a dispenser or any cavity or enclosure in connection with it. A seal is required in any conduit that extends upward beyond the vertical boundary of a Class I, Division 2 area, such as a conduit supplying a lighting standard within a hazardous area. The required seal must be placed so that no coupling or fitting of any kind is between the seal and the boundary line dividing the hazardous and nonhazardous areas. A seal is required at or beyond the horizontal boundary at the point where the conduit emerges from below grade. The seal must be in the first fitting where the conduit emerges into a nonhazardous area. If the conduit does not extend beyond the 20-foot horizontal boundary but emerges and extends upward while still within the hazardous area, the seal must be placed so that no coupling or fitting is between it and the vertical boundary.

Circuits leading to or through a gasoline dispenser enclosure (see Figure 9-20) must be provided with switches or circuit breakers that simultaneously disconnect all circuit conductors, including the neutral, from the source of supply. *NEC® Section 514-5* states that single-pole breakers utilizing handle ties are not permitted for this application. The wiring of bulk storage plants includes locations where gasoline or other volatile, flammable liquids are stored in tanks having an aggregate capacity of one carload or more and from which such products are distributed usually by tank truck.

The degree of hazard or flammability of liquids is commonly measured in terms of the flash point. The flash point is the minimum temperature at which the liquid gives off vapors sufficient to form an ignitable mixture with air near the surface of the liquid when tested under established procedures. Flammable liquids having a flash point less than 100°F require the greatest protection against accidental ignition. Gasoline, for example, has a flash point of about 45°F and is very hazardous material under all

climatic conditions. Diesel fuel has a flash point of 100°F. The flash point of jet fuels ranges from 10°F to 145°F. See *Fire Hazard Properties in Flammable Liquid, Gas and Volatile Solids*, NFPA 325-1994. Adequately ventilated indoor areas containing pumps, bleeders, withdrawal fitting, meters, and similar equipment that is connected to pipelines handling flammable liquids under pressure are considered Class I, Division 2 locations where such spaces are within a 5-foot radius in any direction of an exterior surface of such equipment.

A Class I, Division 2 location also extends 25 feet horizontally from the surface of such equipment and upward to a height of 3 feet above grade. (See Figure 9-21.) Where such spaces are inadequately vented, they are considered Class I, Division 1 locations. Other provisions affecting ventilating requirements are discussed in the flammable liquids code, NFPA 30. Outdoor areas containing the same equipment itemized in the preceding paragraph are to be considered Class I, Division 2 locations if within a 3-foot radius in all directions from the exterior surfaces of such equipment.

The Class I, Division 2 location also extends 10 feet horizontally from any surface of such equipment and extends upward to a height of 18 inches above grade. According to *NEC® Table 515-2* and Figure 9-21, such spaces are considered Class I, Division 1 locations within 3 feet in all directions of the vent pipe or fill pipe opening. Class I, Division 2 locations extend in all directions within the space between a 3-foot and 5-foot radius from the vent pipe or fill pipe opening and horizontally within the radius of 10 feet from the vent or fill pipe opening to a height of 18 inches above grade. (see Figure 9-21). In all indoor areas where volatile, flammable liquids are transferred to individual containers or where fire-enforced ventilation is not reliably maintained, the entire area must be classified as a Class I, Division 1 location (*Table 515-2(b)*).

More limited hazardous locations are also defined for outdoor areas where volatile, flammable liquids are transferred to individual containers as well as for indoor areas that have reliably maintained, fire-enforced ventilation.

In outside locations, the space extending 3 feet in all directions from a vehicle tank when loading through an open dome or from a vent when loading

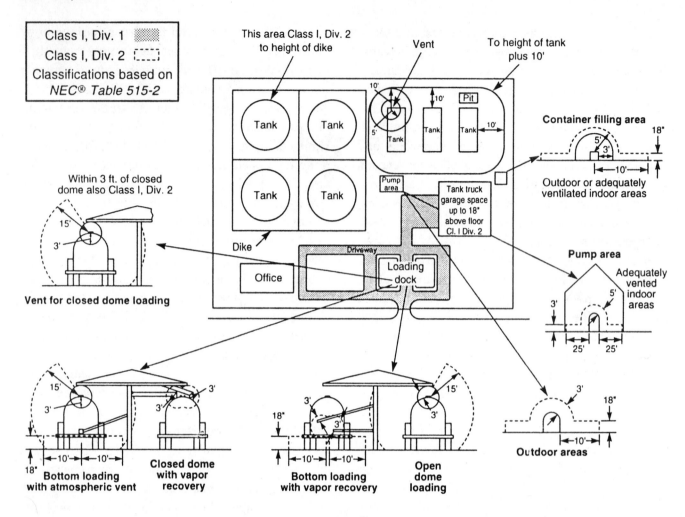

Figure 9-21   Bulk storage facility installation as per *NEC® Article 515* (Courtesy of Crouse-Hinds, Division of Cooper Industries)

through a closed dome having atmospheric venting is classified as a Class I, Division 1 location. Between a 3-foot and 5-foot radius, it is a Class I, Division 2 location. (See Figure 9-21.) Hazardous areas surrounding loading facilities and tank facilities are defined in *NEC® Table 515-2* as spaces within 3 feet in all directions of a fixed connection used in bottom loading or unloading. Loading through a closed dome having atmospheric venting or in loading through a closed dome having a vapor recovery system are considered Class I, Division 2 locations. Where bottom loading is used, the Class I, Division 2 location covers a 10-foot radius from the point of connection to a height of 18 inches above grade. (See Figure 9-21.)

In above-ground tanks the space between the shell and above the roof of floating roof tanks are Class I, Division 1 locations. The area within 10 feet of the shell sides and ends and the roof of an above-ground tank that does not have a floating roof is classified as a Class I, Division 2 location. Where dikes are provided, the area within the dike is also a Class I, Division 2 location. Any point within a 5-foot radius of a vent opening of an above-ground tank is a Class I, Division 1 location. The space between a 5-foot and a 10-foot radius from the vent of an above-ground tank is a Class I, Division 2 location. (See Figure 9-21.) Hazardous locations near fill pipes and vent pipes for underground tanks are defined in *NEC® Table 514-2*. (See Figure 9-21.) Any pit or depression that lies wholly or partly within either a Class I, Division 1 or Division 2 location must be classified entirely as Class I, Division 1 location unless provided with forced draft ventilation. Where such forced draft ventilation is provided, all such spaces may be classified as a Class I, Division 2 locations. Any pit or depression that contains gasoline piping, valves, or fittings is deemed a Class I, Division 2 location when located in areas that would not otherwise be classified as Class I areas, such as where gasoline pipeline or the pit outside the bulk storage plant contains meters or valves.

Storage and repair garages for tank trucks are classified as commercial garages, that is, the Class I, Division 2 area extends up to a height of 18 inches above grade unless the enforcement authority rules that the hazardous area must extend higher than 18 inches. Wiring within hazardous areas is required to comply with the *NEC®* requirements for Class I, Division 1 or Class I, Division 2. Fixed equipment installed above the hazardous areas in bulk storage plants is required to be of the totally enclosed type if capable of producing arcs or sparks unless provided with screens or guards to prevent the escape of the hot particles produced. Underground wiring is required to be installed with threaded rigid metal conduit or threaded steel intermediate conduit, or where buried not less than 2 feet below the earth is permitted to be in rigid nonmetallic conduit or an approved cable.

Where rigid nonmetallic conduit is used, threaded rigid metal conduit or threaded intermediate metal conduit shall be used for the last 2 feet of underground run to emergence or to the point of connection to the above-ground raceway. Where cable is used, it shall be enclosed in threaded rigid metal conduit or threaded steel intermediate metal conduit from the point of the lowest buried cable level to the point of connection above the raceway.

Sealing fittings are required in electrical systems in accordance with the *NEC®* provisions for Class I areas, that is, in all raceways entering or leaving a hazardous location, all raceways entering or leaving an arcing device, and all raceways 2 inches in size or larger entering or leaving a junction box and containing terminals, taps, or splices. Office areas, boiler rooms, and similar locations beyond the limits of the hazardous area are not required to come within the special wiring rules of these articles.

SPRAY APPLICATION, DIPPING, AND COATING PROCESSES

The application of flammable liquids, combustible liquids, and combustible powders by spray operations and the application of flammable liquids or combustible liquids at temperatures above their flash point by dipping, coating, and other means regularly or frequently are covered in *NEC® Article 516.* Additional further information regarding the safeguards for these processes may be found in NFPA 33-1995, NFPA 34-1995, and NFPA 91-1995. These finishes may be applied by dipping, brushing, spraying, pneumatic spraying, fixed electrostatic spraying, detearing, electrostatic hand spraying, electrostatic fluidized beds, or powder spraying with electrostatic power-spraying guns.

Two distinct hazards are present in paint-spraying operations: explosive mixtures in the air from the spray and its vapor, and the combustible residue that may accumulate inside the booth. The interior spray booths and their exhaust ducts and all spaces within 20 feet horizontally in any direction from any spraying operation more extensive than touch-up spraying and not included within spray booths, and all space within 20 feet horizontally in any direction of dip tanks and their drainboards are classified as Class I, Division 1 locations. These particular facilities may contain both Class I and Class II requirements, depending on the material applied.

Interiors of spray booths and rooms, generally the interior of exhaust ducts, any area in the direct path of spray operations, areas for dipping and coating operations, and all space within a 5-foot radial distance from the vapor sources extending from these surfaces to the floor are all considered hazardous areas. Where the vapor source is a liquid exposed in the process, and the drainboard and any dipped or coated object from which it is possible to measure the vapor for concentrations exceeding 25% of the lower flammable limit at a distance of 1 foot in any direction, pits within 25 feet horizontally of the vapor source (if the pit is beyond 25 feet, a vapor stop must be provided or the entire pit is classified), the interior of any enclosed coating or dipping process—these areas are classified as Class I or Class II, Division 1 locations, depending on the material being applied.

The Class I, Division 2 locations include the following spaces: open spraying areas, all space outside but within 20 feet horizontally and 10 feet vertically of the Division 1 location as defined and not separated by partitions (see Figure 9-22), spraying operations conducted within an enclosed-top, open-face, or front

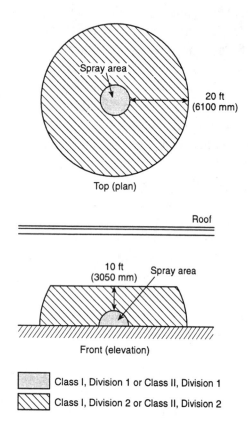

Top (plan)

Roof

10 ft
(3050 mm)   Spray area

Front (elevation)

☐ Class I, Division 1 or Class II, Division 1

▨ Class I, Division 2 or Class II, Division 2

ˣFigure 516-2(b)(1)   Class I or Class II, Division 2 location adjacent to an unenclosed spray operation.

Figure 9-22 (a) Class I or Class II Division 2 locations. Adjacent to an unenclosed spray operation (Reprinted with permission from NFPA 70-1999)

spray booth or room, the space shown in Figure 9-23, and the space within 3 feet in all directions from the openings other than the open face or front.

Class I and II, Division 2 locations shown in Figure 23 extend from the open face or front of the spray booth or room as follows: If the ventilation system is interlocked with the spraying equipment so as to make the spraying equipment inoperative when the ventilation system is not in operation, the space extends 5 feet from the open face or front of the booth or room, as otherwise shown in Figure 9-21b. When the ventilation system is not interlocked with the spraying equipment, the space extends 10 feet from the open face or front of the spray booth. (See Figure 9-23.) For spraying operations conducted within an open-top spray booth, the space 3 feet vertically above the booth and within 3 feet of other

booth openings is considered a Division 2 location. For spraying operations confined to an enclosed booth or room, the space within 3 feet in all directions from any openings must be considered as a Division 2 location, as shown in Figure 9-23.

The dip tanks and drainboards and the space 3 feet above the floor and extending 20 feet horizontally in all directions from the Division 1 location are not considered hazardous where the vapor source area is 5 feet or fewer and where the contents of the open tank, trough, or container do not exceed 5 gallons. In addition, the vapor concentration during operation shutdown periods shall not exceed 25% of the lower flammable limit outside the Class I location specified in *NEC® Section 516-2(a)(4)*.

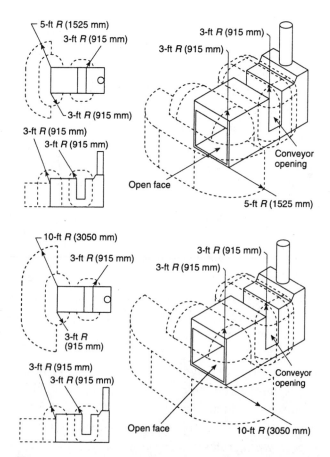

ˣFigure 516-2(b)(2)   Class I or Class II, Division 2 locations adjacent to a closed top, open face, or open front spray booth or room.

Figure 9-23 Class I or Class II Division 2 locations adjacent to a closed top, open face, or open front spray booth or room. (Reprinted with permission from NFPA 70-1999.)

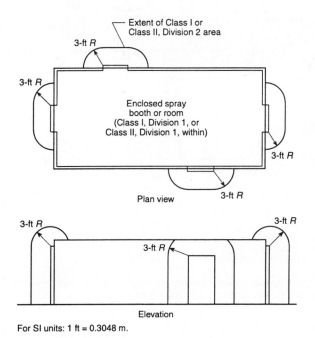

3-ft R

Extent of Class I or Class II, Division 2 area

3-ft R

Enclosed spray booth or room (Class I, Division 1, or Class II, Division 1, within)

3-ft R

Plan view

3-ft R

3-ft R

3-ft R

3-ft R

Elevation

For SI units: 1 ft = 0.3048 m.

*Figure 516-2(b)(4)   Class I (or Class II), Division 2 locations adjacent to an enclosed spray booth or spray room.

Figure 9-24   Class I or Class II Division 2 locations adjacent to openings in an enclosed spray booth or room. (Reprinted with permission from NFPA 70-1999.)

In an enclosed coating and dipping operation, the spaces adjacent to these operations are not considered to be classified except within 3 feet in all directions of any opening or enclosure that is classified as a Division 2 location. Adjacent areas that are cut off by tight partitions without communicating openings and within which hazardous vapors or combustible powders are not likely to be released are also not classified.

Areas utilizing drying, curing, or fusion apparatus provided with positive mechanical ventilation adequate to prevent accumulation of flammable concentrations of vapors and provided with effective interlocks to deenergize all electrical equipment other than the equipment approved for Class I locations is permitted to be classified as nonhazardous where the authority having jurisdiction so judges.

Additional safeguards for these drying ovens

or furnaces may be found in NFPA 86-1995. All electric wiring within the Class I location shall comply with the applicable provisions of *NEC® Article 501*. Additional information regarding fixed electrostatic equipment, electrostatic hand-spraying equipment, and the wiring of equipment for Class II locations and the areas above Class I and Class II locations may be found in *Article 516*.

HEALTH CARE FACILITIES, RESIDENTIAL AND CUSTODIAL FACILITIES, AND LIMITED CARE FACILITIES

The requirements for these areas are found in *NEC® Article 517*. Additional information regarding the installation, testing, and safeguards of these areas may be found in NFPA 99-1996. Specific definitions of areas within these facilities may be found in *Section 517-3*. Health care facilities include the electrical equipment installations in hospitals, nursing homes, extended care facilities, clinics, and dental offices. In addition, *Article 517* includes provisions for electrical installations in areas provided for patients undergoing a type of treatment involving the use of electrical energized equipment connected to the heart.

Requirements for essential and emergency electrical systems in hospitals have also been covered in this *NEC®* article. These specialized systems must meet exacting, detailed requirements, a discussion of which is beyond the scope of this manual. The following comments are limited to the electrical requirements in special hazardous areas in hospitals, usually identified as anesthetizing locations, including hospital operating rooms, delivery rooms, anesthesia rooms, and any corridor, utility room, or other area that may be used for this purpose, or where flammable anesthetics are stored.

The hazardous space and flammable anesthetizing locations extend from the floor to a height of 5 feet and are classified as Class I, Division 1 locations. The entire room where flammable anesthetics are stored is classified as a Class I, Division 1 location. Wiring and equipment within the hazardous space operating at more than 8 volts must comply

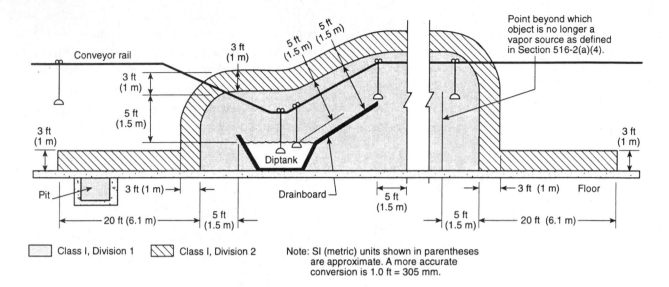

^x**Figure 516-2(b)(5)   Electrical area classification for open processes without vapor containment or ventilation.**

Figure 9-25   The extent of Class I Division 1 and Class I Division 2 hazardous (Classified) locations for open dipping processes. (Reprinted with permission from NFPA 70-1999.)

with the requirements for Class I, Division 1, Group C locations. Receptacles installed above the hazardous space and intended for the use of room equipment and instruments should be of the same type as required within the hazardous space to accommodate plugs on the equipment and to comply with these requirements.

All circuits wholly or partially within the anesthetizing location, except circuits supplying only fixed lighting and other surgical lighting fixtures and approved, permanently installed X-ray equipment, must be supplied from an underground system isolated from a distribution system supplying any other areas. Isolation is provided by one or more bubble-wound transformers, generator sets, or batteries installed in a nonhazardous location. Maximum potential on the primaries or secondaries of isolating transformers is limited to 300 volts. Branch circuits supplying an anesthetizing location may not supply any other location.

Special insulation on circuit conductors is also required to minimize capacitance of circuitry. These circuits are required to be supplied by the un-

grounded side of an isolated transformer. Branch circuits supplying only fixed lighting fixtures in nonhazardous areas other than surgical lighting fixtures or approved, permanently installed X-ray equipment may be supplied by grounded circuits. If supplied by a grounded circuit, the conductors must be in a separate raceway from the conductors of the ungrounded circuit. The lighting fixtures and X-ray equipment must be located 8 feet above the floor, switches controlling the grounded circuits located outside the anesthetizing location. Low-voltage current circuits are required by the *NEC*® for electrical apparatuses and equipment having exposed, current-carrying parts frequently in contact with patients' bodies. Such equipment circuits must be designed and approved for operation of 8 volts or fewer. More detailed and complete information may be found in NFPA 99.

> **NOTE:** The NFPA 99 handbook provides additional assistance for designing or installing a health care facility.

INTRINSICALLY SAFE SYSTEMS—NEC® ARTICLE 504

Intrinsic safety is an explosion-prevention design technique applied to electrical equipment and the wiring of hazardous locations where flammable or combustible material is present. The technique is based on limiting electrical and thermal energy levels below those required to ignite the specific hazardous atmospheric mixture present. Intrinsically safe wiring shall not be capable of releasing sufficient electrical or thermal energy under normal or abnormal conditions to cause ignition to a specific flammable or combustible atmospheric mixture in its most easily ignitable concentration. It is important to define the class and group for which any proposed intrinsically safe electrical circuits are to be installed. Intrinsic safety is a technique for the worst-case hazardous locations; therefore consideration of the division is not necessary. Because intrinsic safety maintains energy levels below those required to ignite specific hazardous mixtures, it is important to know what the energy allowances are for operational and safety considerations. Intrinsically safe systems (*NEC® Article 504*) were introduced as acceptable wiring systems in the 1990 Code. Careful interpretation and application of this article provides safe, hazard-free installations.

The use of intrinsically safe equipment is primarily limited to process control instrumentation, because these electrical systems lend themselves to low-energy requirements. ANSI/UL 913-1997 provides information on the design test and evaluation. Underwriters Laboratories and Factory Mutual list several devices in this catagory. The equipment and its associated wiring must be installed so they are positively separated from the nonintrinsically safe circuits. Induced voltages can defeat the concept of intrinsically safe systems. *NEC® Section 504-1* explains that this article is intended to cover intrinsically safe apparatus and wiring in Class I, II, and III locations. All intrinsically safe apparatuses and associated apparatuses must be approved and accepted as modified in *Article 504*. All applied articles of the Code apply to these systems.

In applying *NEC® Article 504*, the entire intrinsically safe system must be installed in accordance with controlled drawings, as required in *Section 504-10*. The controlled drawing identification must be marked on all equipment and apparatuses. Generally, intrinsically safe systems may be permitted using any wiring method suitable for unclassified locations. However, sealing shall be provided in accordance with *Section 504-70*, and separation of intrinsically safe systems shall be provided in accordance with *Section 504-30*.

Although general wiring methods are permitted, intrinsically safe systems always fail in the safe position. If a wire or cable is crushed or cut accidentally, the operation of the equipment is fail-safe in every instance. No arcing or sparking can occur owing to the design and nature of intrinsically safe systems. Designers generally require physical protection such as metal raceways to protect intrinsically safe conductors and provide protection from costly accidental shutdowns and operational stoppages.

NEC® 504-70 requires that conduits and cables be specified in accordance with the hazardous location in which they are installed (e.g., *Article 501* for Class I hazardous flammable vapors and gases). Sealing of intrinsically safe systems must be provided as specified in *Section 501-5*. For flammable or combustible dust, sealing is required as specified in *NEC® 502-5*. Although this newly accepted system provides safe installations, it is paramount that they be installed precisely as designed and shown on the control drawings and as permitted in *Article 504* of the *National Electrical Code®*. Manufacturers of these systems should provide additional guidance to the installer.

NONINCENDIVE

Nonincendive equipment or circuits are permitted by *NEC® Sections 501-3(b)(1) Exception, Item C* and *501-14(b)(1) Exception, Item C*. They are defined by manufacturers as components having contacts that make or break an incendive circuit and whose contacting mechanism, or the enclosure in which the contacts are housed, is constructed so that the component is not capable of igniting the surrounding specific flammable gas, vapor, or air mixture. The housing of a nonincendive component is not intended to exclude those flammable atmospheres. Although this definition does not appear in the *National Electrical Code*®, equipment may be permitted by the previously mentioned sections, which permit general-purpose enclosures for circuits that under normal conditions do not release sufficient energy to ignite a specific ignitable atmospheric mixture.

> **NOTE:** Wiring methods to these enclosures are not accepted and must comply with the applicable sections listed in *Sections 501-3, 501-4,* and *501-5* of the *National Electrical Code*®, which require wiring and sealing.

REVIEW QUESTIONS

1. In a storage and repair facility for motorcycles ten or more machines are repaired or worked on at the same time, compliance with *NEC*® Section 511-2 is mandated for the purpose of classifying the facility. True or False?

2. *NEC*® Section 514-6(a) requires that the first fitting be a sealoff device where the conduit emerges from the ground or concrete at underground pits near gasoline tanks with submersible pumps. True or False?

3. A business that specializes in oil changes for gasoline-powered vehicles has inspection pits and service bay areas where personnel have access to the underportions of the vehicles. Extension cords, portable lights, and portable power tools are used in these service bays and also in the pits. Are the pits classified? If so, what classification would these pits have, and under what conditions?

4. I want to install paddle fans in an aircraft hangar. The installation will be above 5 feet and within 10 feet of the wings and power plants of the aircraft housed within. Is this permissible? If not, where could these paddle fans be located without violating the requirements of the *National Electrical Code*®?

5. Exposed metal raceway surfaces in general patient care areas in a health care facility are not permitted to exceed what differences at frequencies of 1000 Hertz or less measured across a 1000-ohm resistance?

6. An underground gasoline tank remote pump from the dispenser is listed for a Class I, Division 1 location, with the pump having a factory seal provided between the terminal housing and the motor. Is a listed sealoff fitting required by *NEC*® 501-5(a)(1) for this underground gasoline pump necessary?

7. A commercial garage has an attached car showroom. There is only one doorway between them and no fire separation. May the showroom be wired in accordance with the first three

chapters of the Code, such as in EMT nonmetallic sheath cable or Type AC cable? What if the garage portion of the building has four air changes per hour?

8. A diesel dispenser is mounted adjacent to a gasoline dispenser on a service station island. A sealing fitting is installed in the conduit entering the gasoline dispenser. Is a sealing fitting required in the conduit entering the diesel fuel dispenser?

9. A rigid metal conduit runs under a commercial garage floor and extends up through the 18-inch hazardous area with no couplings or fittings in that area. Is a sealoff fitting required in this conduit?

10. What wiring methods are approved for use in a Class I, Division 1 location?

Appendix

Codes and Standards

Figure A-1 Allowable Ampacity of Multiple Conductors in Raceway

Note: for other Amperages see *NEC $^{\circledR}$ Section 310–15*

Number of Conductors in Raceway

1 to 3	4 to 6	7 to 9	10 to 20	21 to 30	31 to 40	41 and above
15	12	10.5	7.5	6.75	6	5.25
20	16	14	10	9	8	7
25	20	17.5	12.5	11.25	10	8.75
30	24	21	15	13.5	12	10.5
35	28	24.5	17.5	15.75	14	12.25
40	32	28	20	18	16	14.00
45	36	31.5	22.5	20.25	18	15.75
50	40	35	25	22.5	20	17.50
55	44	38.5	27.5	24.75	22	19.25
60	48	42	30	27	24	21.00
65	52	45.5	32.5	29.25	26	22.75
70	56	49	35	31.5	28	24.50
75	60	52.5	37.5	33.75	30	26.25
80	64	56	40	36.00	32	28.00
85	68	59.5	42.5	38.25	34	29.75
90	72	63	45	40.50	36	31.50
95	76	66.5	47.5	42.75	38	33.25
100	80	70	50	45.00	40	35.00
110	88	77	55	49.50	44	38.50
120	96	84	60	54.00	48	42.00
125	100	87.5	62.5	56.25	50	43.75
150	120	105	75	67.50	60	52.50
175	140	122.5	87.5	78.75	70	61.25
200	160	140	100	90.00	80	70.00
225	180	157.5	112.5	101.25	90	78.75
250	200	175	125	112.50	100	87.50
275	220	192.5	137.5	123.75	110	96.25

Figure A-1   Allowable Ampacity of Multiple Conductors in Raceway
(continued)

Note: for other Amperages see *NEC* ®*Section 310–15*   +

Number of Conductors in Raceway

1 to 3	4 to 6	7 to 9	10 to 20	21 to 30	31 to 40	41 and above
300	240	210	150	135.00	120	105.00
325	260	227.5	162.5	146.25	130	113.75
350	280	245	175	157.50	140	122.50
375	300	262.5	187.5	168.75	150	131.25
400	320	280	200	180.00	160	140.00
425	340	297.5	212.5	191.25	170	148.75
450	360	315	225	202.50	180	157.50
475	380	332.5	237.5	213.75	190	166.25
500	400	350	250	225.00	200	175.00

Example: What size Type THW copper conductor is required for a 65-ampere load where the conductor is in a raceway having 3, 6, 9, 18, 30, 39, or 45 conductors?

Solution:

No. of conductors in a raceway	Derating factor, percent	Min. ampacity before derating	Type THW conductor size required to supply 65 ampere load
3	100	65	6
6	80	85	4
9	70	100	3
18	50	130	1
30	45	150	1/0
39	40	175	2/0
45	35	200	3/0

See *Sections 310-10* and *310-15* for additional derating requirements that may apply.

Figure A-2

Table B
Impedance Factors for Finding Voltage Drop in Copper Conductors
(Courtesy of American Iron and Steel Institute)
Single-phase AC, Power factor

Wire size, AWG	100	90	80	70	60	DC
14	.5880	.5360	.4790	.4230	.3650	.5880
12	.3690	.3330	.3030	.2680	.2320	.3690
10	.2320	.2150	.1935	.1718	.1497	.2320
8	.1462	.1373	.1248	.1117	.0981	.1462
6	.0918	.0882	.0812	.0734	.0653	.0918
4	.0578	.0571	.0533	.0489	.0440	.0578
2	.0367	.0379	.0361	.0337	.0309	.0363
1	.0291	.0311	.0299	.0284	.0264	.0288
1/0	.0233	.0257	.0252	.0241	.0227	.0229
2/0	.0187	.0213	.0212	.0206	.0196	.0181
3/0	.0149	.0179	.0181	.0177	.0171	.0144
4/0	.0121	.0152	.0156	.0155	.0151	.0114

Wire size, kcmil	100	90	80	70	60	DC
250	.0102	.0136	.0143	.0143	.0141	.0097
300	.0086	.0121	.0128	.0131	.0130	.0081
350	.0074	.0109	.0118	.0122	.0122	.0069
400	.0066	.0101	.0111	.0115	.0116	.0060
500	.0054	.0089	.0099	.0105	.0108	.0048
600	.0047	.0083	.0093	.0099	.0103	.0040
700	.0041	.0077	.0088	.0094	.0098	.0034
750	.0039	.0075	.0086	.0093	.0097	.0032
800	.0037	.0073	.0084	.0091	.0095	.0030
900	.0034	.0069	.0081	.0088	.0093	.0027
1000	.0031	.0067	.0079	.0086	.0091	.0021

Voltage drop in two wires of a single-phase AC or DC circuit is equal to current × length of run in hundreds of feet × impedance factor.
Impedance factors in table multiplied by 0.86 will give phase-to-phase voltage drop for 3-phase AC circuits.

Example: Find voltage drop in a 300-foot circuit using No. 4/0 Type THWN copper conductors in a 3-phase circuit. Allowable ampacity (*NEC* ® *Table 310-16*) is 230; power factor, 80 percent.

Solution:
Voltage drop = 230 × 3.00 × 0.0156 × 0.86 = 9.3 volts.

Figure A-3 General Lighting Loads
(Reprinted with permission from NFPA 70–1999.)

Table 220-11. Lighting Load Demand Factors

Type of Occupancy	Portion of Lighting Load to Which Demand Factor Applies (Volt-Amperes)	Demand Factor (Percent)
Dwelling units	First 3000 or less at From 3001 to 120,000 at Remainder over 120,000 at	100 35 25
Hospitals*	First 50,000 or less at Remainder over 50,000 at	40 20
Hotels and motels, including apartment houses without provision for cooking by tenants*	First 20,000 or less at From 20,001 to 100,000 at Remainder over 100,000 at	50 40 30
Warehouses (storage)	First 12,500 or less at Remainder over 12,500 at	100 50
All others	Total volt-amperes	100

*The demand factors of this table shall not apply to the computed load of feeders or services supplying areas in hospitals, hotels, and motels where the entire lighting is likely to be used at one time, as in operating rooms, ballrooms, or dining rooms.

Table 220-3(a). General Lighting Loads by Occupancies

Type of Occupancy	Unit Load per Square Foot (Volt-Amperes)
Armories and auditoriums	1
Banks	3½[b]
Barber shops and beauty parlors	3
Churches	1
Clubs	2
Court rooms	2
Dwelling units[a]	3
Garages — commercial (storage)	½
Hospitals	2
Hotels and motels, including apartment houses without provision for cooking by tenants[a]	2
Industrial commercial (loft) buildings	2
Lodge rooms	1½
Office buildings	3½[b]
Restaurants	2
Schools	3
Stores	3
Warehouses (storage)	¼
In any of the above occupancies except one-family dwellings and individual dwelling units of two-family and multifamily dwellings:	
Assembly halls and auditoriums	1
Halls, corridors, closets, stairways	½
Storage spaces	¼

Note: For SI units, 1 ft^2 = 0.093 m^2.

[a]See Section 220-3(b)(10).

[b]In addition, a unit load of 1 volt-ampere per square foot shall be included for general-purpose receptacle outlets where the actual number of general-purpose receptacle outlets is unknown.

Figure A-4   USES FOR RACEWAY AND OTHER WIRING METHODS

CHAPTER 3, NEC®

Installation conditions

NEC ARTICLE	TYPE	For conductors over 600 volts	Inside Buildings	Outside Buildings	Under Ground	Cinder Fill	Embedded in Concrete	Wet locations	Dry Locations	Corrosive Locations	Severe Corrosive Locations	Hazardous Locations	Mechanical Injury	Severe Mechanical injury	Exposed work	Concealed work
330	TYPE MI MINERAL INSULATED-SHEATHED CABLE	X	P	P	C	X	P	P	P	C	X	P	P	P	P	P
331	ELECTRICAL NONMETALLIC TUBING*	X	C	X	X	X	C	P	P	P	C	X	X	X	C	C
333	TYPE AC ARMORED CABLE	X	P	X	X	X	X	X	P	X	X	X	X	X	C	P
334	TYPE MC METAL CLAD CABLE	C	C	C	C	X	C	C	P	X	X	C	X	X	P	P
336	NONMETALLIC SHEATHED CABLE ROMEX TYPE NM	X	P	C	X	X	X	X	P	X	X	X	X	X	P	P
338	TYPE SE SERVICE-ENTRANCE CABLE	X	C	P	X	X	X	C	C	X	X	X	X	X	C	C
388	TYPE USE SERVICE-ENTRANCE CABLE	X	X	C	P		X	C	X	X	X	X	X	X	X	X,
339	TYPE UF UNDERGROUND FEEDER AND BRANCH CIRCUIT CABLE	X	P	P	P		X	P	P	X	X	X	X	X	C	C
343	NONMETALLIC UNDERGROUND CONDUIT WITH CONDUCTORS	P	X	-	P	P		C	X	C	C	X	-	-	-	-
345	IMC INTERMEDIATE METAL CONDUIT	P	P	P	P	C	P	P	P	P	C	P	P	P	P	P
345	IMC INTERMEDIATE COATED METAL CONDUIT	P	C	P	P	P	P	P	P	P	P	P	P	P	P	P
346	RIGID METAL CONDUIT (STEEL)	P	P	P	P	C	P	P	P	P	C	P	P	P	P	P
346	RIGID COATED METAL CONDUIT	P	C	P	P	P	P	P	P	P	P	P	P	P	P	P
346	RIGID METAL CONDUIT (ALUMINUM)	P	P	P	X	X	X	P	P	X	X	P	P	C	P	P
347	RIGID NONMETALLIC CONDUIT O (SCHEDULE 40) *	C	P	P	P	P	P	P	P	P	P	X	X	X	C	P
347	RIGID NONMETALLIC CONDUIT (SCHEDULE 80) *	C	P	P	P	P	P	P	P	P	P	X	P	P	P	P
348	EMT ELECTRICAL METALLIC TUBING	C	P	P	P	C	P	P	P	C	P	C	P	X	P	P
349	FLEXIBLE METALLIC TUBING	C	C	X	X	X	X	X	C	X	X	X	X	X	C	C
350	FLEXIBLE METAL CONDUIT STEEL AND ALUMINUM	X	P	C	X	X	X	C	P	X	X	X	X	X	P	P
351A	LIQUIDTIGHT FLEXIBLE METAL CONDUIT	X	P	P	C	C	C	C	P	X	X	C	X	X	P	P
351b	LIQUIDTIGHT FLEXIBLE NONMETALLIC CONDUIT *	X	P	C	C	C	C	C	P	P	P	C	X	X	P	P
352A	SURFACE METAL RACEWAYS—STRUT TYPE CHANNEL RACEWAYS	C	P	C	-	-	-	C	P	C	C	C	C	C	P	-
352B	SURFACE NONMETALLIC RACEWAYS *	C	P	C	-	-	-	C	P	C	C	C	C	X	P	-
354	UNDERFLOOR RACEWAYS	C	P	-	-	-	C	X	P	X	X	C	C	C	-	P
356	CELLULAR METAL FLOOR RACEWAYS	C	P	-	-	-	P	X	P	-	-	—	-	-	-	-
362A	WIREWAYS METAL	C	P	C	X	X	X	P	P	X	X	C	P	X	P	X
362B	WIREWAYS NONMETALLIC	C	C	X	X	X	X	C	C	C	C	X	C	X	P	X
364	BUSWAYS	C	P	C	X	X	X	C	P	X	X	X	P	X	P	C

P - GENERALLY PERMITTED   C - CONDITIONAL   X - NOT PERMITTED   — CONDITIONS DO NOT APPLY

Wiring methods covered by *Chapter 3, NEC®* are limited to systems utilizing voltages not exceeding 600 volts unless higher voltages are specifically permitted. *NEC® 300–21 & 22* may further limit some wiring methods.

* - Where subject to chemicals for which the raceway is specifically approved

* - (FPN) Extreme cold may cause some nonmetallic conduits to become brittle and therefore more susceptible to damage from physical contact

* - Except where the enclosed conductors' ambient temperature would exceed those for which the conduit is approved

Figure A-5   Load and Circuit Chart for Residential Electrical Systems
(Courtesy of American Iron and Steel Institute)

	Typical connected watts	Preferred circuit, kW	Volts	Wires*	Circuit breaker or fuse amperes	Outlets on Circuit	Comments
Kitchen area							
Range	12000	10	120/240	3/6	50	1	Use of more than one outlet is not recommended.
Oven, built in	4500	6	120/240	3/10	30	1	May be direct connected
Range top	6000	6	120/240	3/10	30	1	May be direct connected.
Range top	3300	4	120/240	3/12	20	1	May be direct connected.
Dishwasher	1200	2	120	2/12	20	1	Appliance may be direct connected on a single circuit.
Waste disposer	300	2	120	2/12	20	1	Appliance may be direct connected on a single circuit.
Broiler	1500	2	120	2/12	20	1 or more	Heavy-duty appliance regularly used at one location should have a separate circuit. Only one such unit should be attached to a single circuit at the same time.
Fryer	1300	2	120	2/12	20	1 or more	Heavy-duty appliance regularly used at one location should have a separate circuit. Only one such unit should be attached to a single circuit at the same time.
Coffeemaker	1000	2	120	2/12	20	1 or more	Heavy-duty appliance regularly used at one location should have a separate circuit. Only one such unit should be attached to a single circuit at the same time.
Refrigerator	300	2	120	2/12	20	1 or more	Separate circuit serving only refrigerator is recommended.
Freezer	350	2	120	2/12	20	1 or more	Separate circuit serving only refrigerator is recommended.
Laundry area							
Washer	1200	2	120	2/12	20	1 or more	
Dryer	5000	6	120/240	3/10	30	1	
Hand Iron	1650	2	120	2/12	20	1 or more	
Water heater	1000	2	120	2/12	20	1 or more	Consider possible use in other locations.
Water heater	4500	6	120/240	2/10	30	1	
Living areas							
Workshop	1500	2	120	2/12	20	1 or more	Separate circuits recommended.
Portable heater	1300	2	120	2/12	20	1	Should not be connected to circuit serving heavy-duty loads.
Television	300	2	120	2/12	20	1 or more	Should not be connected to circuit serving appliances.
Portable lighting	1200	2	120	2/12	20	1 or more	Provide one circuit for each 500 sq. ft. Divided receptacle may be switch controlled.
Fixed utilities							
Fixed lighting	1200	2	120	2/12	20	1 or more	Provide at least one circuit for each 1200 watts of fixed lighting.
Air conditioner, 3/4 hp	1200	2	120	2/12	20	1	Consider 4 kW circuits to all window or console type air conditioners. Outlets may then be adapted to individual 120/240-volt machines. Connection to general purpose or appliance circuits is not recommended.
1 1/2 hp	2400	4	120/240	3/12	20	1	
Central air conditioner	5000	6	120/240				Consult manufacturer for recommended connections.
Sump pump	300	2	120	2/12	20	1 or more	May be direct connected.
Heating plant	600	2	120	2/12	20	1	Direct connected. Some local codes require separate circuit.
Fixed bathroom heater	1500	2	120	2/12	20	1	Direct connected.
Attic fan	300	2	120	2/12	20	1 or more	May be direct connected. Individual circuit is recommended.

NOTE: (Number of wires)/(wire size, no.). Number of wires does not include required equipment-grounding wire.

Figure A-6 Basic Electrical Formulas
(Courtesy of American Iron and Steel Institute)

DC Circuit Characteristics

Ohm's Law:

$$E = IRI = \frac{E}{R}R = \frac{E}{I}$$

E = voltage impressed on circuit (volts)

I = current flowing in circuit (amperes)

R = circuit resistance (ohms)

Resistances in Series:

$$R_t = R_1 + R_2 + R_3 + \ldots$$

R_T = total resistance (ohms)

R_1, R_2 etc. = individual resistances (ohms)

Resistances in Parallel:

$$R_t = \frac{1}{\frac{1}{R_1} + \frac{1}{R_2} + \frac{1}{R_3} + \ldots}$$

Formulas for the conversion of electrical and mechanical power:

$$HP = \frac{watts}{746} (watts \times .00134)$$

$$HP = \frac{kilowatts}{.746} (kilowatts \times 1.34)$$

Kilowatts = HP × .746

Watts = HP × 746

HP = Horsepower

In direct-current circuits, electrical power is equal to the product of the voltage and current:

$$P = EI = I_2R = \frac{E_2}{R}$$

P = power (watts)

E = voltage (volts)

I = current (amperes)

R = resistance (ohms)

Solving the basic formula for I, E, and R gives

$$I = \frac{P}{E} = \sqrt{\frac{P}{R}}; \quad E = \frac{P}{I} = \sqrt{RP}; R = \frac{E2}{P} = \frac{P}{T^2}$$

Energy

Energy is the capacity for doing work. Electrical energy is expressed in kilowatt-hours (kWhr), one kilowatt-hour representing the energy expended by a power source of 1 kW over a period of 1 hour.

Efficiency

Efficiency of a machine, motor or other device is the ratio of the energy output (useful energy delivered by the machine) to the energy input (energy delivered to the machine), usually expressed as a percentage:

$$Efficiency = \frac{output}{input} \times 100\%$$

$$or \ Output = imput \times \frac{efficiency}{100\%}$$

Torque

Torque may be described as a force tending to cause a body to rotate. It is expressed in pound-feet or pounds of force acting at a certain radius:

Torque (pound-feet) = force tending to produce rotation (pounds) × distance from center of rotation to point at which force is applied (feet).

Relations between torque and horsepower:

$$Torque = \frac{33,000 \times HP}{6.28 \times rpm}$$

$$HP = \frac{6.28 \times rpm \ time \ torque}{33,000}$$

rpm = speed of rotating part (revolutions per minute)

AC Circuit Characteristics

The instantaneous values of an alternating current or voltage vary from zero to maximum value each half cycle. In the practical formula that follows, the "effective value" of current and voltage is used, defined as follows:

Effective value = 0.707 × maximum instantaneous value

Inductances in Series and Parallel:

The resulting circuit inductance of several inductances in series or parallel is determined exactly as the sum of resistances in series or parallel as described under DC circuit characteristics.

Impedance:

Impedance is the total opposition to the flow of alternating current. It is a function of resistance, capacitive reactance and inductive reactance. The following formulae relate these circuit properties:

$$X_L = 2\pi HzL \quad Xc = \frac{1}{2\pi HzC} \quad Z = \sqrt{R^2 + (X_L - X_C^2)}$$

X_L = inductive reactance (ohms)

X_c = capacitive reactance (ohms)

Z = impedance (ohms)

Hz - (Hertz) cycles per second

C = capacitance (farads)

L - inductance (henrys)

R = resistance (ohms)

π = 3.14

In circuits where one or more of the properties L, C, or R is absent, the impedance formula is simplified as follows:

Resistance only:	Inductance only:	Capacitance only:
Z = R	$Z = X_L$	Z = XC
Resistance and Inductance only:	Resistance and Capacitance only:	Inductance and Capacitance only:
$Z = \sqrt{R^2 + X_L^2}$	$Z = \sqrt{R^2 + X_C^2}$	$Z = \sqrt{X_L - X_C^2}$

Ohm's law for AC circuits:

$$E = 1 \times Z \qquad I = \frac{E}{Z} \qquad Z = \frac{E}{I}$$

Capacitances in Parallel:

$$C_t = C_1 + C_2 \ C_3 + \ldots$$

C_t = total capacitance (farads)

$C_1 C_2 C_3 \ldots$ = individual cpacitances (farads)

Capacitances in series:

$$C_t = \frac{1}{\dfrac{1}{C_1} + \dfrac{1}{C_2} + \dfrac{1}{C_3} + \cdots}$$

Phase Angle

An alternating current through an inductance lags the voltage across the inductance by an angle computed as follows:

$$\text{Tangent of angle of lag} = \frac{X_L}{R}$$

An alternating current through a capacitance leads the voltage across the capacitance by an angle computed as follows:

$$\text{Tangent of angle of lead} = \frac{X_C}{R}$$

The resultant angle by which a current leads or lags the voltage in an entire circuit is called the phase angle and is computed as follows:

$$\text{Cosine of phase angle} = \frac{R \text{ of circuit}}{Z \text{ of circuit}}$$

Power Factor

Power factor of a circuit or system is the ratio of actual power (watts) to apparent power (volt-amperes), and is equal to the cosine of the phase angle of the circuit:

$$PF = \frac{\text{actual power}}{\text{apparent power}} = \frac{\text{watts}}{\text{volts} \times \text{amperes}} = \frac{kW}{kVA} = \frac{R}{Z}$$

KW = kilowatts

kVA = kilowatt-amperes = volt-amperes + 1,000

PF = power factor (expressed as decimal or percent)

Single-Phase Circuits

$$kVA = \frac{EI}{1,000} = \frac{kW}{PF} \quad kW = kVA \times PF$$

$$I = \frac{P}{E \times PF} \quad E = \frac{P}{I \times PF} \quad PF = \frac{P}{E \times I}$$

$$P = E \times I \times PF$$

P = power (watts)

Two-Phase Circuits

$$I = \frac{P}{2 \times E \times PF} \quad E = \frac{P}{2 \times I \times PF} \quad PF = \frac{P}{E \times I}$$

$$kVA = \frac{2 \times E \times I}{1000} = \frac{kW}{PF} \quad kW = kVA \times PF$$

$$P = 2 \times E \times 1 \times PF$$

E = phase voltage (volts

Three-Phase Circuits, Balanced Star or Wye

$$I_N = 0 \quad I = I_P \quad E = \sqrt{3}E_P = 1.73 E_P$$

$$E_P = \frac{E}{\sqrt{3}} = \frac{E}{1.73} = 0.577E$$

I_N = current in neutral (amperes)

I = line current per phase (amperes)

I_P = current in each phase winding (amperes)

E = voltage, phase to phase (volts)

E_P = voltage, phase to neutral (volts)

Three-Phase Circuits, Balanced Delta

$$I = 1.732 \times I_P \quad I_P = \frac{1}{\sqrt{3}} = 0.577 \times I$$

$$E = E_P$$

Power:

Balanced 3-Wire, 3-Phase Circuit, Delta or Wye

For unit power factor (PF = 1.0):

$$P = 1.732 \times E \times I$$

$$I = \frac{P}{\sqrt{3}} \quad E = 0.577\frac{P}{E} \qquad E = \frac{P}{\sqrt{3}} \times I = 0.577\frac{P}{I}$$

P = total power (watts)

For any load:

$$P = 1.732 \times E \times I \times PF \quad VA = 1.732 \times E \times I$$

$$E = \frac{P}{PF \times 1.73 \times I} = 0.577 \times \frac{P}{PF \times I}$$

$$I = \frac{P}{PF \times 1.73 \times E} = 0.577 \times \frac{P}{I} \times E$$

$$PF = \frac{P}{1.73 \times I \times E} = \frac{0.577 \times P}{I \times E}$$

VA = apparent power (volt-amperes)

P = actual power (watts)

E = line voltage (volts)

I = line current (amperes)

Power Loss:

Any AC or DC Circuit

$$P = I_2 RI = \sqrt{\frac{P}{R}} \quad R = \frac{P}{I^2}$$

P = power heat loss in circuit (watts)

I = effective current in conductor (amperes)

R = conductor resistance (ohms)

Load Calculations

Branch Circuits—Lighting & Appliance 2-Wire:

$$I = \frac{\text{total connected load (watts)}}{\text{line voltage (volts)}}$$

I = current load on conductor (amperes)

3-Wire:

Apply same formula as for 2–wire branch circuit, considering each line to neutral separately. Use line-to-neutral voltage; result gives current in line conductors.

Figure A-7
(Courtesy of American Iron and Steel Institute)

Electrical Wiring Symbols

Selected from American National Standard Graphic for
Electrical Wiring and Layout Diagrams Used in Architecture and Building Construction
ANSI Y32.9-1972

1. Lighting Outlets

Ceiling   *Wall*

1.1 Surface or pendant incandescent, mercury-vapor, or similar lamp fixture

1.2 Recessed incandescent, mercury-vapor or similar lamp fixture

1.3 Surface of pendant individual fluorescent fixture

1.4 Recessed individual fluorescent fixture

1.5 Surface or pendant continuous-row fluorescent fixture

1.6 Recessed continuous-row fluorescent fixture

1.8 Surface or pendant exit light

1.9 Recessed exit light

1.10 Blanked outlet

1.11 Junction box

1.12 Outlet controlled by low-voltage switching when relay is installed in outlet box

2. Receptacle Outlets

Grounded   *Ungrounded*

2.1 Single receptacle outlet

2.2 Duplex receptacle outlet

2.3 Triplex receptacle outlet

2.4 Quadrex receptacle outlet

2.5 Duplex receptacle outlet – split wired

2.6 Triplex receptacle outlet – split wired

2.7 Single special-purpose receptacle outlet – split wired

NOTE 2.7A: Use numeral or letter as a subscript alongside the symbol, keyed to explanation in the drawing list of symbols, to indicate type of receptacle or usage.

2.8 Duplex special-purpose receptacle outlet
See note 2.7A

2.9 Range outlet (typical)
See note 2.7A

2.10 Special-purpose connection or provision for connection
Use subscript letters to indicate function (SW – dishwasher; CD – clothes dryer, etc).

DW   UNG DW

2.12 Clock hanger receptacle

©   UNG

2.13 Fan hanger receptacle

Ⓕ   UNG

2.14 Floor single receptacle outlet

UNG

2.15 Floor duplex receptacle outlet

UNG

2.16 Floor special-purpose outlet
See note 2.7A

UNG

2.17 Floor telephone outlet – public

2.18 Floor telephone outlet – private

2.19 Underfloor duct and junction box for triple, double, or single duct system (as indicated by the number of parallel lines)

2.20 Cellular floor header duct

3. Switch Outlets

3.1 Single-pole switch

S

3.2 Double-pole switch

S_2

3.3 Three-way switch

S_3

3.4 Four-way switch

S_4

3.5 Key-operated switch

S_K

3.6 Switch and pilot lamp

S_P

3.7 Switch for low-voltage switching system

S_L

3.8 Master switch for low-voltage switching system

S_{LM}

3.9 Switch and single receptacle

S

3.10 Switch and double receptacle

S

3.11 Door switch

S_D

3.12 Time switch

S_T

3.13 Circuit breaker switch

S_{CB}

3.14 Momentary contact switch or pushbutton
for other than signalling system

SMC

3.15 Ceiling pull switch

Ⓢ

5.13 Radio outlet

R

5.14 Television outlet

TV

6. Panelboards, switchboards, and related equipment

6.1 Flush-mounted panel board and cabinet
NOTE 6.1A: Identify by notation or schedule

6.2 Surface-mounted panel board and cabinet
See note 6.1A

6.3 Switchboard, power control center, unit
substations (should be drawn to scale)
See note 6.1A

6.4 Flush-mounted terminal cabinet
See note 6.1A

NOTE 6.4A: In small-scale drawings theTC may be
indicated alongside the symbol

TC

6.5 Surface-mounted terminal cabinet
See note 6.1A and 6.4A

TC

6.6 Pull box
Identify in relation to wiring system section and size

6.7 Motor or other power controller
See note 6.1A

MC

6.8 Externally operated disconnection switch
See note 6.1A

6.9 Combination controller and disconnection means
See note 6.1A

7. Bus Ducts and Wireways

7.1 Trolley duct
See note 6.1A

| T | | T | | T | |

7.2 Busway (service, feeder, or plug-in)
See note 6.1A

| B | | B | | B | |

7.3 Cable through, ladder, or channel
See note 6.1A

| BP | | BP | | BP | |

7.4 Wireway
See note 6.1A

| W | | W | | W | |

9. Circuiting

Wiring method indentification by notation on
drawing or in specifications.

9.1 Wiring concealed in ceiling or wall

NOTE 9.1A: Use heavy weight line to identify service
and feed runs.

9.2 Wiring concealed in floor
See note 9.1A

9.2 Wiring exposed
See note 9.1A

— — — — — — — — — —

9.4 Branch circuit home run to panel board
Number of arrows indicates number of circuits. (A numeral at each arrow may be used to identify circuit number.)

———————————→ 2   → 1

NOTE: Any circuit without further identification indicates a 2-wire circuit. For a greater number of wires, indicate with cross lines (see 9.4.1, Applications)

9.4.1 Applications:

——— /// ——— 3 wires;

——— //// ——— 4 wires, etc

Unless indicated otherwise, the wire size of the circuit is the minimum size required by the specification. Indicate size in inches and identify different functions of wiring system, such as signaling, by notation or other means.

9.6 Wiring turned up

————————————○

9.7 Wiring turned down

————————————●

Figure A-8
(Courtesy of American Iron and Steel Institute)

Table F
Approximate Diameters for Conduit, Locknuts, and Bushings

Component	Nominal or trade size of conduit, inches											
Approximate diameter, inches	$1/2$	$3/4$	1	$1\,1/4$	$1\,1/2$	2	$2\,1/2$	3	$3\,1/2$	4	5	6
Conduit	$7/8$	$1\,1/16$	$1\,3/8$	$1\,11/16$	$1\,15/16$	$2\,3/8$	$2\,7/8$	$3\,1/2$	4	$4\,1/2$	$5\,9/16$	$6\,5/8$
Locknut	$1\,1/8$	$1\,3/8$	$1\,11/16$	$2\,3/16$	$2\,7/16$	3	$3\,7/16$	$4\,3/16$	$5\,3/8$	$6\,11/16$	$7\,15/16$	
Bushing	1	$1\,1/4$	$1\,1/2$	$1\,7/8$	$2\,1/8$	$2\,5/8$	$3\,3/16$	$3\,7/8$	$4\,7/16$	5	$6\,1/4$	$7\,3/8$

Figure A-9
(Courtesy of American Iron and Steel Institute)

Table G
**Recommended Minimum Spacing of Rigid Steel Conduit
and EMT at Junction and Pull Boxes**

Nominal or trade size inches	Distances between centers, inches											
	$1/2$	$3/4$	1	$1\,1/4$	$1\,1/2$	2	$2\,1/2$	3	$3\,1/2$	4	5	6
$1/2$	$1\,3/8$	—	—	—	—	—	—	—	—	—	—	—
$3/4$	$1\,1/2$	$1\,5/8$	—	—	—	—	—	—	—	—	—	—
1	$1\,3/4$	$1\,7/8$	2	—	—	—	—	—	—	—	—	—
$1\,1/4$	2	$2\,1/8$	$2\,1/4$	$2\,1/2$	—	—	—	—	—	—	—	—
$1\,1/2$	$2\,1/8$	$2\,1/4$	$2\,3/8$	$2\,5/8$	$2\,3/4$	—	—	—	—	—	—	—
2	$2\,3/8$	$2\,1/2$	$2\,3/4$	3	$3\,1/8$	$3\,3/8$	—	—	—	—	—	—
$2\,1/2$	$2\,5/8$	$2\,3/4$	3	$3\,1/4$	$3\,3/8$	$3\,5/8$	4	—	—	—	—	—
3	3	$3\,1/8$	$3\,3/8$	$3\,5/8$	$3\,3/4$	4	$4\,3/8$	$4\,3/4$	—	—	—	—
$3\,1/2$	$3\,3/8$	$3\,1/2$	$3\,5/8$	$3\,7/8$	4	$4\,3/8$	$4\,5/8$	5	$4\,3/4$	—	—	—
4	$3\,11/16$	$3\,7/8$	4	$4\,1/4$	$4\,3/8$	$4\,3/4$	5	$5\,3/8$	$5\,5/8$	6	—	—
5	$4\,3/8$	$4\,1/2$	$4\,5/8$	$4\,7/8$	5	$5\,3/8$	$5\,5/8$	6	$6\,1/4$	$6\,5/8$	$7\,1/4$	—
6	5	$5\,1/8$	$5\,1/4$	$5\,1/2$	$5\,5/8$	6	$6\,1/4$	$6\,5/8$	7	$7\,1/4$	8	$8\,5/8$

Figure A-10   I.D. and O.D. Type AC Cable

(Courtesy of Allied Tube & Conduit; Carlon, a Lamson & Sessions Company; and AFC/A Monogram Company)

WEIGHTS AND DIMENSIONS FOR GALVANIZED RIGID CONDUIT

Trade Size, Inches	Approx. Wt. per 100 ft. (30.5m)		Nominal Outside Dia.[1]		Nominal Wall Thickness[2]		Length of Finished Conduit[3]		Quantity In Primary Bundle		Quantity In Master Bundle		Approx. Wt. of Master Bundle		Volume of Master Bundle	
	lb.	kg	in.	mm	in.	mm	ft.	m	ft.	m	ft.	m	lb.	kg	cu ft.	cu m
½	80	36.29	0.840	21.3	0.104	2.6	9'11¼"	3.03	100	30.48	2500	762	2000	907	20.8	0.59
¾	109	49.44	1.050	26.7	0.107	2.7	9'11¼"	3.03	50	15.24	2000	610	2180	989	24.3	0.69
1	165	74.84	1.315	33.4	0.126	3.2	9'11"	3.02	50	15.24	1250	381	2063	936	21.7	0.61
1¼	215	97.52	1.660	42.2	0.133	3.4	9'11"	3.02	30	9.14	900	274	1935	878	23.3	0.66
1½	258	117.03	1.900	48.3	0.138	3.5	9'11"	3.02	—	—	800	244	2064	936	27.8	0.79
2	352	159.67	2.375	60.3	0.146	3.7	9'11"	3.02	—	—	600	183	2112	958	33.8	0.96
2½	567	257.19	2.875	73.0	0.193	4.9	9'10½"	3.01	—	—	370	113	2098	952	29.2	0.83
3	714	323.87	3.500	88.9	0.205	5.2	9'10½"	3.01	—	—	300	91	2142	972	31.3	0.89
3½	860	390.10	4.000	101.6	0.215	5.5	9'10¼"	3.00	—	—	250	76	2150	975	34.7	0.98
4	1000	453.60	4.500	114.3	0.225	5.7	9'10¼"	3.00	—	—	200	61	2000	907	33.7	0.95
5	1320	598.75	5.563	141.3	0.245	6.2	9'10"	3.00	—	—	150	46	1980	898	41.3	1.17
6	1785	809.68	6.625	168.3	0.266	6.8	9'10"	3.00	—	—	100	30	1785	810	38.9	1.10

[1]Outside diameter tolerances: +/- .015 in. (.38mm) for trade sizes ½ in.
through 2 in. +/- .025 in. (.64mm) for trade sizes 2½ in. through 4 in. +/- 1%
for trade sizes 5 in. and 6 in.
[2]For more information only; not a spec requirement.
[3]Without Coupling Length Tolerances: +/- .25 in (6.35mm).

WEIGHTS AND DIMENSIONS FOR INTERMEDIATE METAL CONDUIT

Trade Size, Inches	Approx. Wt. per 100 ft. (30.5m)		Nominal Outside Dia.[1]		Minimum Wall Thickness[2]		Length of Finished Conduit[3]		Quantity In Primary Bundle		Quantity In Master Bundle		Approx. Wt. of Master Bundle		Volume of Master Bundle	
	lb.	kg	in.	mm	in.	mm	ft.	m	ft.	m	ft.	m	lb.	kg	cu ft.	cu m
½	60	27.22	0.815	20.7	0.070	1.8	9'11¼"	3.03	100	30.48	3500	1067	2233	1013	26.7	0.76
¾	82	37.29	1.029	26.1	0.075	1.9	9'11¼"	3.03	50	15.24	2500	762	2078	943	30.7	0.87
1	116	52.62	1.290	32.8	0.085	2.2	9'11"	3.02	50	15.24	1700	518	2035	923	30.7	0.87
1¼	150	68.04	1.638	41.6	0.085	2.2	9'11"	3.02	—	—	1350	411	2209	1002	36.3	1.03
1½	182	82.55	1.883	47.8	0.090	2.3	9'11"	3.02	—	—	1100	335	2122	963	38.2	1.08
2	242	109.77	2.360	59.9	0.095	2.4	9'11"	3.02	—	—	800	244	2096	951	45.8	1.30
2½	428	194.14	2.857	72.6	0.140	3.5	9'10½"	3.01	—	—	370	113	1652	749	29.2	0.83
3	526	238.59	3.476	88.3	0.140	3.5	9'10½"	3.01	—	—	300	91	1618	734	31.3	0.89
3½	612	277.60	3.971	100.9	0.140	3.5	9'10¼"	3.00	—	—	240	73	1576	715	34.7	0.98
4	682	309.35	4.466	113.4	0.140	3.5	9'10¼"	3.00	—	—	240	73	1809	821	42.8	1.21

[1]Outside diameter tolerances: +/- .005 in. (.13mm) for trade sizes ½"
through 1". +/- .0075 in. (.19mm) for trade size 1¼". through 2 in. +/- .010 in.
(.25mm) for trade size 2½" through 4 in.
[2]Wall thickness tolerances: + .015 in. (.38mm) and - .000 for trade sizes ½"
through 2 in. + .20 in. (.51 mm) and -.000 for trade sizes 2½" through 4 in.
Without Coupling Length Tolerances: +/- .25 in (6.35mm).

Figure A-10 *(continued)*

WEIGHTS AND DIMENSIONS FOR ELECTRICAL METALLIC TUBING

Trade Size, Inches	Approx. Wt. per 100 ft. (30.5m)		Nominal Outside Dia.[1]		Nominal Wall Thickness		Length of Finished Conduit[2]		Quantity In Primary Bundle		Quantity In Master Bundle		Approx. Wt. of Master Bundle		Volume of Master Bundle	
	lb.	kg	in.	mm	in.	mm	ft.	m	ft.	m	ft.	m	lb.	kg	cu ft.	cu m
½	29	13.15	0.706	17.9	0.042	1.067	10	3.05	100	30.48	7000	2134	2037	924	28.7	0.81
¾	45	20.41	0.922	23.4	0.049	1.245	10	3.05	100	30.48	5000	1524	2175	987	35.6	1.01
1	65	29.48	1.163	29.5	0.057	1.448	10	3.05	50	15.24	3000	914	1905	864	33.7	0.95
1¼	96	43.55	1.510	38.4	0.065	1.651	10	3.05	50	15.24	2000	610	1894	859	35.0	0.99
1½	111	50.35	1.740	44.2	0.065	1.651	10	3.05	50	15.24	1500	457	1692	767	34.2	0.97
2	141	63.96	2.197	55.8	0.065	1.651	10	3.05	—	—	1200	366	1693	768	46.7	1.32
2½	215	97.52	2.875	73.0	0.072	1.829	10	3.05	—	—	610	186	1412	640	41.5	1.18
3	260	117.94	3.500	88.9	0.072	1.829	10	3.05	—	—	510	155	1429	648	48.9	1.38
3½	325	147.42	4.000	101.6	0.083	2.108	10	3.05	—	—	370	113	1248	566	48.6	1.38
4	390	176.90	4.500	114.3	0.083	2.108	10	3.05	—	—	300	91	1134	514	48.3	1.37

[1]Outside diameter tolerances: +/- .005 in. (.13mm) for trade sizes ½" through 2". +/- .010 in. (.25mm) for trade size 2½". +/- .015 in. (.38mm) for trade size 3". +/- .020 in. (.51mm) for trade sizes 3½" and 4".
[2] Length tolerances: +/-.25" (6.35mm).

PLUS 40® Heavy Wall

Nom. Size	Part No.	O.D.	I.D.	Wall	Wt. Per 100 Feet	Feet Per Bundle
½	49005	.840	.622	.109	18	100
¾	49007	1.050	.824	.113	23	100
1	49008	1.315	1.049	.133	35	100
1¼	49009	1.660	1.380	.140	48	50
1½	49010	1.900	1.610	.145	57	50
2	49011	2.375	2.067	.154	76	50
2½	49012	2.875	2.469	.203	125	10
3	49013	3.500	3.066	.216	164	10
3½	49014	4.000	3.548	.226	198	10
4	49015	4.500	4.026	.237	234	10
5	49016	5.563	5.047	.258	317	10
6	49017	6.625	6.065	.280	412	10

Rigid nonmetallic conduit is normally supplied in standard 10' lengths, with one belled end per length. For specific requirements. it may be produced in lengths shorter or longer than 10', with or without belled ends.

PLUS 80® Extra Heavy Wall

Nom. Size	Part No.	O.D.	I.D.	Wall	Wt. Per 100 Feet	Feet Per Bundle
½	49405	.840	.546	.147	21	100
¾	49407	1.050	.742	.154	29	100
1	49408	1.315	.957	.179	43	100
1¼	49409	1.660	1.278	.191	60	50
1½	49410	1.900	1.500	.200	72	50
2	49411	2.375	1.939	.218	100	10
2½	49412	2.875	2.323	.276	153	10
3	49413	3.500	2.900	.300	212	10
4	49415	4.500	3.826	.337	310	10
5	49416	5.563	4.813	.375	431	10
6	49417	6.625	6.193	.432	592	10

Rigid nonmetallic conduit is normally supplied in standard 10' lengths, with one belled end per length. For specific requirements, it may be produced in lengths shorter or longer than 10', with or without belled ends.

Figure A-10 *(continued)*

METAL-CLAD CABLE WEIGHTS, DIMENSIONS & PACKAGING

	With Insulated Green Ground				
Product Code	AWG Size	Nominal Weight Per 1000 ft.	Nominal Outer Diameter in Inches	Coil Length in Feet	Reel Length in Feet
1701	14-2 Solid	175	.450	250	1000
1702	14-3 Solid	205	.480	250	1000
1703	14-4 Solid	230	.510	250	1000
1704	12-2 Solid	215	.495	250	1000
1705	12-3 Solid	255	.530	250	1000
1706	12-4 Solid	295	.565	250	1000
1707	10-2 Solid	285	.560	125	1000
1708	10-3 Solid	340	.600	125	1000
1709	10-4 Solid	395	.645	125	1000
1718	8-2 Stranded	450	.710	200	500
1716	8-3 Stranded	545	.770	200	500
1717	8-4 Stranded	645	.835	125	500
1719	6-2 Stranded	590	.795	125	500
1720	6-3 Stranded	720	.865	125	500
1721	6-4 Stranded	860	.945	100	500
1724	4-2 Stranded	785	.945	100	500
1725	4-3 Stranded	985	1.035	100	500
1728	4-4 Stranded	1195	1.135	100	500
1729	2-2 Stranded	1045	1.075	100	500
1726	2-3 Stranded	1340	1.180	100	500
1727	2-4 Stranded	1640	1.295	100	500
1737	1-3 Stranded	1530	1.185	100	500
	1-4 Stranded	1860	1.325	100	500
1750	1/0-3 Stranded	1780	1.270	50	500
	1/0-4 Stranded	2220	1.390	50	500
	2/0-3 Stranded	2095	1.350	50	500
	2/0-4 Stranded	2650	1.520	50	500
	3/0-3 Stranded	2580	1.520	50	500
	3/0-4 Stranded	3245	1.690	50	500
	4/0-3 Stranded	3040	1.580	50	500
	4/0-4 Stranded	3855	1.770		500
	250-3 Stranded	3535	1.720		500

Note: Grounding conductors are sized in accordance with Table 5.3 of U.L. Standard 1569. All dimensions and weights are subject to normal manufacturing tolerances.

Figure A-10 *(continued)*

INTERNAL DIAMETER OF ARMOR

Type of Armored Cable	Type of Circuit Conductor	AWG Size of Circuit Conductors	MINIMUM INTERNAL DIAMETER OF ARMOR IN INCHES							
			Cable with Two Circuit Conductors and No Grounding Conductor		Cable with Three Circuit Conductors and No Grounding Conductor and Cable with Two Circuit Conductors and a Grounding Conductor		Cable with Four Circuit Conductors and No Grounding Conductor and Cable with Three Circuit Conductors and a Grounding Conductor		Cable with Four Circuit Conductors and a Grounding Conductor	
			Solid	Stranded	Solid	Stranded	Solid	Stranded	Solid	Stranded
ACTHH	THHN	14	0.260		0.280		0.320		0.345	
		12	0.270		0.295		0.335		0.370	
		10	0.316		0.340		0.381		0.427	
		8	0.410	0.444	0.441	0.477	0.494	0.535	0.554	0.599
		6		0.520		0.559		0.627		0.702
		4		0.656		0.705		0.790		0.886
		2		0.776		0.834		0.935		1.048

EXTERNAL DIAMETER OF ARMOR

Type of Armored Cable	Type of Circuit Conductor	AWG Size of Circuit Conductors	MINIMUM EXTERNAL DIAMETER OF ARMOR IN INCHES							
			Cable with Two Circuit Conductors and No Grounding Conductor		Cable with Three Circuit Conductors and No Grounding Conductor and Cable with Two Circuit Conductors and a Grounding Conductor		Cable with Four Circuit Conductors and No Grounding Conductor and Cable with Three Circuit Conductors and a Grounding Conductor		Cable with Four Circuit Conductors and a Grounding Conductor	
			Solid	Stranded	Solid	Stranded	Solid	Stranded	Solid	Stranded
ACTHH	THHN	14	0.433		0.453		0.486		0.522	
		12	0.467		0.489		0.520		0.545	
		10	0.476		0.500		0.541		0.587	
		8	0.570	0.604	0.601	0.637	0.654	0.695	0.714	0.759
		6		0.700		0.739		0.807		0.882
		4		0.836		0.885		0.970		1.066
		2		0.956		1.014		1.115		1.228

Figure A-11 Kwik Fittings Conduit and Tubing
(Courtesy of Allied Tube & Conduit)

Weights and Dimensions for Kwik-Couple RIGID

Trade Size, Inches	Approx. Wt. per 100 Ft. (30.5m) (with couplings attached)		Nominal Outside Dia.[1]		Nominal Wall Thickness[2]		Length of Finished Conduit (without coupling[3])		Quantity in Master Bundle		Approx. Wt. of Master Bundle		Volume of Master Bundle	
	lb	kg	in	mm	in	mm	ft and in	m	ft	m	lb	kg	cu ft	cu m
2½	564	255.83	2.875	73.0	0.193	4.9	9'10½"	3.01	400	122	2,256	1,023	28.5	.8065
3	711	322.51	3.500	88.9	0.205	5.2	9'10½"	3.01	300	91	2,133	968	30.7	.8688
3½	849	385.11	4.000	101.6	0.215	5.5	9'10¼"	3.00	250	76	2,123	963	35.5	1.005
4	998	452.69	4.500	114.3	0.225	5.7	9'10¼"	3.00	200	61	1,996	905	32.8	.9283

[1]Outside diameter tolerances: +/- .025 in. (.64 mm).   [2]For information only; not a spec requirement.   [3]Length tolerances +/- .25 in. (6.35 mm).

Weights and Dimensions for Kwik-Couple IMC

Trade Size, Inches	Approx. Wt. per 100 Ft. (30.5m) (with couplings attached)		Nominal Outside Dia.[1]		Wall Thickness[2]		Length of Finished Conduit (without coupling[3])		Quantity in Master Bundle		Approx. Wt. of Master Bundle		Volume of Master Bundle	
	lb	kg	in	mm	in	mm	ft and in	m	ft	m	lb	kg	cu ft	cu m
2½	444	201.40	2.857	72.6	0.140	3.6	9'10½"	3.01	400	122	1,776	806	28.5	.8065
3	537	243.58	3.476	88.3	0.140	3.6	9'10½"	3.01	300	91	1,611	731	30.7	.8688
3½	648	293.93	3.971	100.9	0.140	3.6	9'10¼"	3.00	250	76	1,620	735	35.5	1.005
4	742	336.57	4.466	113.4	0.140	3.6	9'10¼"	3.00	200	61	1,484	673	32.8	.9283

[1]Outside diameter tolerances: +/- .010 in. (.25 mm). [2]Wall thickness tolerances: + .020 in. (.51 mm) and - .000. [3]Length tolerance +/- .25 in. (6.35 mm).

Weights and Dimensions for KWIK-FIT EMT

Trade Size, Inches	Approx. Wt. Per 100 Ft. (30.5m)		Nominal Outside Dia.[1]		Nominal Wall Thickness		Length of Finished Conduit[2]		Quantity in Master Bundle		Approx. Wt. of Master Bundle		Volume of Master Bundle	
	lb	kg	in	mm	in	mm	ft	m	ft	m	lb	kg	cu ft	cu m
2½	215	97.52	2.875	73.0	0.072	1.829	10	3.05	350	107	752	341	29.2	.83
3	260	117.94	3.500	88.9	0.072	1.829	10	3.05	300	91	780	354	33.5	.95
3½	325	147.42	4.000	101.6	0.083	2.108	10	3.05	250	76	812	368	43.4	1.23
4	390	176.90	4.500	114.3	0.083	2.108	10	3.05	250	76	975	442	52.5	1.48

[1]Outside diameter tolerances: +/- .010 in. (.25 mm) for trade size 2½ in.; +/- .015 in. (.38mm) for trade size 3 in.; +/- .020 in. (.51 mm) for trade sizes 3½ in. and 4 in.   [2]Length tolerance: +/- .25 in. (6.35 mm).

Figure A-12 Environmental Resistance Ratings—Nonmetallic Wiring Methods
(Courtesy of Carlon, A Lamson & Sessions Company)

These environmental resistance ratings are based upon tests where the specimens were placed in complete submergence in the reagent listed. In many applications Carflex* conduit can be used in process areas where these chemicals are manufactured or used because worker safety requirements dictate that any air presence or splashing be at a very low level. Most liquidtight conduit is located in areas suitable for worker access.
If there are any questions for specific suitability in a given environment, prototype samples should be tested under actual conditions.

RATING CODE

A-Excellent service.	**B-Good service life.**	**C-Fair or limited service.**	**D-Unsatisfactory service.**
No harmful effect to reduce service life. Suitable for continuous service.	Moderate to minor effect. Good for intermittent service. Generally suitable for continuous service.	Depends on operating conditions. Generally suitable for intermittent continuous service.	Not recommended.

* These ratings apply only to Carflex* conduit & tubing.

Chemical	Concentration	Resistance	Chemical	Concentration	Resistance	Chemical	Concentration	Resistance
Acetate Solvents		D	Creosote		D	Methyl Acetate		D
Acetic Acid	10%	B	Cresol		C	Methyl Alcohol		C
Acetic Acid (Glacial)		C	Cresylic Acid		D	Methyl Bromide		D
Acetone		D	Cyclohexane		B	Methyl Ethyl Ketone		D
Acrylonitrile		A	DDT Weed Killer		A	Methylene Chloride		D
Alcohols (Aliphatic)		C	Dibutyl Phthalate		D	Mineral Oil		A
Aluminum Chloride		A	Diesel Oils		C	Monochlorobenzene		D
Aluminum Sulfate (Alums)		A	Diethylene Glycol		B	Muriatic Acid (See Hydrochloric Acid)		—
Ammonia (Anhydrous Liquids)		D	Diethyl Ether		A	Naphtha		C
Ammonia (Aqueous)		A	Di-isodecyl Phthalate		D	Naphthalene		D
Ammoniated Latex		A	Dioctyl Phthalate		D	Nitric Acid	10%	A
Ammonium Chloride		A	Dow General Weed Killer (Phenol)		D	Nitric Acid	35%	A
Ammonium Hydroxide		A	Dow General Weed Killer (H₂O)		B	Nitric Acid	70%	D
Amyl Acetate		D	Ethyl Alcohol		C	Oleic Acid		A
Aniline Oils		D	Ethylene Dichloride		D	Oleum		D
Aromatic Hydrocarbons		D	Ethylene Glycol		B	Oxalic Acid		A
Asphalt		D	Ferric Chloride		A	Pentachlorophenol in Oil		B
ASTM Fuel A		C	Ferric Sulfate		A	Pentane		C
ASTM Fuel B		D	Ferrous Chloride		A	Perchloroethylene		D
ASTM #1 Oil		B	Ferrous Sulfate		A	Petroleum Ether		C
ASTM #3 Oil		C	Formaldehyde		D	Phenol		B
Barium Chloride		A	Fuel Oil		B	Phosphoric Acid	85%	A
Barium Sulfide		A	Furfural		C	Pitch		B
Barium Hydroxide		A	Gallic Acid		A	Potassium Hydroxide		A
Benzene (Benzol)		D	Gasoline (Hi Test)		C	Propyl Alcohol		B
Benzine (Petroleum Ether)		C	Glycerine		A	Ritchfield "A" Weed Killer		C
Black Liquor		A	Grease		A	Sea Water		A
Bordeaux Mixture		A	Green Sulfate Liquor		A	Sodium Hydroxide	10%	A
Boric Acid		A	Heptachlor in Petroleum Solvents		A	Sodium Hydroxide	50%	C
Butyl Acetate		D	Heptane		C	Soybean Oil		C
Butyl Alcohol		B	Hexane		C	Sodium Cyanide		A
Calcium Hydroxide		A	Hydrobromic Acid		A	Stoddard Solvent		D
Calcium Hypochlorite		A	Hydrochloric Acid	10%	A	Styrene		D
Carbolic Acid (Phenol)		B	Hydrochloric Acid	40%	C	Sulfur Dioxide (liquid)		D
Carbon Dioxide		A	Hydrofluoric Acid	70%	D	Sulfuric Acid	50%	A
Carbon Disulfide		D	Hydrofluoroboric Acid		A	Sulfuric Acid	98%	D
Carbon Tetrachloride		D	Hydrofluorosilicic Acid		A	Sulfurous Acid		B
Carbonic Acid		A	Hydrogen Peroxide	10%	A	Tall Oil		D
Casein		A	Iso-octane		C	Tannic Acid		A
Caustic Soda		A	Isopropyl Acetate		D	Toluene		D
Chlorine Gas (wet)		D	Isopropyl Alcohol		B	Trichlorethylene		D
Chlorine Gas (dry)		D	Jet Fuels (JP-3, 4, and 5)		C	Triethanol Amine		C
Chlorine (water solution)		C	Kerosene		C	Tricresyl Phosphate (Skydrol)		D
Chlorobenzene		D	Ketones		D	Turpentine		C
Chlorinated Hydrocarbons		D	Linseed Oil		A	Vinegar		A
Chromic Acid	10%	B	Lubricating Oils		A	Vinyl Chloride		D
Citric Acid		A	Magnesium Chloride		A	Water		A
Coal Tar		D	Magnesium Hydroxide		A	White Liquor		A
Coconut Oil		C	Magnesium Sulfate		A	Xylene		D
Corn Oil		A	Malathion 50 in Aromatics		D	Zinc Chloride		A
Cottonseed Oil		C	Malic Acid		A	Zinc Sulfate		A

Figure A-13   Chemical Resistance Charts—Coated Conduit (Metal)
(Courtesy of Robroy Industries)

Urethane Interior Coating Chemical Resistance Chart

Solutions	Conc.	Temp.	Splashing	Liquid	Fumes
Acetic Acid	10%	75	yes	yes	yes
Acid Copper Plating Solution		75	yes	yes	yes
Alkaline Cleaners		75	yes	yes	yes
Aluminum Chloride	Sat'd	75	yes	yes	yes
Aluminum Sulfate	Sat'd	75	yes	yes	yes
Alums	Sat'd	75	yes	yes	yes
Ammonium Chloride	Sat'd	75	yes	yes	yes
Ammonium Hydroxide	28%	75	yes	yes	yes
Ammonium Hydroxide	10%	75	yes	yes	yes
Ammonium Sulfate	Sat'd	75	yes	yes	yes
Ammonium Thiocyanate	Sat'd	75	yes	yes	yes
Amyl Alcohol	Any	75	?	no	?
Arsenic Acids	Any	75	yes	yes	yes
Barium Sulfide	Sat'd	75	yes	yes	yes
Black Liquor	Sat'd	75	yes	yes	yes
Benzoic Acid	Sat'd	75	yes	yes	yes
Brass Plating Solution	Any	75	yes	yes	yes
Bromine Water	Sat'd	75	yes	no	yes
Butyl Alcohol	Any	75	?	no	?
Cadmium Plating Solution	Any	75	yes	yes	yes
Calcium Bisulfite	Any	75	yes	yes	yes
Calcium Chloride	Sat'd	75	yes	yes	yes
Calcium Hypochlorite	Sat'd	75	yes	yes	yes
Carbonic Acid	Sat'd	75	yes	yes	yes
Casein	Sat'd	75	yes	yes	yes
Castor Oil	Any	75	yes	yes	yes
Caustic Soda	35%	75	yes	yes	yes
Caustic Soda	10%	75	yes	yes	yes
Caustic Potash	35%	75	yes	yes	yes
Caustic Potash	10%	75	yes	yes	yes
Chlorine Water	Sat'd	75	yes	yes	yes
Chromium Plating Solution	Any	75	yes	yes	yes
Citric Acid	Sat'd	75	yes	yes	yes
Copper Chloride (Cupric)	Sat'd	75	yes	yes	yes
Copper CyanidePlating Sol	Any	75	yes	yes	yes
(High Speed)	Any	75	yes	yes	yes
(with Alkali Cyanides)	Sat'd	75	yes	yes	yes
Copper Sulfate	Sat'd	75	yes	yes	yes
Cocoanut Oil	Sat'd	75	yes	yes	yes
Cottonseed Oil	Sat'd	75	yes	yes	yes
Disodium Phosphate	Sat'd	75	yes	yes	yes
Ethyl Alcohol	Any	75	?	no	?
Ethylene Glycol	Any	75	?	no	?
Ferric Chloride	45%	75	yes	yes	yes
Ferrous Sulfate	Sat'd	75	yes	yes	yes
Fluoboric Acid	Any	75	yes	yes	yes
Formaldehyde	37%	75	yes	yes	yes
Formic Acid	85%	75	yes	no	yes
Gallic Acid	Sat'd	75	yes	yes	yes
Glucose	Any	75	yes	yes	yes
Glue	Any	75	yes	yes	yes
Glycerine	Any	75	yes	yes	yes
Gold Plating Solution	Any	75	yes	yes	yes
Hydrochloric Acid	10%	75	yes	yes	yes
Hydrochloric Acid	21.5%	75	yes	yes	yes
Hydrochloric Acid	37.5%	75	yes	no	yes
Hydrofluoric Acid	4%	75	yes	yes	yes
Hydrofluoric Acid	10%	75	yes	yes	yes
Hydrofluoric Acid	48%	75	yes	no	yes
Hydrogen Peroxide	30%	75	yes	yes	yes
Hydrogen Sulfide	Sat'd	75	yes	yes	yes
Hydroquinone	Any	75	yes	yes	yes
Indium Plating Solution	Any	75	yes	yes	yes
Lactic Acid	50%	75	yes	yes	yes
Lactic Acid	Any	75	yes	yes	yes
Lead Plating Solution	Any	75	yes	yes	yes
Malic acid	Any	75	yes	yes	yes
Methyl Alcohol	Any	75	?	no	?
Mineral Oils	Any	75	yes	yes	yes
Nickel Acetate	Sat'd	75	yes	yes	yes
Nickel Plating Solution		75	yes	yes	yes
Nickel Salts	Sat'd	75	yes	yes	yes
Nitric Acid	35%	75	yes	yes	yes
Nitric Acid	40%	75	yes	yes	yes
Nitric Acid	60%	75	yes	no	yes
Nitric Acid/	15 %				
Hydrofluoric Acid	4%	75	yes	yes	yes
Nitric Acid/	16 %				
Sodium Dichromate	13%	75	yes	yes	yes
Water	71%				
Oleic Acid	Any	75	yes	yes	yes
Oxalic Acid	Sat'd	75	yes	yes	yes
	Any	75	yes	yes	yes
Phenol	Sat'd	75	yes	no	yes
Phosphoric Acid	75%	75	yes	yes	yes
Phosphoric Acid	85%	75	yes	yes	yes
Potassium Acid Sulfate	Sat'd	75	yes	yes	yes
Potassium Antimonate	Sat'd	75	yes	yes	yes
Potassium Bisulfite	Sat'd	75	yes	yes	yes
Potassium Chloride	Sat'd	75	yes	yes	yes
Potassium Cuprocyanide	Sat'd	75	yes	yes	yes
Potassium Cyanide	Sat'd	75	yes	yes	yes
Potassium Diachromate	Sat'd	75	yes	yes	yes
Potassium Hypochlorite	Sat'd	75	yes	?	yes
Potassium Sulfide	Sat'd	75	yes	yes	yes
Potassium Thiosulfate	Sat'd	75	yes	yes	yes
Propyl Alcohol	Sat'd	75	?	no	?
Rhodium Plating Solution	Sat'd	75	yes	yes	yes
Silver Plating Solution	Sat'd	75	yes	yes	yes
Soaps	Any	75	yes	yes	yes
Sodium Acid Sulfate	Sat'd	75	yes	yes	yes
Sodium Antimonate	Sat'd	75	yes	yes	yes
Sodium Bicarbonate	Sat'd	75	yes	yes	yes
Sodium Bisulfite	Sat'd	75	yes	yes	yes
Sodium Chloride	Sat'd	75	yes	yes	yes
Sodium Cyanide	Sat'd	75	yes	yes	yes
Sodium Dichromate	Sat'd	75	yes	yes	yes
Sodium Hydroxide	10%	75	yes	yes	yes
Sodium Hydroxide	35%	75	yes	yes	yes
Sodium Hydroxide	73%	75	no	no	yes
Sodium Hypochlorite	Sat'd	75	yes	?	yes
Sodium Hypochlorite	15%	75	yes	?	yes
Sodium Sulfide	Sat'd	75	yes	yes	yes
Sodium Thiosulfate	Sat'd	75	yes	yes	yes
Sulfuric Acid	15%	75	yes	yes	yes
Sulfuric Acid	50%	75	yes	yes	yes
Sulfuric Acid	70%	75	yes	yes	yes
Sulfuric Acid	98%	75	no	no	yes
Sulfurous Acid	2%	75	yes	yes	yes
Sulfurous Acid	6%	75	yes	no	yes
Tannic Acid	Sat'd	75	yes	yes	yes
Tartaric Acid	Sat'd	75	yes	yes	yes
Tin Chloride Aqueous	Sat'd	75	yes	yes	yes
Tin Plating Solution	Sat'd	75	yes	yes	yes
Triethaneolamine	Sat'd	75	yes	yes	yes
Trisodium Phosphate	Sat'd	75	yes	yes	yes
Water	Sat'd	75	yes	yes	yes
White Liquor		75	yes	yes	yes
Zinc Plating Solution		75	yes	yes	yes
Zinc Sulfate	Sat'd	75	yes	yes	yes

Figure A-13 *(continued)*

PVC Coating Chemical Resistance Chart

Solutions	Conc.	Temp.	Splashing	Liquid	Fumes
Acetic Acid	10%	75	yes	yes	yes
Acid Copper Plating Solution		75	yes	yes	yes
Alkaline Cleaners		75	yes	yes	yes
Aluminum Chloride	Sat'd	75	yes	yes	yes
Aluminum Sulfate	Sat'd	75	yes	yes	yes
Alums	Sat'd	75	yes	yes	yes
Ammonium Chloride	Sat'd	75	yes	yes	yes
Ammonium Hydroxide	28%	75	yes	yes	yes
Ammonium Hydroxide	10%	75	yes	yes	yes
Ammonium Sulfate	Sat'd	75	yes	yes	yes
Ammonium Thiocyanate	Sat'd	75	yes	yes	yes
Amyl Alcohol	Any	75	?	no	?
Arsenic Acids	Any	75	yes	yes	yes
Barium Sulfide	Sat'd	75	yes	yes	yes
Black Liquor	Sat'd	75	yes	yes	yes
Benzoic Acid	Sat'd	75	yes	yes	yes
Brass Plating Solution	Any	75	yes	yes	yes
Bromine Water	Sat'd	75	yes	no	yes
Butyl Alcohol	Any	75	?	no	?
Cadmium Plating Solution	Any	75	yes	yes	yes
Calcium Bisulfite	Any	75	yes	yes	yes
Calcium Chloride	Sat'd	75	yes	yes	yes
Calcium Hypochlorite	Sat'd	75	yes	yes	yes
Carbonic Acid	Sat'd	75	yes	yes	yes
Casein	Sat'd	75	yes	yes	yes
Castor Oil	Any	75	yes	yes	yes
Caustic Soda	35%	75	yes	yes	yes
Caustic Soda	10%	75	yes	yes	yes
Caustic Potash	35%	75	yes	yes	yes
Caustic Potash	10%	75	yes	yes	yes
Chlorine Water	Sat'd	75	yes	yes	yes
Chromium Plating Solution	Any	75	yes	yes	yes
Citric Acid	Sat'd	75	yes	yes	yes
Copper Chloride (Cupric)	Sat'd	75	yes	yes	yes
Copper CyanidePlating Sol	Any	75	yes	yes	yes
(High Speed)	Any	75	yes	yes	yes
(with Alkali Cyanides)	Sat'd	75	yes	yes	yes
Copper Sulfate	Sat'd	75	yes	yes	yes
Cocoanut Oil	Sat'd	75	yes	yes	yes
Cottonseed Oil	Sat'd	75	yes	yes	yes
Disodium Phosphate	Sat'd	75	yes	yes	yes
Ethyl Alcohol	Any	75	?	no	?
Ethylene Glycol	Any	75	?	no	?
Ferric Chloride	45%	75	yes	yes	yes
Ferrous Sulfate	Sat'd	75	yes	yes	yes
Fluoboric Acid	Any	75	yes	yes	yes
Formaldehyde	37%	75	yes	yes	yes
Formic Acid	85%	75	yes	no	yes
Gallic Acid	Sat'd	75	yes	yes	yes
Glucose	Any	75	yes	yes	yes
Glue	Any	75	yes	yes	yes
Glycerine	Any	75	yes	yes	yes
Gold Plating Solution	Any	75	yes	yes	yes
Hydrochloric Acid	10%	75	yes	yes	yes
Hydrochloric Acid	21.5%	75	yes	yes	yes
Hydrochloric Acid	37.5%	75	yes	no	yes
Hydrofluoric Acid	4%	75	yes	yes	yes
Hydrofluoric Acid	10%	75	yes	yes	yes
Hydrofluoric Acid	48%	75	yes	no	yes
Hydrogen Peroxide	30%	75	yes	yes	yes
Hydrogen Sulfide	Sat'd	75	yes	yes	yes
Hydroquinon	Any	75	yes	yes	yes
Indium Plating Solution	Any	75	yes	yes	yes
Lactic Acid	50%	75	yes	yes	yes
Lactic Acid	Any	75	yes	yes	yes
Lead Plating Solution	Any	75	yes	yes	yes
Malic acid	Any	75	yes	yes	yes
Methyl Alcohol	Any	75	?	no	?
Mineral Oils	Any	75	yes	yes	yes
Nickel Acetate	Sat'd	75	yes	yes	yes
Nickel Plating Solution		75	yes	yes	yes
Nickel Salts	Sat'd	75	yes	yes	yes
Nitric Acid	35%	75	yes	yes	yes
Nitric Acid	40%	75	yes	yes	yes
Nitric Acid	60%	75	yes	no	yes
Nitric Acid/ Hydrofluoric Acid	15 % / 4%	75	yes	yes	yes
Nitric Acid/ Sodium Dichromate Water	16 % / 13% / 71%	75	yes	yes	yes
Oleic Acid	Any	75	yes	yes	yes
Oxalic Acid	Sat'd	75	yes	yes	yes
	Any	75	yes	yes	yes
Phenol	Sat'd	75	yes	no	yes
Phosphoric Acid	75%	75	yes	yes	yes
Phosphoric Acid	85%	75	yes	yes	yes
Potassium Acid Sulfate	Sat'd	75	yes	yes	yes
Potassium Antimonate	Sat'd	75	yes	yes	yes
Potassium Bisulfite	Sat'd	75	yes	yes	yes
Potassium Chloride	Sat'd	75	yes	yes	yes
Potassium Cuprocyanide	Sat'd	75	yes	yes	yes
Potassium Cyanide	Sat'd	75	yes	yes	yes
Potassium Diachromate	Sat'd	75	yes	yes	yes
Potassium Hypochlorite	Sat'd	75	yes	?	yes
Potassium Sulfide	Sat'd	75	yes	yes	yes
Potassium Thiosulfate	Sat'd	75	yes	yes	yes
Propyl Alcohol	Sat'd	75	?	no	?
Rhodium Plating Solution	Sat'd	75	yes	yes	yes
Silver Plating Solution	Sat'd	75	yes	yes	yes
Soaps	Any	75	yes	yes	yes
Sodium Acid Sulfate	Sat'd	75	yes	yes	yes
Sodium Antimonate	Sat'd	75	yes	yes	yes
Sodium Bicarbonate	Sat'd	75	yes	yes	yes
Sodium Bisulfite	Sat'd	75	yes	yes	yes
Sodium Chloride	Sat'd	75	yes	yes	yes
Sodium Cyanide	Sat'd	75	yes	yes	yes
Sodium Dichromate	Sat'd	75	yes	yes	yes
Sodium Hydroxide	10%	75	yes	yes	yes
Sodium Hydroxide	35%	75	yes	yes	yes
Sodium Hydroxide	73%	75	no	no	yes
Sodium Hypochlorite	Sat'd	75	yes	?	yes
Sodium Hypochlorite	15%	75	yes	?	yes
Sodium Sulfide	Sat'd	75	yes	yes	yes
Sodium Thiosulfate	Sat'd	75	yes	yes	yes
Sulfuric Acid	15%	75	yes	yes	yes
Sulfuric Acid	50%	75	yes	yes	yes
Sulfuric Acid	70%	75	yes	yes	yes
Sulfuric Acid	98	75	no	no	yes
Sulfurous Acid	2%	75	yes	yes	yes
Sulfurous Acid	6%	75	yes	no	yes
Tannic Acid	Sat'd	75	yes	yes	yes
Tartaric Acid	Sat'd	75	yes	yes	yes
Tin Chloride Aqueous	Sat'd	75	yes	yes	yes
Tin Plating Solution	Sat'd	75	yes	yes	yes
Triethaneolamine	Sat'd	75	yes	yes	yes
Trisodium Phosphate	Sat'd	75	yes	yes	yes
Water	Sat'd	75	yes	yes	yes
White Liquor		75	yes	yes	yes
Zinc Plating Solution		75	yes	yes	yes
Zinc Sulfate	Sat'd	75	yes	yes	yes

Recommended Exposure: Splashing, Liquid, Fumes

Figure A-14   Hazardous Substances Flash Point, Ignition Temperatures–Gases and Vapors

Table 1

Class I* Group	Substance	Auto-* Ignition Temp.		Flash** Point		Explosive Limits** Per Cent by Volume		Vapor** Density (Air Equals 1.0)
		°F	°C	°F	°C	Lower	Upper	
C	Acetaldehyde	347	175	-38	-39	4.0	60	1.5
D	Acetic Acid	867	464	103	39	4.0	19.9 @ 200° F	2.1
D	Acetic Anhydride	600	316	120	49	2.7	10.3	3.5
D	Acetone	869	465	-4	-20	2.5	13	2.0
D	Acetone Cyanohydrin	1270	688	165	74	2.2	12.0	2.9
D	Acetonitrile	975	524	42	6	3.0	16.0	1.4
A	Acetylene	581	305	gas	gas	2.5	100	0.9
B(C)	Acrolein (inhibited)[1]	455	235	-15	-26	2.8	31.0	1.9
D	Acrylic Acid	820	438	122	50	2.4	8.0	2.5
D	Acrylonitrile	898	481	32	0	3.0	17	1.8
D	Adiponitrile	—	—	200	93	—	—	—
C	Allyl Alcohol	713	378	70	21	2.5	18.0	2.0
D	Allyl Chloride	905	485	-25	-32	2.9	11.1	2.6
B(C)	Allyl Glycidyl Ether[1]	—	—	—	—	—	—	—
D	Ammonia[2]	928	498	gas	gas	15	28	0.6
D	n-Amyl Acetate	680	360	60	16	1.1	7.5	4.5
D	sec-Amyl Acetate	—	—	89	32	—	—	4.5
D	Aniline	1139	615	158	70	1.3	11	3.2
D	Benzene	928	498	12	-11	1.3	7.9	2.8
D	Benzyl Chloride	1085	585	153	67	1.1	—	4.4
B(D)	1,3-Butadiene[1]	788	420	gas	gas	2.0	12.0	1.9
D	Butane	550	288	gas	gas	1.6	8.4	2.0
D	1-Butanol	650	343	98	37	1.4	11.2	2.6
D	2-Butanol	761	405	75	24	1.7 @ 212° F	9.8 @ 212° F	2.6
D	n-Butyl Acetate	790	421	72	22	1.7	7.6	4.0
D	iso-Butyl Acetate	790	421	—	—	—	—	—
D	sec-Butyl Acetate	—	—	88	31	1.7	9.8	4.0
D	t-Butyl Acetate	—	—	—	—	—	—	—
D	n-Butyl Acrylate (inhibited)	559	293	118	48	1.5	9.9	4.4
C	n-Butyl Formal	—	—	—	—	—	—	—
B(C)	n-Butyl Glycidyl Ether[1]	—	—	—	—	—	—	—
C	Butyl Mercaptan	—	—	35	2	—	—	3.1
C	t-Butyl Toluene	—	—	—	—	—	—	—
D	Butylamine	594	312	10	-12	1.7	9.8	2.5
D	Butylene	725	385	gas	gas	1.6	10.0	1.9
C	n-Butyraldehyde	425	218	-8	-22	1.9	12.5	2.5
D	n-Butyric Acid	830	443	161	72	2.0	10.0	3.0
—[3]	Carbon Disulfide	194	90	-22	-30	1.3	50.0	2.6
C	Carbon Monoxide	1128	609	gas	gas	12.5	74.0	1.0
C	Chloroacetaldehyde	—	—	—	—	—	—	—
D	Chlorobenzene	1099	593	82	28	1.3	9.6	3.9
C	1-Chloro-1-Nitropropane	—	—	144	62	—	—	4.3
D	Chloroprene	—	—	-4	-20	4.0	20.0	3.0
D	Cresol	1038-1110	559-599	178-187	81-86	1.1-1.4	—	—
C	Crotonaldehyde	450	232	55	13	2.1	15.5	2.4
D	Cumene	795	424	96	36	0.9	6.5	4.1
D	Cyclohexane	473	245	-4	-20	1.3	8.0	2.9
D	Cyclohexanol	572	300	154	68	—	—	3.5
D	Cyclohexanone	473	245	111	44	1.1 @ 212° F	9.4	3.4
D	Cyclohexene	471	244	<20	<-7	—	—	2.8
D	Cyclopropane	938	503	gas	gas	2.4	10.4	1.5
D	p-Cymene	817	436	117	47	0.7 @ 212° F	5.6	4.6
C	n-Decaldehyde	—	—	—	—	—	—	—
D	n-Decanol	550	288	180	82	—	—	5.5
D	Decene	455	235	<131	<55	—	—	4.84
D	Diacetone Alcohol	1118	603	148	64	1.8	6.9	4.0
D	o-Dichlorobenzene	1198	647	151	66	2.2	9.2	5.1
D	1,1-Dichloroethane	820	438	22	-6	5.6	—	—
D	1,2-Dichloroethylene	860	460	36	2	5.6	12.8	3.4
C	1,1-Dichloro-1-Nitroethane	—	—	168	76	—	—	5.0
D	1,3-Dichloropropene	—	—	95	35	5.3	14.5	3.8
C	Dicyclopentadiene	937	503	90	32	—	—	—
D	Diethyl Benzene	743-842	395-450	133-135	56-57	—	—	4.6
C	Diethyl Ether	320	160	-49	-45	1.9	36.0	2.6
C	Diethylamine	594	312	-9	-23	1.8	10.1	2.5
C	Diethylaminoethanol	—	—	—	—	—	—	—
C	Diethylene Glycol Monobutyl Ether	442	228	172	78	0.85	24.6	5.6
C	Diethylene Glycol Monomethyl Ether	465	241	205	96	—	—	—
D	Di-isobutyl Ketone	745	396	120	49	0.8 @ 200° F	7.1 @ 200° F	4.9
D	Di-isobutylene	736	391	23	-5	0.8	4.8	3.9
C	Di-isopropylamine	600	316	30	-1	1.1	7.1	3.5
C	N-N-Dimethyl Aniline	700	371	145	63	—	—	4.2
D	Dimethyl Formamide	833	455	136	58	2.2 @ 212° F	15.2	2.5
D	Dimethyl Sulfate	370	188	182	83	—	—	4.4
C	Dimethylamine	752	400	gas	gas	2.8	14.4	1.6
C	1,4-Dioxane	356	180	54	12	2.0	22	3.0
D	Dipentene	458	237	113	45	0.7 @ 302° F	6.1 @ 302° F	4.7
C	Di-n-propylamine	570	299	63	17	—	—	3.5
C	Dipropylene Glycol Methyl Ether	—	—	185	85	—	—	5.11
D	Dodecene	491	255	—	—	—	—	—

Figure A-14 *(continued)*

Table 1 *(continued)*

Class I Group	Substance	Auto-Ignition Temp. °F	°C	Flash Point °F	°C	Explosive Limits Per Cent by Volume Lower	Upper	Vapor Density (Air Equals 1.0)
C	Epichlorohydrin	772	411	88	31	3.8	21.0	3.2
D	Ethane	882	472	gas	gas	3.0	12.5	1.0
D	Ethanol	685	363	55	13	3.3	19	1.6
D	Ethyl Acetate	800	427	24	-4	2.0	11.5	3.0
D	Ethyl Acrylate (inhibited)	702	372	50	10	1.4	14	3.5
D	Ethyl sec-Amyl Ketone	—	—	—	—	—	—	—
D	Ethyl Benzene	810	432	70	21	1.0	6.7	3.7
D	Ethyl Butanol	—	—	—	—	—	—	—
D	Ethyl Butyl Ketone	—	—	115	46	—	—	4.0
D	Ethyl Chloride	966	519	-58	-50	3.8	15.4	2.2
D	Ethyl Formate	851	455	-4	-20	2.8	16.0	2.6
D	2-Ethyl Hexanol	448	231	164	73	0.88	9.7	4.5
D	2-Ethyl Hexyl Acrylate	485	252	180	82	—	—	—
C	Ethyl Mercaptan	572	300	<0	<-18	2.8	18.0	2.1
C	n-Ethyl Morpholine	—	—	—	—	—	—	—
C	2-Ethyl-3-Propyl Acrolein	—	—	155	68	—	—	4.4
D	Ethyl Silicate	—	—	125	52	—	—	7.2
D	Ethylamine	725	385	<0	<-18	3.5	14.0	1.6
C	Ethylene	842	450	gas	gas	2.7	36.0	1.0
D	Ethylene Chlorohydrin	797	425	140	60	4.9	15.9	2.8
D	Ethylene Dichloride	775	413	56	13	6.2	16	3.4
C	Ethylene Glycol Monobutyl Ether	460	238	143	62	1.1 @ 200° F	12.7 @ 275° F	4.1
C	Ethylene Glycol Monobutyl Ether Acetate	645	340	160	71	0.88 @ 200° F	8.54 @ 275° F	—
C	Ethylene Glycol Monoethyl Ether	455	235	110	43	1.7 @ 200° F	15.6 @ 200° F	3.0
C	Ethylene Glycol Monoethyl Ether Acetate	715	379	124	52	1.7	—	4.72
D	Ethylene Glycol Monomethyl ether	545	285	102	39	1.8 @ STP	14 @ STP	2.6
B(C)	Ethylene Oxide[1]	804	429	-20	-28	3.0	100	1.5
D	Ethylenediamine	725	385	93	34	4.2	14.4	2.1
C	Ethylenimine	608	320	12	-11	3.6	46.0	1.5
C	2-Ethylhexaldehyde	375	191	112	44	0.85 @ 200° F	7.2 @ 275° F	4.4
B	Formaldehyde (Gas)	795	429	gas	gas	7.0	73	1.0
D	Formic Acid (90%)	813	434	122	50	18	57	—
D	Fuel Oils	410-765	210-407	100-336	38-169	0.7	5	—
C	Furfural	600	316	140	60	2.1	19.3	3.3
C	Furfuryl Alcohol	915	490	167	75	1.8	16.3	3.4
D	Gasoline	536-880	280-471	-36 to -50	-38 to -46	1.2-1.5	7.1-7.6	3-4
D	Heptane	399	204	25	-4	1.05	6.7	3.5
D	Heptene	500	260	<32	<0	—	—	3.39
D	Hexane	437	225	-7	-22	1.1	7.5	3.0
D	Hexanol	—	—	145	63	—	—	3.5
D	2-Hexanone	795	424	77	25	—	8	3.5
D	Hexenes	473	245	<20	<-7	—	—	3.0
D	sec-Hexyl Acetate	—	—	—	—	—	—	—
C	Hydrazine	74-518	23-270	100	38	2.9	9.8	1.1
B	Hydrogen	968	520	gas	gas	4.0	75	0.1
C	Hydrogen Cyanide	1000	538	0	-18	5.6	40.0	0.9
C	Hydrogen Selenide	—	—	—	—	—	—	—
C	Hydrogen Sulfide	500	260	gas	gas	4.0	44.0	1.2
D	Isoamyl Acetate	680	360	77	25	1.0 @ 212° F	7.5	4.5
D	Isoamyl Alcohol	662	350	109	43	1.2	9.0 @ 212° F	3.0
D	Isobutyl Acrylate	800	427	86	30	—	—	4.42
C	Isobutyraldehyde	385	196	-1	-18	1.6	10.6	2.5
C	Isodecaldehyde	—	—	185	85	—	—	5.4
D	Iso-octyl Alcohol	—	—	180	82	—	—	—
C	Iso-octyl Aldehyde	387	197	—	—	—	—	—
D	Isophorone	860	460	184	84	0.8	3.8	—
D	Isoprene	428	220	-65	-54	1.5	8.9	2.4
D	Isopropyl Acetate	860	460	35	2	1.8 @ 100° F	8	3.5
D	Isopropyl Ether	830	443	-18	-28	1.4	7.9	3.5
C	Isopropyl Glycidyl Ether	—	—	—	—	—	—	—
D	Isopropylamine	756	402	-35	-37	—	—	2.0
D	Kerosene	410	210	110-162	43-72	0.7	5	—
D	Liquefied Petroleum Gas	761-842	405-450	—	—	—	—	—
B	Manufactured Gas (containing more than 30% H₂ by volume)	—	—	—	—	1.4	7.2	3.4
D	Mesityl Oxide	652	344	87	31	1.4	7.2	3.4
D	Methane	999	630	gas	gas	5.0	15.0	0.6
D	Methanol	725	385	52	11	6.0	36	1.1
D	Methyl Acetate	850	454	14	-10	3.1	16	2.8
D	Methyl Acrylate	875	468	27	-3	2.8	25	3.0
D	Methyl Amyl Alcohol	—	—	106	41	1.0	5.5	—
D	Methyl n-Amyl Ketone	740	393	102	39	1.1 @ 151° F	7.9 @ 250° F	3.9
C	Methyl Ether	662	350	gas	gas	3.4	27.0	1.6
D	Methyl Ethyl Ketone	759	404	16	-9	1.7 @ 200° F	11.4 @ 200° F	2.5
D	2-Methyl-5-Ethyl Pyridine	—	—	155	68	1.1	6.6	4.2
C	Methyl Formal	460	238	—	—	—	—	—
D	Methyl Formate	840	449	-2	-19	4.5	23	2.1

Figure A-14 *(continued)*

Table 1 *(continued)*

Class I* Group	Substance	Auto-* Ignition Temp. °F	Auto-* Ignition Temp. °C	Flash** Point °F	Flash** Point °C	Explosive Limits** Per Cent by Volume Lower	Explosive Limits** Per Cent by Volume Upper	Vapor** Density (Air Equals 1.0)
D	Methyl Isobutyl Ketone	840	440	64	18	1.2 @ 200° F	8.0 @ 200° F	**3.5**
D	Methyl Isocyanate	994	534	19	-7	5.3	26	**1.97**
C	Methyl Mercaptan	—	—	—	—	3.9	21.8	**1.7**
D	Methyl Methacrylate	792	422	50	10	1.7	8.2	**3.6**
D	2-Methyl-1-Propanol	780	416	82	28	1.7 @ 123° F	10.6 @ 202° F	**2.6**
D	2-Methyl-2-Propanol	892	478	52	11	2.4	8.0	**2.6**
D	alpha-Methyl Styrene	1066	574	129	54	1.9	6.1	—
C	Methylacetylene	—	—	gas	gas	1.7	—	1.4
C	Methylacetylene-Propadiene (stabilized)	—	—	—	—	—	—	—
D	Methylamine	806	430	gas	gas	4.9	20.7	1.0
D	Methylcyclohexane	482	250	25	-4	1.2	6.7	3.4
D	Methylcyclohexanol	565	296	149	65	—	—	3.9
D	o-Methylcyclohexanone	—	—	118	48	—	—	3.9
D	Monoethanolamine	770	410	185	85	—	—	2.1
D	Monoisopropanolamine	705	374	171	77	—	—	2.6
C	Monomethyl Aniline	900	482	185	85	—	—	3.7
C	Monomethyl Hydrazine	382	194	17	-8	2.5	92	1.6
C	Morpholine	590	310	98	37	1.4	11.2	3.0
D	Naphtha (Coal Tar)	531	277	107	42	—	—	—
D	Naphtha (Petroleum)⁴	550	288	<0	<-18	1.1	5.9	2.5
D	Nitrobenzene	900	482	190	88	1.8 @ 200° F	—	4.3
C	Nitroethane	778	414	82	28	3.4	—	2.6
C	Nitromethane	785	418	95	35	7.3	—	2.1
C	1-Nitropropane	789	421	96	36	2.2	—	3.1
C	2-Nitropropane	802	428	75	24	2.6	11.0	3.1
D	Nonane	401	205	88	31	0.8	2.9	4.4
D	Nonene	—	—	78	26	—	—	4.35
D	Nonyl Alcohol	—	—	165	74	0.8 @ 212° F	6.1 @ 212° F	5.0
D	Octane	403	206	56	13	1.0	6.5	3.9
D	Octene	446	230	70	21	—	—	3.9
D	n-Octyl Alcohol	—	—	178	81	—	—	4.5
D	Pentane	470	243	<-40	<-40	1.5	7.8	2.5
D	1-Pentanol	572	300	91	33	1.2	10.0 @ 212° F	3.0
D	2-Pentanone	846	452	45	7	1.5	8.2	3.0
D	1-Pentene	527	275	0	-18	1.5	8.7	2.4
D	Phenylhydrazine	—	—	190	88	—	—	—
D	Propane	842	450	gas	gas	2.1	9.5	1.6
D	1-Propanol	775	413	74	23	2.2	13.7	2.1
D	2-Propanol	750	399	53	12	2.0	12.7 @ 200° F	2.1
D	Propiolactone	—	—	165	74	2.9	—	2.5
C	Propionaldehyde	405	207	-22	-30	2.6	17	2.0
D	Propionic Acid	870	466	126	52	2.9	12.1	2.5
D	Propionic Anhydride	545	285	145	63	1.3	9.5	4.5
D	n-Propyl Acetate	842	450	55	13	1.7 @ 100° F	8	3.5
C	n-Propyl Ether	419	215	70	21	1.3	7.0	3.53
B	Propyl Nitrate	347	175	68	20	2	100	—
D	Propylene	851	455	gas	gas	2.0	11.1	1.5
D	Propylene Dichloride	1035	557	60	16	3.4	14.5	3.9
B(C)	Propylene Oxide¹	840	449	-35	-37	2.3	36	2.0
D	Pyridine	900	482	68	20	1.8	12.4	2.7
D	Styrene	914	490	88	31	1.1	7.0	3.6
C	Tetrahydrofuran	610	321	6	-14	2.0	11.8	2.5
D	Tetrahydronaphthalene	725	385	160	71	0.8 @ 212° F	5.0 @ 302° F	4.6
C	Tetramethyl Lead	—	—	100	38	—	—	6.5
D	Toluene	896	480	40	4	1.2	7.1	3.1
D	Tridecene	—	—	—	—	—	—	—
C	Triethylamine	480**	249**	16	-9	1.2	8.0	3.5
D	Triethylbenzene	—	—	181	83	—	—	5.6
D	Tripropylamine	—	—	105	41	—	—	4.9
D	Turpentine	488	253	95	35	0.8	—	—
D	Undecene	—	—	—	—	—	—	—
C	Unsymmetrical Dimethyl Hydrazine (UDMH)	480	249	5	-15	2	95	2.0
C	Valeraldehyde	432	222	54	12	—	—	3.0
D	Vinyl Acetate	756	402	18	-8	2.6	13.4	3.0
D	Vinyl Chloride	882	472	gas	gas	3.6	33.0	2.2
D	Vinyl Toluene	921	494	120	49	—	11.0	4.1
D	Vinylidene Chloride	1058	570	-19	-28	6.5	15.5	3.4
D	Xylenes	867-984	464-529	81-90	27-32	1.0-1.1	7.0	3.7

¹ If equipment is isolated by sealing all conduit ½ in. or larger, in accordance with Section 501-5(a) of NFPA 70, *National Electrical Code*, equipment for the group classification shown in parentheses is permitted.

² For classification of areas involving Ammonia, see *Safety Code for Mechanical Refrigeration*, ANSI/ASHRAE 15, and *Safety Requirements for the Storage and Handling of Anhydrous Ammonia*, ANSI/CGA G2.1.

³ Certain chemicals may have characteristics that require safeguards beyond those required for any of the above groups. Carbon disulfide is one of these chemicals because of its low autoignition temperature and the small joint clearance to arrest its flame propagation.

⁴ Petroleum Naphtha is a saturated hydrocarbon mixture whose boiling range is 20° to 135° C. It is also known as benzine, ligroin, petroleum ether, and naphtha.

* Data from NFPA 497M-1986, *Classification of Gases, Vapors and Dusts for Electrical Equipment in Hazardous (Classified) Locations*.

**Data from NFPA 325M-1984, *Fire Hazard Properties of Flammable Liquids, Gases and Volatile Solids*.

(Courtesy of Crouse Hinds Division of Cooper Industries)

Figure A-14   Hazardous Substances Flash Point, Ignition Temperatures—Dust

Table 2

Class II, Group E	Minimum Cloud or Layer Ignition Temp.		
Material[2]	°F		°C
Aluminum, atomized collector fines	1022	Cl	550
Aluminum, A422 flake	608		320
Aluminum — cobalt alloy (60-40)	1058		570
Aluminum — copper alloy (50-50)	1526		830
Aluminum — lithium alloy (15% Li)	752		400
Aluminum — magnesium alloy (Dowmetal)	806	Cl	430
Aluminum — nickel alloy (58-42)	1004		540
Aluminum — silicon alloy (12% Si)	1238	NL	670
Boron, commercial-amorphous (85% B)	752		400
Calcium Silicide	1004		540
Chromium, (97%) electrolytic, milled	752		400
Ferromanganese, medium carbon	554		290
Ferrosilicon (88%, 9% Fe)	1472		800
Ferrotitanium (19% Ti, 74.1% Fe, 0.06% C)	698	Cl	370
Iron, 98%, H_2 reduced	554		290
Iron, 99%, Carbonyl	590		310
Magnesium, Grade B, milled	806		430
Manganese	464		240
Silicon, 96%, milled	1436	Cl	780
Tantalum	572		300
Thorium, 1.2%, O_2	518	Cl	270
Tin, 96%, atomized (2% Pb)	806		430
Titanium, 99%	626	Cl	330
Titanium Hydride, (95% Ti, 3.8% H_2)	896	Cl	480
Vanadium, 86.4%	914		490
Zirconium Hydride, (93.6% Zr, 2.1% H_2)	518		270

Class II, Group G			
AGRICULTURAL DUSTS			
Alfalfa Meal	392		200
Almond Shell	392		200
Apricot Pit	446		230
Cellulose	500		260
Cherry Pit	428		220
Cinnamon	446		230
Citrus Peel	518		270
Cocoa Bean Shell	698		370
Cocoa, natural, 19% fat	464		240
Coconut Shell	428		220
Corn	482		250
Corncob Grit	464		240
Corn Dextrine	698		370
Cornstarch, commercial	626		330
Cornstarch, modified	392		200
Cork	410		210
Cottonseed Meal	392		200
Cube Root, South Amer.	446		230
Flax Shive	446		230
Garlic, dehydrated	680	NL	360
Guar Seed	932	NL	500
Gum, Arabic	500		260
Gum, Karaya	464		240
Gum, Manila (copal)	680	Cl	360
Gum, Tragacanth	500		260
Hemp Hurd	428		220
Lycopodium	590		310
Malt Barley	482		250
Milk, Skimmed	392		200
Pea Flour	500		260
Peach Pit Shell	410		210
Peanut Hull	410		210
Peat, Sphagnum	464		240
Pecan Nut Shell	410		210
Pectin	392		200
Potato Starch, Dextrinated	824	NL	440
Pyrethrum	410		210
Rauwolfia Vomitoria Root	446		230
Rice	428		220
Rice Bran	914	NL	490
Rice Hull	428		220
Safflower Meal	410		210
Soy Flour	374		190
Soy Protein	500		260
Sucrose	662	Cl	350
Sugar, Powdered	698	Cl	370

Class II, Group G (cont'd)	Minimum Cloud or Layer Ignition Temp.		
	°F		°C
AGRICULTURAL DUSTS			
Tung, Kernels, Oil-Free	464		240
Walnut Shell, Black	428		220
Wheat	428		220
Wheat Flour	680		360
Wheat Gluten, gum	968	NL	520
Wheat Starch	716	NL	380
Wheat Straw	428		220
Woodbark, Ground	482		250
Wood Flour	500		260
Yeast, Torula	500		260
CARBONACEOUS DUSTS[3]			
Asphalt, (Blown Petroleum Resin)	950	Cl	510
Charcoal	356		180
Coal, Kentucky Bituminous	356		180
Coal, Pittsburgh Experimental	338		170
Coal, Wyoming	—		—
Gilsonite	932		500
Lignite, California	356		180
Pitch, Coal Tar	1310	NL	710
Pitch, Petroleum	1166	NL	630
Shale, Oil	—		—
CHEMICALS			
Acetoacetanilide	824	M	440
Acetoacet-p-phenetidide	1040	NL	560
Adipic Acid	1022	M	550
Anthranilic Acid	1076	M	580
Aryl-nitrosomethylamide	914	NL	490
Azelaic Acid	1130	M	610
2,2-Azo-bis-butyronitrile	662		350
Benzoic Acid	824	M	440
Benzotriazole	824	M	440
Bisphenol-A	1058	M	570
Chloroacetoacetanilide	1184	M	640
Diallyl Phthalate	896	M	480
Dicumyl Peroxide (suspended on $CaCO_3$), 40-60	356		180
Dicyclopentadiene Dioxide	788	NL	420
Dihydroacetic Acid	806	NL	430
Dimethyl Isophthalate	1076	M	580
Dimethyl Terephthalate	1058	M	570
3,5 - Dinitrobenzoic Acid	860	NL	460
Dinitrotoluamide	932	NL	500
Diphenyl	1166	M	630
Ditertiary Butyl Paracresol	878	NL	470
Ethyl Hydroxyethyl Cellulose	734	NL	390
Fumaric Acid	968	M	520
Hexamethylene Tetramine	770	S	410
Hydroxyethyl Cellulose	770	NL	410
Isotoic Anhydride	1292	NL	700
Methionine	680		360
Nitrosoamine	518	NL	270
Para-oxy-benzaldehyde	716	Cl	380
Paraphenylene Diamine	1148	M	620
Paratertiary Butyl Benzoic Acid	1040	M	560
Pentaerythritol	752	M	400
Phenylbetanaphthylamine	1256	NL	680
Phthalic Anydride	1202	M	650
Phthalimide	1166	M	630
Salicylanilide	1130	M	610
Sorbic Acid	860		460
Stearic Acid, Aluminum Salt	572		300
Stearic Acid, Zinc Salt	950	M	510
Sulfur	428		220
Terephthalic Acid	1256	NL	680
DRUGS			
2-Acetylamino-5-nitrothiazole	842		450
2-Amino-5-nitrothiazole	860		460
Aspirin	1220	M	660
Gulasonic Acid, Diacetone	788	NL	420
Mannitol	860	M	460
Nitropyridone	806	M	430
1-Sorbose	698	M	370
Vitamin B1, mononitrate	680	NL	360
Vitamin C (Ascorbic Acid)	536		280

Figure A-14 *(continued)*

Table 2 *(continued)*

Class II, Group G (cont'd)	Minimum Cloud or Layer Ignition Temp.[1]		
	°F		°C
DYES, PIGMENTS, INTERMEDIATES			
Beta-naphthalene-azo-Dimethylaniline	347		175
Green Base Harmon Dye	347		175
Red Dye Intermediate	347		175
Violet 200 Dye	347		175
PESTICIDES			
Benzethonium Chloride	716	CI	380
Bis(2-Hydroxy-5-chlorophenyl) methane	1058	NL	570
Crag No. 974	590	CI	310
Dieldrin (20%)	1022	NL	550
2, 6-Ditertiary-butyl-paracresol	788	NL	420
Dithane	356		180
Ferbam	302		150
Manganese Vancide	248		120
Sevin	284		140
∝,∝ - Trithiobis (N,N-Dimethylthio-formamide)	446		230
THERMOPLASTIC RESINS AND MOLDING COMPOUNDS			
Acetal Resins			
Acetal, Linear (Polyformaldehyde)	824	NL	440
Acrylic Resins			
Acrylamide Polymer	464		240
Acrylonitrile Polymer	860		460
Acrylonitrile - Vinyl Pyridine Copolymer	464		240
Acrylonitrile-Vinyl Chloride- Vinylidene Chloride Copolymer (70-20-10)	410		210
Methyl Methacrylate Polymer	824	NL	440
Methyl Methacrylate - Ethyl Acrylate Copolymer	896	NL	480
Methyl Methacrylate-Ethyl Acrylate-Styrene Copolymer	824	NL	440
Methyl Methacrylate-Styrene- Butadiene-Acrylonitrile Copolymer	896	NL	480
Methacrylic Acid Polymer	554		290
Cellulosic Resins			
Cellulose Acetate	644		340
Cellulose Triacetate	806	NL	430
Cellulose Acetate Butyrate	698	NL	370
Cellulose Propionate	860	NL	460
Ethyl Cellulose	608	CI	320
Methyl Cellulose	644		340
Carboxymethyl Cellulose	554		290
Hydroxyethyl Cellulose	644		340
Chlorinated Polyether Resins			
Chlorinated Polyether Alcohol	860		460
Nylon (Polyamide) Resins			
Nylon Polymer (Polyhexa-methylene Adipamide)	806		430
Polycarbonate Resins			
Polycarbonate	1310	NL	710
Polyethylene Resins			
Polyethylene, High Pressure Process	716		380
Polyethylene, Low Pressure Process	788	NL	420
Polyethylene Wax	752	NL	400
Polymethylene Resins			
Carboxypolymethylene	968	NL	520

Class II, Group G (cont'd)	Minimum Cloud or Layer Ignition Temp.		
	°F		°C
THERMOPLASTIC RESINS AND MOLDING COMPOUNDS			
Polypropylene Resins			
Polypropylene (No Antioxidant)	788	NL	420
Rayon Resins			
Rayon (Viscose) Flock	482		250
Styrene Resins			
Polystyrene Molding Cmpd.	1040	NL	560
Polystyrene Latex	932		500
Styrene-Acrylonitrile (70-30)	932	NL	500
Styrene-Butadiene Latex (> 75% Styrene; Alum Coagulated)	824	NL	440
Vinyl Resins			
Polyvinyl Acetate	1022	NL	550
Polyvinyl Acetate/Alcohol	824		440
Polyvinyl Butyral	734	NL	390
Vinyl Chloride - Acrylonitrile Copolymer	878		470
Polyvinyl Chloride - Dioctyl Phthalate Mixture	608	NL	320
Vinyl Toluene - Acrylonitrile Butadiene Copolymer	936	NL	530
THERMOSETTING RESINS AND MOLDING COMPOUNDS			
Allyl Resins			
Allyl Alcohol Derivative (CR-39)	932	NL	500
Amino Resins			
Urea Formaldehyde Molding Compound	860	NL	460
Urea Formaldehyde - Phenol Formaldehyde Molding Compound (Wood Flour Filler)	464		240
Epoxy Resins			
Epoxy	1004	NL	540
Epoxy - Bisphenol A	950	NL	510
Phenol Furfural	590		310
Phenolic Resins			
Phenol Formaldehyde	1076	NL	580
Phenol Formaldehyde Molding Cmpd (Wood Flour Filler)	932	NL	500
Phenol Formaldehyde, Polyalkylene- Polyamine Modified	554		290
Polyester Resins			
Polyethylene Terephthalate	932	NL	500
Styrene Modified Polyester- Glass Fiber Mixture	680		360
Polyurethane Resins			
Polyurethane Foam, No Fire Retardant	824		440
Polyurethane Foam, Fire Retardant	734		390
SPECIAL RESINS AND MOLDING COMPOUNDS			
Alkyl Ketone Dimer Sizing Compound	320		160
Cashew Oil, Phenolic, Hard	356		180
Chlorinated Phenol	1058	NL	570
Coumarone-Indene, Hard	968	NL	520
Ethylene Oxide Polymer	662	NL	350
Ethylene-Maleic Anhydride Copolymer	1004	NL	540
Lignin, Hydrolized, Wood-Type, Fines	842	NL	450
Petrin Acrylate Monomer	428	NL	220
Petroleum Resin (Blown Asphalt)	932		500
Rosin, DK	734	NL	390
Rubber, Crude, Hard	662	NL	350
Rubber, Synthetic, Hard (33% S)	608	NL	320
Shellac	752	NL	400
Sodium Resinate	428		220
Styrene — Maleic Anhydride Copolymer	878	CI	470

[1]Normally, the minimum ignition temperature of a layer of a specific dust is lower than the minimum ignition temperature of a cloud of that dust. Since this is not universally true, the lower of the two minimum ignition temperatures is listed. If no symbol appears between the two temperature columns, then the layer ignition temperature is shown. "CI" means the cloud ignition temperature is shown. "NL" means that no layer ignition temperature is available and the cloud ignition temperature is shown. "M" signifies that the dust layer melts before it ignites; the cloud ignition temperature is shown. "S" signifies that the dust layer sublimes before it ignites; the cloud ignition temperature is shown.

[2]Certain metal dusts may have characteristics that require safeguards beyond those required for atmospheres containing the dusts of aluminum, magnesium, and their commercial alloys. For example, zirconium, thorium, and uranium dusts have extremely low ignition temperatures (as low as 20° C) and minimum ignition energies lower than any material classified in any of the Class I or Class II groups.

[3]The 1987 NEC classifies carbonaceous dusts as Group F, some of which may be conductive.

(Courtesy of Crouse Hinds Division of Cooper Industries)

Figure A-15   American Wire Gauge Table

AMERICAN WIRE GAUGE TABLE								
B & S Gauge No.	Diam. in Mils	Area in Circular Mils	Ohms per 1 000 Ft. (ohms per 100 m)			Pounds per 1 000 Ft. (kg per 100 m)		
			Copper* 68°F (20°C)	Copper* 167°F (75°C)	Aluminum 68°F (20°C)	Copper		Aluminum
0000	460	211 600	0.049 (0.016)	0.0596 (0.0195)	0.0804 (0.0263)	640	(95.2)	195	(29.0)
000	410	167 800	0.0618 (0.020)	0.0752 (0.0246)	0.101 (0.033)	508	(75.5)	154	(22.9)
00	365	133 100	0.078 (0.026)	0.0948 (0.031)	0.128 (0.042)	403	(59.9)	122	(18.1)
0	325	105 500	0.0983 (0.032)	0.1195 (0.0392)	0.161 (0.053)	320	(47.6)	97	(14.4)
1	289	83 690	0.1239 (0.0406)	0.151 (0.049)	0.203 (0.066)	253	(37.6)	76.9	(11.4)
2	258	66 370	0.1563 (0.0512)	0.190 (0.062)	0.526 (0.084)	201	(29.9)	61.0	(9.07)
3	229	52 640	0.1970 (0.0646)	0.240 (0.079)	0.323 (0.106)	159	(23.6)	48.4	(7.20)
4	204	41 740	0.2485 (0.0815)	0.302 (0.099)	0.408 (0.134)	126	(18.7)	38.4	(5.71)
5	182	33 100	0.3133 (0.1027)	0.381 (0.125)	0.514 (0.168)	100	(14.9)	30.4	(4.52)
6	162	26 250	0.395 (1.29)	0.481 (0.158)	0.648 (0.212)	79.5	(11.8)	24.1	(3.58)
7	144	20 820	0.498 (0.163)	0.606 (0.199)	0.817 (0.268)	63.0	(9.37)	19.1	(2.84)
8	128	16 510	0.628 (0.206)	0.764 (0.250)	1.03 (0.338)	50.0	(7.43)	15.2	(2.26)
9	114	13 090	0.792 (0.260)	0.963 (0.316)	1.30 (0.426)	39.6	(5.89)	12.0	(1.78)
10	102	10 380	0.999 (0.327)	1.215 (0.398)	1.64 (0.538)	31.4	(4.67)	9.55	(1.42)
11	91	8 234	1.260 (0.413)	1.532 (0.502)	2.07 (0.678)	24.9	(3.70)	7.57	(1.13)
12	81	6 530	1.588 (0.520)	1.931 (0.633)	2.61 (0.856)	19.8	(2.94)	6.00	(0.89)
13	72	5 178	2.003 (0.657)	2.44 (0.80)	3.29 (1.08)	15.7	(2.33)	4.8	(0.71)
14	64	4 107	2.525 (0.828)	3.07 (1.01)	4.14 (1.36)	12.4	(1.84)	3.8	(0.56)
15	57	3 257	3.184 (1.043)	3.98 (1.27)	5.22 (1.71)	9.86	(1.47)	3.0	(0.45)
16	51	2 583	4.016 (1.316)	4.88 (1.60)	6.59 (2.16)	7.82	(1.16)	2.4	(0.36)
17	45.3	2 048	5.06 (1.66)	6.16 (2.02)	8.31 (2.72)	6.20	(0.922)	1.9	(0.28)
18	40.3	1 624	6.39 (2.09)	7.77 (2.55)	10.5 (3.44)	4.92	(0.731)	1.5	(0.22)
19	35.9	1 288	8.05 (2.64)	9.79 (3.21)	13.2 (4.33)	3.90	(0.580)	1.2	(0.18)
20	32.0	1 022	10.15 (3.33)	12.35 (4.05)	16.7 (5.47)	3.09	(0.459)	0.94	(0.14)
21	28.5	810	12.8 (4.2)	15.6 (5.11)	21.0 (6.88)	2.45	(0.364)	0.745	(0.110)
22	25.4	642	16.1 (5.3)	19.6 (6.42)	26.5 (8.69)	1.95	(0.290)	0.591	(0.09)
23	22.6	510	20.4 (6.7)	24.8 (8.13)	33.4 (10.9)	1.54	(0.229)	0.468	(0.07)
24	20.1	404	25.7 (8.4)	31.2 (10.2)	42.1 (13.8)	1.22	(0.181)	0.371	(0.05)
25	17.9	320	32.4 (10.6)	39.4 (12.9)	53.1 (17.4)	0.97	(0.14)	0.295	(0.04)
26	15.9	254	40.8 (13.4)	49.6 (16.3)	67.0 (22.0)	0.77	(0.11)	0.234	(0.03)
27	14.2	202	51.5 (16.9)	62.6 (20.5)	84.4 (27.7)	0.61	(0.09)	0.185	(0.03)
28	12.6	160	64.9 (21.3)	78.9 (25.9)	106 (34.7)	0.48	(0.07)	0.147	(0.02)
29	11.3	126.7	81.8 (26.8)	99.5 (32.6)	134 (43.9)	0.384	(0.06)	0.117	(0.02)
30	10.0	100.5	103.2 (33.8)	125.5 (41.1)	169 (55.4)	0.304	(0.04)	0.092	(0.01)
31	8.93	79.7	130.1 (42.6)	158.2 (51.9)	213 (69.8)	0.241	(0.04)	0.073	(0.01)
32	7.95	63.2	164.1 (53.8)	199.5 (65.4)	269 (88.2)	0.191	(0.03)	0.058	(0.01)
33	7.08	50.1	207 (68)	252 (82.6)	339 (111)	0.152	(0.02)	0.046	(0.01)
34	6.31	39.8	261 (86)	317 (104)	428 (140)	0.120	(0.02)	0.037	(0.01)
35	5.62	31.5	329 (108)	400 (131)	540 (177)	0.095	(0.01)	0.029	
36	5.00	25.0	415 (136)	505 (165)	681 (223)	0.076	(0.01)	0.023	
37	4.45	19.8	523 (171)	636 (208)	858 (281)	0.0600	(0.01)	0.0182	
38	3.96	15.7	660 (216)	802 (263)	1080 (354)	0.0476	(0.01)	0.0145	
39	3.53	12.5	832 (273)	1012 (332)	1360 (446)	0.0377	(0.01)	0.0115	
40	3.15	9.9	1049 (344)	1276 (418)	1720 (564)	0.0299	(0.01)	0.0091	
41									
42	2.50	6.3							
43									
44	1.97	3.9							

*Resistance figures are given for standard annealed copper. For hard-drawn copper add 2%.

(*Courtesy Loper and Tedsen*, Direct Current Fundamentals, *3rd Edition, Delmar Publishers Inc., 1986.*)

Figure A-16   Maximum length of Electrical Metallic Tubing that may safely be used as an equipment grounding circuit conductor. Based on a ground-fault current of 400% of the overcurrent device rating. Circuit 120 volts to ground; 40 volts drop at the point of the fault. Ambient Temperature 25˚C

EMT Size Inches	Conductors AWG No.	Overcurrent Device Rating Amps. 75° C*	Fault Clearing Current 400% O.C. Device Rating Amps.	Maximum Length of EMT Run in Ft.
1/2	3-#12	20	80	395
	4-#10	30	120	358
3/4	4-#10	30	120	404
	4-#8	50	200	332
1	4-#8	50	200	370
	3-#4	85	340	365
1 1/4	3-#2	115	460	391
1 1/2	3-#1	130	520	407
	3-#2/0	175	700	364
2	3-#3/0	200	800	390
	3-#4/0	230	920	367
2 1/2	3-250 kcm	255	1020	406
3	3-350 kcm	310	1240	404
	3-500 kcm	380	1520	370
	3-600 kcm	420	1680	353
4	3-900 kcm	520	2080	353
	3-1000 kcm	545	2180	347

*60° C for 20- and 30-ampere devices.

Based on 1994 Georgia Tech Model

Appendices A-16 through A-21 are tables derived from software (GEMI) and testing developed at Georgia Institute of Technology and sponsored by the Steel Tube Institute of North America (STI), producers of steel, EMT, IMC and Rigid Steel Conduit

Figure A-17 Maximum length of Intermediate Metal Conduit that may safely be used as an equipment grounding circuit conductor. Based on a ground-fault current of 400% of the overcurrent device rating. Circuit 120 volts to ground; 40 volts drop at the point of the fault. Ambient Temperature 25°C

IMC Size Inches	Conductors AWG No.	Overcurrent Device Rating Amps. 75° C*	Fault Clearing Current 400% O.C. Device Rating Amps.	Maximum Length of IMC Run in Ft.
1/2	3-#12	20	80	398
	4-#10	30	120	383
3/4	4-#10	30	120	399
	4-#8	50	200	350
1	4-#8	50	200	362
	3-#4	85	340	382
1 1/4	3-#2	115	460	392
1 1/2	3-#1	130	520	402
	3-#2/0	175	700	377
2	3-#3/0	200	800	389
	3-#4/0	230	920	375
2 1/2	3-250 kcm	255	1020	368
3	3-350 kcm	310	1240	367
	3-500 kcm	380	1520	338
	3-600 kcm	420	1680	325
4	3-900 kcm	520	2080	320
	3-1000 kcm	545	2180	314

*60° C for 20- and 30-ampere devices.

Based on 1994 Georgia Tech Model

Figure A-18   Maximum length of Galvanized Rigid Conduit that may safely be used as an equipment grounding circuit conductor. Based on a ground-fault current of 400% of the overcurrent device rating. Circuit 120 volts to ground; 40 volts drop at the point of the fault. Ambient Temperature 25°C

GRC Size Inches	Conductors AWG No.	Overcurrent Device Rating Amps. 75° C*	Fault Clearing Current 400% O.C. Device Rating Amps.	Maximum Length of GRC Run in Ft.
1/2	3-#12	20	80	384
	4-#10	30	120	364
3/4	4-#10	30	120	386
	4-#8	50	200	334
1	4-#8	50	200	350
	3-#4	85	340	357
1 1/4	3-#2	115	460	365
1 1/2	3-#1	130	520	377
	3-#2/0	175	700	348
2	3-#3/0	200	800	363
	3-#4/0	230	920	347
2 1/2	3-250 kcm	255	1020	356
3	3-350 kcm	310	1240	355
	3-500 kcm	380	1520	327
	3-600 kcm	420	1680	314
4	3-900 kcm	520	2080	310
	3-1000 kcm	545	2180	304

*60° C for 20- and 30-ampere devices.

Based on 1994 Georgia Tech Model

Figure A-19 Maximum length of Equipment Grounding Conductor that may safely be used as an equipment grounding circuit conductor. Based on a ground-fault current of 400% of the overcurrent device rating. Circuit 120 volts to ground; 40 volts drop at the point of fault. Ambient Temperature 25°C

Copper Equipment Grounding Conductor Size***	Copper Circuit Conductors	Maximum Length of Run (in ft.) using Copper Equipment Ground Conductor	Aluminum Equipment Grounding Conductor Size***	Aluminum Circuit Conductors	Maximum Length of Run (in ft.) using Aluminum Equipment Ground Conductor	Overcurrent Device Rating Amps. 75°C**	Fault Clearing Current 400% O.C. Device Rating Amps.
#14	#14	253	#12	#12	244	15	60
#12	#12	300	#10	#12	226	20	80
#10	#10	319	#8	#8	310	30	120
#10	#8	294	#8	#8	232	40	160
#10	#6	228	#8	#4	221	60	240
#8	#3	229	#6	#1	222	100	400
#6	#3/0	201	#4	250kcm	195	200	800
#4	350 kcm	210	#2	500kcm	204	300	1200
#3	600 kcm	195	#1	900kcm	192	400	1600
#2	2-#4/0	160	#1/0	2-400kcm	163	500	2000
#1	2-300kcm	160	#2/0	2-500kcm	161	600	2400
#1/0	3-300kcm	134	#3/0	3-400kcm	131	800	3200
#2/0	4-250kcm	114	#4/0	4-400kcm	115	1000	4000
#3/0	4-300kcm	106	250 kcm	4-500kcm	107	1200	4800
#4/0	4-600kcm	93	350 kcm	4-900kcm	97	1600	6400
250 kcm	5-600kcm	78	400 kcm	5-800kcm	79	2000	8000
350 kcm	6-600kcm	*	600 kcm	6-900kcm	*	2500	10000
400 kcm	8-500kcm	*	600 kcm	8-750kcm	*	3000	15000
500 kcm	8-1000kcm	*	800 kcm	8-1500kcm	*	4000	16000
700 kcm	10-1000kcm	*	1200 kcm	10-1500kcm	*	5000	20000
800 kcm	12-1000kcm	*	1200 kcm	12-1500kcm	*	6000	24000

*Calculations Necessary
**60°C for 20- and 30-ampere devices.
***Based on NEC Chapter 9 Table 8

Based on 1994 Georgia Tech Model

Examples of Maximum Length Equipment Grounding Conductor (Steel EMT, IMC, GRC, and Copper, Copperclad or Aluminum Wire) Computed As A Safe Return Fault Path To Overcurrent Device Based on 1994 Georgia Tech Software (GEMI 2.3) With an Arc Voltage of 40 and 4 IP at 25° C Ambient 120 Volts to Ground

Figure A-20A

Overcurrent Device Rating Amperes (75°C)	400% (4IP) overcurrent Device Rating Amperes	Circuit Conductor Size AWG-kcmil Copper or Aluminum	EMT, IMC GRC Trade Size	(1) Equipment Grounding Conductor Size Copper or Aluminum	Length of EMT Run Computed Maximum (In Feet)	Length of IMC Run Computed Maximum (In Feet)	Length of GRC Run Computed Maximum (In Feet)	(1) Copper Grounding Conductor Max Run (In Feet)	(1) Aluminum or Copperclad Grounding Conductor Max Run (In Feet)
20	80	12	1/2	---	395	398	384	---	---
20	80	12	---	12	---	---	---	300	---
20	80	10 AL	---	10 AL	---	---	---	---	293
30	120	10	1/2	---	---	383	---	---	---
30	120	10	3/4	---	404	399	386	---	---
30	120	10	---	10	---	---	---	319	---
30	120	8 AL	---	8 AL	---	---	---	---	310
40	160	8	3/4	---	---	414	---	---	---
40	160	8	1	---	447	431	418	---	---
40	160	8	---	10	---	---	---	294	---
40	160	8 AL	---	8 AL	---	---	---	---	232
60	240	6	1	---	404	400	382	---	---
60	240	6	---	10	---	---	---	228	---
60	240	4 AL	---	8 AL	---	---	---	---	221
100	400	3	1 1/4	---	402	397	373	---	---
100	400	3	---	8	---	---	---	229	---
100	400	1 AL	---	6 AL	---	---	---	---	222
200	800	3/0	2	---	390	389	363	---	---
200	800	3/0	---	6	---	---	---	201	---
200	800	250 AL	---	4 AL	---	---	---	---	195

(1) Per NEC Table 250-95.
Applicable to non-metallic conduit runs.

*NEC Wire Fill Table Permits Smaller Conduit Size

Note: Software is not limited to above examples

Examples of Maximum Length Equipment Grounding Conductor (Steel EMT, IMC, GRC, and Copper, Copperclad or Aluminum wire) Computed As A Safe Return Fault Path To Overcurrent Device
Based on 1994 Georgia Tech Software (GEMI 2.3)
With an Arc Voltage of 40 and 4 IP at 25° C Ambient
277 Volts to Ground

Figure A-20B

Overcurrent Device Rating Amperes (75°C)	400% (4IP) overcurrent Device Rating Amperes	Circuit Conductor Size AWG-kcmil Copper or Aluminum	EMT, IMC GRC Trade Size	(1) Equipment Grounding Conductor Size Copper or Aluminum	Length of EMT Run Computed Maximum (In Feet)	Length of IMC Run Computed Maximum (In Feet)	Length of GRC Run Computed Maximum (In Feet)	(1) Copper Grounding Conductor Max Run (In Feet)	(1) Aluminum or Copperclad Grounding Conductor Max Run (In Feet)
20	80	12	1/2	---	1170	1179	1140	---	---
20	80	12	---	12	---	---	---	890	---
20	80	10 AL	---	10 AL	---	---	---	---	870
30	120	10	1/2	---	---	1135	---	---	---
30	120	10	3/4	---	1199	1182	1143	---	---
30	120	10	---	10	---	---	---	946	---
30	120	8 AL	---	8 AL	---	---	---	---	920
40	160	8	3/4	---	---	1228	---	---	---
40	160	8	1	---	1326	1276	1239	---	---
40	160	8	---	10	---	---	---	871	---
40	160	8 AL	---	8 AL	---	---	---	---	690
60	240	6	1	---	1197	1186	1131	---	---
60	240	6	---	10	---	---	---	676	---
60	240	4 AL	---	8 AL	---	---	---	---	657
100	400	3	1 1/4	---	1192	1176	1107	---	---
100	400	3	---	8	---	---	---	680	---
100	400	1 AL	---	6 AL	---	---	---	---	659
200	800	3/0	2	---	1157	1155	1077	---	---
200	800	3/0	---	6	---	---	---	598	---
200	800	50 A	---	4 AL	---	---	---	---	578

(1) Per NEC Table 250-95.
Applicable to non-metallic conduit runs.

*NEC Wire Fill Table Permits Smaller Conduit Size

Note: Software is not limited to above examples

Figure A21
Article 505 Class I Zone 0, 1, and 2 locations

A new *Article 505* was added to the 1996 *NEC*® introducing the European zone concept as a new optional method for engineers for what has been the accepted Class I areas. It provides a new level of the classification of Class I hazardous (Classified) locations. This alternate method uses the Zone concept to meet the requirements for electrical equipment and wiring for all voltages in locations where fire or explosive hazards may exist due to flammable gases or vapors and flammable liquids.

The new article modifies the general rules of the code as it applies to the electric wiring and equipment in locations as Class I, Zone 0, Zone 1, Zone 2.

A-21A Proponents of this new article presented this comparison chart to explain the impact of adding *Article 505* to the 1996 *NEC*®

Comparison of the Two Systems for Hazardous Locations

◆ Current Class – Division System

◆ Proposed Zone Alternative

Class I Division 1

Class I Zone 0

Class I Zone 1

Class I Division 2 ——— Class I Zone 2

A-21B This is an example of how an area may be used as an alternative to classifying an area under the zone concept Article 505.

ALTERNATIVE AREA CLASSIFICATION

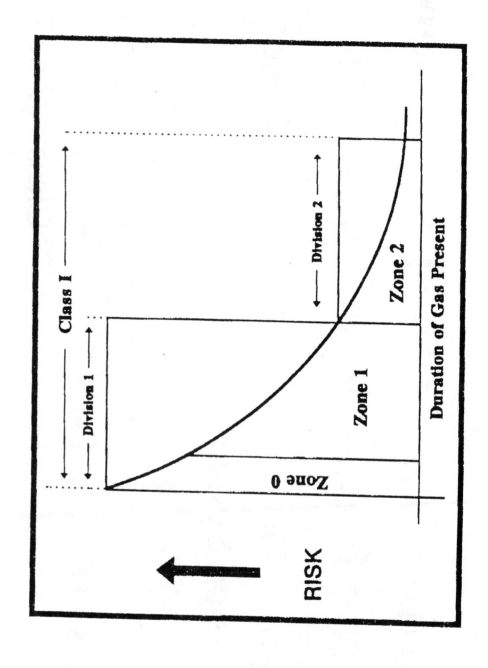

Index

Note: Page numbers in **bold type** reference non-text material